The series "Studies in Computational Intelligence" (SCI) publishes new developments and advances in the various areas of computational intelligence—quickly and with a high quality. The intent is to cover the theory, applications, and design methods of computational intelligence, as embedded in the fields of engineering, computer science, physics and life sciences, as well as the methodologies behind them. The series contains monographs, lecture notes and edited volumes in computational intelligence spanning the areas of neural networks, connectionist systems, genetic algorithms, evolutionary computation, artificial intelligence, cellular automata, self-organizing systems, soft computing, fuzzy systems, and hybrid intelligent systems. Of particular value to both the contributors and the readership are the short publication timeframe and the world-wide distribution, which enable both wide and rapid dissemination of research output.

Indexed by SCOPUS, DBLP, WTI Frankfurt eG, zbMATH, SCImago.

All books published in the series are submitted for consideration in Web of Science.

Park Gyei-Kark · Dipak Kumar Jana · Prabir Panja · Mohd Helmy Abd Wahab

Editors

Engineering Mathematics and Computing

 Springer

Editors
Park Gyei-Kark
Division of Maritime Transportation
Mokpo National Maritime University
Mokpo, Korea (Republic of)

Dipak Kumar Jana
Department of Applied Science
Haldia Institute of Technology
Haldia, India

Prabir Panja
Department of Applied Science
Haldia Institute of Technology
Haldia, India

Mohd Helmy Abd Wahab
Department of Electronic Engineering
Universiti Tun Hussein Onn Malaysia
Batu Pahat, Malaysia

ISSN 1860-949X   ISSN 1860-9503 (electronic)
Studies in Computational Intelligence
ISBN 978-981-19-2302-9   ISBN 978-981-19-2300-5 (eBook)
https://doi.org/10.1007/978-981-19-2300-5

Mathematics Subject Classification: 58J05, 90B60, 76W05, 68T40, 97N10, 94C12, 90B50, 68T07, 51K05

This Springer imprint is published by the registered company Springer Nature Singapore Pte Ltd.
The registered company address is: 152 Beach Road, #21-01/04 Gateway East, Singapore 189721, Singapore

Preface

The Third International Conference on Engineering Mathematics and Computing (ICEMC 2020) was held at the Haldia Institute of Technology, Haldia, West Bengal, India, during 5–7 February 2020. Haldia is a city and a municipality in Purba Medinipur in the Indian state of West Bengal, and Haldia Institute of Technology is a premier institution training engineers and computer scientists for the past several years. It has gained its reputation through its institutional dedication to teaching and research.

In response to the call for papers for ICEMC 2020, 70 papers were submitted for presentation and inclusion in the proceedings of the conference. The papers were evaluated and ranked on the basis of their significance, novelty, and technical quality by at least two reviewers per paper. After a careful blind refereeing process, 19 papers were selected for inclusion in the conference proceedings. The papers cover current research in intelligent systems, soft computing, machine learning, and natural language processing, image and video processing, computer network and security, cryptography, and data hiding, rough set, fuzzy logic, operations research, optimization, uncertain theory and applications, etc. ICEMC 2020 has eminent personalities both from India and abroad (Serbia) who delivered invited talks. The speakers from India are recognized leaders in government, industry, and academic institutions like Indian Statistical Institute, Kolkata, IIT Bombay, Vidyasagar University, etc. All of them are involved in research dealing with the current issues of interest related to the theme of the conference. The conference was one key-note talk by Prof. Siddhartha Bhattacharyya, Christ University, Bangalore, India, and 10 invited talks.

A conference of this kind would not be possible to organize without the full support from different people across different committees. All logistics and general organizational aspects are looked after by the organizing committee members who spent their time and energy in making the conference a reality. We also thank all the technical program committee members and external reviewers for thoroughly reviewing the papers submitted for the conference and sending their constructive suggestions within the deadlines. Our hearty thanks to Springer for agreeing to publish the proceedings in the Studies in Computational Intelligence series.

We are indebted to Haldia Institute of Technology, Haldia, India, for sponsoring the event. Their support has significantly helped in raising the profile of the conference.

Last but not the least; our sincere thanks go to all the authors who submitted papers to ICEMC 2020 and to all speakers and participants. We sincerely hope that the readers will find the proceedings stimulating and inspiring.

Mokpo, Korea (Republic of)	Park Gyei-Kark
Haldia, India	Dipak Kumar Jana
Haldia, India	Prabir Panja
Batu Pahat, Malaysia	Mohd Helmy Abd Wahab

Committees

Patron

Lakshman Seth, Chairman, Haldia Institute of Technology, Haldia, India

Organizing Committee

Asish Lahiri, Haldia Institute of Technology, India
Sayantan Seth, Haldia Institute of Technology, India
M. N. Bandyopadhyay, Haldia Institute of Technology, India
Asit Kumar Saha, Haldia Institute of Technology, India
Anjan Mishra, Haldia Institute of Technology, India
Sudipta Kumar Basu, Haldia Institute of Technology, India
Debasis Das, Haldia Institute of Technology, India
Sk. Sahnawaj, Haldia Institute of Technology, India
Sarbari Samanta, Haldia Institute of Technology, India
Mihir Baran Bera, Haldia Institute of Technology, India
Joyeeta Majumdar, Haldia Institute of Technology, India
Apratim Mitra, Haldia Institute of Technology, India
Subhankar Joardar, Haldia Institute of Technology, India
Prabir Panja, Haldia Institute of Technology, India
Anupam De, Haldia Institute of Technology, India
Shuvendu Chakraborty, Haldia Institute of Technology, India
Manoj Mondal, Haldia Institute of Technology, India
Somashri Karan, Haldia Institute of Technology, India
Nabin Sen, Haldia Institute of Technology, India
Biplab Sinha Mahapatra, Haldia Institute of Technology, India
Sumana Mandal, Haldia Institute of Technology, India
Bidesh Chakraborty, Haldia Institute of Technology, India

Subhabrata Barman, Haldia Institute of Technology, India
Snehasish Kumar Karan, Haldia Institute of Technology, India
Palash Roy, Haldia Institute of Technology, India
Dipak Kumar Jana (Organizing Chair), Haldia Institute of Technology, India
Petr Dostál, Brno University of Technology, Czech Republic, UK
Ivana Štajner-Păpuga, University of Novi Sad, Serbia
Moti Zwilling, Ariel University, Samaria
Doskočil Radek, Brno University, Czech Republic, UK
Roman Šeneřřk, Tomas Bata University, Zlin
Zuzana Janková, Brno University Technology, Czech Republic, UK
Jin Peng, Huanggang Normal University, China
Waichon Lio, Beihang University, China
Samarjit Kar, National Institute of Technology, Durgapur, India
Tanmoy Som, Indian Institute of Technology BHU, India.

Technical Program Committee

Sofía Estelles-Miguel, Universitat Politècnica de València, Valencia, Spain
Palash Dutta, Dibrugarh University, Assam, India
Riccardo Zamolo, University of Trieste, Italy
Angeles Martinez Calomardo, University of Trieste, Italy
Vishal Gupta, Maharishi Markandeshwar, Deemed to be University Mullana, Haryana, India
Stojan Radenović, University of Belgrade, Serbia
Mauparna Nandan, Brainware University, Kolkata, India
Luca Iocchi, Università di Roma "La Sapienza", Italy
Sapan Kumar Das, NIT Jamshedpur, India
Shuping Wan, Jiangxi University of Finance and Economics, China
Andreja Tepavcevic, University of Novi Sad, Serbia
Avishek Chakraborty, Narula Institute of Technology, Kolkata, India
Yanbin Sang, North University of China, China
Guiwu Wei, Sichuan Normal University, P.R. China
Sanjay Kumar, G. B. Pant University of Agriculture and Technology, India
Ahona Ghosh, Maulana Abul Kalam Azad University of Technology, West Bengal, India
Shibaprasad Sen, Future Institute of Engineering and Management, India
Swarnali Sadhukhan, St. Mary's Technical Campus Kolkata, India
Sabah M. Alturfi, University of Kerbala, Kerbala, Iraq
Kushal Pokhrel, Sikkim Manipal Institute of Technology, India
Mohmed A. Hassan, Ain Shams University, Cairo, Egypt
Sutapa Pramanik, Vidyasagar University, India
Shyamal Kumar Mondal, Vidyasagar University, India
Amalesh Kumar Manna, The University of Burdwan, India

Asif khan, Aligarh Muslim University, U.P. India
Prashant Patel, Xavier College, Ahmedabad, India
M. Nasiruzzaman, University of Tabuk, Saudi Arabia
S. Sindu Devi, SRMIST, Ramapuram, Chennai, India
Nabin Sen, Haldia Institute of Technology, India
Sankar Prasad Mondal, Maulana Abul Kalam Azad University of Technology, India
Krit Somkantha, Udon-Thani Rajabhat University, Thailand
Chanchal Kundu, Rajiv Gandhi Institute of Petroleum Technology, India
S. M. Sunoj, Cochin University of Science and Technology, India
Hemen Dutta, Gauhati University, India
S. H. Dong, Instituto Politécnico Nacional, Mexico
Ujjwal Laha, National Institute of Technology, Jamshedpur, India
Peng Zhu, Nanjing University of Science and Technology, China
Ahmed Ismail Ebada, Mansoura University, Mansoura, Egypt
Santosh Chapaneri, University of Mumbai, India
Utpal Nandi, Vidyasagar University, India
Satyabrata Maity, University of Calcutta, India
Kalipada Maity, Mugberia Gangadhar Mahavidyalaya, India
Anupam Mukherjee, BITS Pilani Goa, India
Sujit Samanta, National Institute of Technology, Raipur, India
U. K. Misra, Berhampur University, Odisha, India
Asish Bera, Edge Hill University, UK
Jagannath Samanta, Haldia Institue of Technology, India
Banibrata Bag, Haldia Institue of Technology, India
Ajitha Soundararaj, National Institute of Technology Tiruchirappalli, India
Pradipta Banerjee, Sidho-Kanho-Birsha University, India
Debanjan Nag, ICFAI University Tripura, India.

Contents

About the Editors

Park Gyei-Kark is Full Professor in the Maritime Transportation of Mokpo National Maritime University (MMU), South Korea. He has been elected as President of the Korea Government Institute of Aids to Navigation for three years since January 2018. He is graduated from the Department of Nautical Science, Korea National Maritime University in Busan and earned his Ph.D. in Systems Science from the Tokyo Institute of Technology, Japan, in 1993. He also is a master in international economics in which he did his second Ph.D. from Chonnam National University, South Korea, in 2010. His fields of interest are maritime information systems and port transportation systems by using fuzzy theory and uncertainty theory. He worked as President of Korea intelligent Systems in 2012 and President of the Korea Association of Port Economics from 2017 to 2018. He has published more than 200 scientific research articles in the field of intelligent systems and maritime economics.

Dipak Kumar Jana is Professor and Head of the Department, School of Applied Science and Humanities, Haldia Institute of Technology, Haldia, West Bengal, India. He did his Ph.D. from the Indian Institute of Engineering Science and Technology, Shibpur, Kolkata, West Bengal, India, and his M.Sc. in applied mathematics with specialization in operations research from Vidyasagar University, West Bengal. As qualified National Eligibility Test (NET-CSIR) for Junior Research Fellow (JRF) and GATE, he has been teaching mathematics to both undergraduate and postgraduate students. He is Member of Operational Research Society of India, Indian Science Congress Association, and Calcutta Mathematical Society.

Professor Jana has published more than 98 papers in several international journals of repute: Journal of Information Science, Journal of Cleaner Production, Applied Soft Computing, Computers and Industrial Engineering, Separation and Purification Technology, International Journal of Uncertainty, Fuzziness and Knowledge-Based Systems, International Journal of Computing Science and Mathematics, Journal of Official Statistics, International Journal of Operations Research, International Journal of Advanced Operations Management, International Journal of Computer Applications, OPESARCH, and Journal of Transportation Security. He has authored eight books including A Textbook of Engineering Operations Research, GATE Mathematics, Advanced Engineering Mathematics, Advanced Numerical Methods, and Basic Engineering Mathematics. He has also been working as Scholar of Tsinghua University, Beijing.

Prabir Panja is Assistant Professor in the Department of Applied Science and Humanities, Haldia Institute of Technology, Haldia, West Bengal, India. He has completed his Ph.D. degree in mathematical biology from Vidyasgar University, West Bengal, India, in 2017, and M.Sc. in applied mathematics from Vidyasagar University, West Bengal, in 2012, with specialization in operations research. He teaches operations research, numerical analysis, engineering mathematics, discrete mathematics to the undergraduate and postgraduate students. His research areas are mathematical modeling of ecological and epidemiological problems.

Dr. Panja has published more than 25 papers in different reputed international journals, such as Nonlinear Dynamics, Chaos, Solitons and Fractals, Computational and Applied Mathematics, International Journal of Biomathematics, Theory in Biosciences, International Journal of Nonlinear Sciences, and Numerical Simulation. He has contributed two chapters in the book, Mathematical Modeling and Soft Computing in Epidemiology. He is also a reviewer for Applied Mathematics and Computation, Mathematical Biosciences, Nonlinear Dynamics, and International Journal of Biomathematics. He has organized two international conferences: International Conference on Information Technology and Applied Mathematics (in 2019) and International Conference on Engineering Mathematics and Computing (in 2020).

Mohd Helmy Abd Wahab is Senior Lecturer and former Head of Multimedia Engineering Lab and Intelligent System Lab in the Department of Computer Engineering, Faculty of Electrical and Electronic Engineering, Universiti Tun Hussein Onn Malaysia (UTHM) in 2014 and 2009, respectively. He also is Principle Research Fellow, Faculty of Electrical and Electronic Engineering, Advanced Telecommunication Research Center, Japan. Earlier, he was a visiting research fellow at the Center of Excellence on Geopolymer and Green Technology under cluster green ICT, Universiti Malaysia Perlis (UniMAP) from 2018 to 2020. He received Bachelor of Information Technology with Honors from Universiti Utara Malaysia and Master of Science (Intelligent System) from the same university in 2002 and 2004, respectively. He has authored/co-authored two books on database and WAP application and published more than 100 technical papers in conferences and peer-reviewed journals. He also has served as Editor-in-Chief in Advances in Computing and Intelligent System Journal, and the Guest Editor for special issue in "Wireless and Mobile Networks" in the International Journal of Advanced Computer Science and Applications, and Deputy Editor-in-Chief for the International Journal of Software Engineering and Computing. He has also been active committee member on various international conferences, editorial team and has been active manuscript reviewer.

Dr. Wahab has completed several research grants and won several medals in research and innovation showcases. He also has received several teaching awards. He was awarded the 2nd Runner Up for Grand Prize Award Category Innovation

Video Challenge for the project "Online Learning System with Authentication and Identification Mechanism using Neuro-Fuzzy Algorithm," and for the same project, he also received the 1st Grand Prize Award for FB Post and Contest as well as gold medals, respectively at Asia International Innovation Exhibition 2020 (AIINEX2020) organized by Connection Asia and UNIKL. He also received an international award (gold medal) for the project, "Portable Dwi-Switch Home Control Appliance for Elderly and Disabled People," in 2020, in Bangkok, International Intellectual Property, Invention, Innovation and Technology Exposition (IPITEx 2020) in conjunction with Thailand Inventors' Day 2020 from February 2 to 6, 2020.

Multilevel Meshfree RBF-FD Method for Elliptic Partial Differential Equations

Nikunja Bihari Barik and T. V. S. Sekhar

Abstract In this work, two multilevel augmented local radial basis function (RBF-FD) meshless algorithms have been developed to enhance the convergence rate of the RBF-FD scheme. The developed algorithm is tested with the 2D Poisson equation and convection-diffusion equation. The developed schemes save 60.1% and 79.9% of the CPU time for the Poisson equation and 76.4% and 76.4% of the CPU time for the convection-diffusion equation with respect to the usual local RBF method by the v-cycle and bootstrap multilevel RBF-FD methods, respectively. The iteration matrix of the enhanced local RBF method satisfies the necessary and sufficient conditions for convergence.

Keywords Multilevel local RBF method · Elliptic partial differential equations · Convection diffusion equation · CPU time

1 Introduction

It is well known that meshfree methods like RBF methods handle complex geometries efficiently. If the domain is large, the global RBF collocation methods may fail to give a converged solution due to ill-conditioning of the resulting algebraic system of equations. Sometimes, convergence can be achieved at the cost of instability. The RBF-based local method proposed by [1–3] is a better conditioned linear system and offers more flexibility for nonlinearities. Chandini and Sanyasiraju [4] applied this method for solving non-linear convection-diffusion equations. Generally, RBF

N. B. Barik (✉)
Department of Mathematics, VIT-AP University, Amaravati, Vijayawada 522237,
Andhra Pradesh, India
e-mail: nikunja.b@vitap.ac.in

T. V. S. Sekhar
School of Basic Sciences, Indian Institute of Technology Bhubaneswar,
Bhubaneswar 752050, India
e-mail: sekhartvs@iitbbs.ac.in

© The Author(s), under exclusive license to Springer Nature Singapore Pte Ltd. 2023
P. Gyei-Kark et al. (eds.), *Engineering Mathematics and Computing*,
Studies in Computational Intelligence 1042,
https://doi.org/10.1007/978-981-19-2300-5_1

methods are not computationally efficient when compared to grid-based methods such as finite difference methods, finite volume methods and finite element methods. Meshfree methods involve the computation of derivatives which requires weights of supporting nodes as well as square root calculations which reduces computational efficiency. The calculation of weights in turn requires matrix inversion. More nodes have to be used to improve the accuracy of the solution, which further increases the bandwidth of matrices of the system of equations that correspond to the governing partial differential equations. Therefore, the solution procedure slows down due to the relatively dense matrix equations further reducing computational efficiency. To make the RBF numerical scheme efficient, Ding et al. [5] combined the conventional FD scheme with meshfree least square-based finite differences (MLSFD). In a similar manner, Javed et al. [6] used a hybrid scheme which combines RBF-FD with conventional FD schemes. However, the generation of an efficient and effective mesh is a significantly tedious and time-consuming task although such a mesh ensures accurate results. Some researchers are developing efficient meshfree schemes by combining other meshfree methods and mesh-based methods which removes the original advantage of the meshfree method. Recently, without merging with other numerical schemes, an efficient local RBF method based on Hyman [7] was developed by Barik and Sekhar [8–10]. Therefore, an appropriate extension of the local RBF meshless scheme is devised in such a way that the convergence rate of the scheme is enhanced. The aim of the paper is to develop multilevel local RBF methods without the help of other meshfree methods in order to get minimized CPU time with the same order of accuracy as the usual local RBF method.

2 Multilevel RBF-FD Formulation

The RBF-FD interpolation for a set of m distinct data points $\mathbf{x}_j$, and corresponding data values f_j, $j = 1, 2, \ldots, m$ is given by

$$s(\mathbf{x}) = \sum_{j=1}^{m} \lambda_j \phi(\|\mathbf{x} - \mathbf{x}_j\|) + \sum_{k=1}^{N} \alpha_k p_k(\mathbf{x}), \ \mathbf{x} \in \mathbb{R}^d \tag{1}$$

where $\{p_k(\mathbf{x})\}_{k=1}^{N}$ is a basis for $\Pi_n(\mathbb{R}^d)$ (space of all d-variate polynomials with degree less than or equal to n) and $s(\mathbf{x}_j) = f(\mathbf{x}_j)$. N extra conditions are required to solve the linear system formed by the above interpolation. The orthogonal conditions are chosen by taking the expansion coefficient vector $\check{} \in \mathbb{R}^m$ and $\Pi_n(\mathbb{R}^d)$ to get the extra conditions.

That is, $\sum_{j=1}^{m} \lambda_j p_k(\mathbf{x}_j) = 0, \ k = 1, 2, \ldots, N$.

By using all supporting node points, we will get the following symmetric linear system:

$$\left(\begin{array}{c|c} B & p \\ \hline p^T & 0 \end{array}\right) \left(\begin{array}{c} \lambda \\ \alpha \end{array}\right) = \left(\begin{array}{c} \mathbf{f} \\ 0 \end{array}\right) \qquad (2)$$

B is the $m \times m$ matrix with elements $b_{ij} = \phi(\|\mathbf{x}_i - \mathbf{x}_j\|)$, $j = 1, \ldots, m$, $i = 1, \ldots, m$ and p is the $m \times N$ matrix with elements $p_k(\mathbf{x}_j)$ for $k = 1, 2, \ldots, N$.

The derivatives of the function at a given point are calculated by using Lagrange form of RBF interpolant. The interpolant is given by

$$s(\mathbf{x}) = \sum_{j=1}^{m} \psi_j(\mathbf{x}) u(\mathbf{x_j}) \qquad (3)$$

where $\psi_j(\mathbf{x})$ satisfies the cardinal conditions

$$\psi_j(\mathbf{x}_k) = \delta_{jk} = \begin{cases} 1, & \text{if } j = k \\ 0, & \text{if } j \neq k \end{cases} \quad k = 1, 2, \ldots, m \qquad (4)$$

Closed form representation for $\psi_j(\mathbf{x})$ can be obtained by considering that the right-hand side vector of (6) stems from each ψ_j's. Then by Cramer's rule on (6)–(3) gives

$$\psi_j(\mathbf{x}) = \frac{det(Q_j(\mathbf{x}))}{det(Q)} \qquad (5)$$

where

$$Q = \left(\begin{array}{c|c} B & p \\ \hline p^T & 0 \end{array}\right) \qquad (6)$$

and $Q_j(\mathbf{x})$ is the same as Q, except that the jth row is replaced by the vector $A(\mathbf{x})$

$$A(\mathbf{x}) = [\phi_1 \ \phi_2 \ \cdots \ \phi_m \mid p_1(\mathbf{x}) \ p_2(\mathbf{x}) \ \cdots \ p_n(\mathbf{x})] \qquad (7)$$

Here, ϕ_j is the RBF interpolation for jth node, and $p_j(\mathbf{x}) \in \Pi_m(\mathbb{R}^d)$.

To approximate the derivative of a function at a given point, the derivation from (3)–(7) can be used. The linear differential operator of a function u at a given point $\mathbf{x_i}$ is $l(u(\mathbf{x_i}))$ can be calculated by using the neighborhood points of $\mathbf{x_i}$ (say n_i nodes $(\mathbf{x_1}, \ \mathbf{x_2}, \ \ldots, \ \mathbf{x_{m_i}})$. Then

$$l(u(\mathbf{x_i})) \approx \sum_{j=1}^{m_i} c_{ij} u(\mathbf{x_j}), \ i = 1, 2, \ldots, m \qquad (8)$$

By applying Lagrange RBF interpolation (3),

$$l(u(\mathbf{x_i})) \approx l(s(\mathbf{x_i})) = \sum_{j=1}^{m_i} l(\psi_j(\mathbf{x_i})) u(\mathbf{x_j}) \qquad (9)$$

From (8) and (9),

$$c_{ij} = l(\psi_j(\mathbf{x_i})), \quad j = 1, 2, \ldots, m_i, \ i = 1, 2, \ldots, m$$

The weights are computed by solving the linear system:

$$\left(\frac{B \mid p}{p^T \mid 0} \right)_i \left(\frac{C}{\mu} \right)_i = \left(\frac{(l(A(\mathbf{x})))^T}{0} \right)_i \tag{10}$$

where B is the part of the coefficient matrix of Eq. (2), $A(\mathbf{x})$ is the row vector in (7) and μ is a vector related to α in (1) and $C = [c_1, c_2, \ldots, c_{mi}]'$. By using the values of C in (8), we will get an equation on $u_j, \ j = 1, 2, \ldots, m_i$. These m_i points are some points from $u_i, \ i = 1, 2, \ldots, m$ which are nearer to the ith point. The method is tested by considering $m_i = 5$ points. By applying similar procedure for all internal points, we will get a system of equations

$$SU = F \tag{11}$$

Here, we discuss two types of local RBF multilevel methods. We have taken the Fortran platform to work out the numerical examples.

2.1 Bootstrap Local Refinement Multilevel RBF-FD Method

1. Solve the algebraic system of linear equations $S^1 U^1 = F^1$, obtained by discretizing the governing equations using the RBF-FD method until convergence, in the coarsest set of nodes (L^1) using an iterative technique such as Gauss-Seidel.
2. Prolongate the coarsest set solution to the next finer set, i.e. $\widehat{U}^2 = P(U^1)$.
3. The solution at additional points can be obtained by RBF-FD interpolation. Using this as the starting solution, achieve a convergent solution in the finer set, i.e. $S^2 U^2 = F^2$.
4. Repeat the above procedure for the next finer set and so on until the finest set and achieve a convergent solution in the finest set.

2.2 Residue-Based V-Cycle Multilevel RBF-FD Method

1. Smooth on the finest set of nodes for a few iterations (5 or 7 iterations) to get $\widehat{U}^m$ by using the iterative technique such as Gauss-Seidel.
2. Calculate residue $r^m = F^m - S\widehat{U}^m$ in the finest level.
3. By applying restriction operator repeatedly, restrict the residue r^m from the finest level to the coarsest level which is denoted as r^1.

4. Solve the error equation in the coarsest set $S^1 e^1 = r^1$ by the Gauss-Seidel method until convergence.
5. Prolongate the error from the coarsest level (e^1) to the finest level $(\widehat{e}^m)$ by repeatedly applying the prolongation operator.
6. The solution at additional points can be obtained by RBF-FD interpolation.
7. Add this error to get $U^m : U^m = \widehat{U}^m + e^m$.
8. Repeat the above v-cycle multiple times to achieve convergent solution in the desired finest level.

3 Numerical Examples

3.1 Poisson Equation

Consider the two-dimensional Poisson equation

$$u_{xx} + u_{yy} = f(x, y), \ 0 \le x, y \le 1$$

with exact solution $u(x, y) = \sin \pi x \sin \pi y$. The forcing term and boundary condition are calculated by using an analytical solution.

The multiquadric RBF interpolant is taken for computational work. Central type supporting points (Fig. 1) are considered for discretization of the PDE. That is, the five neighborhood points are chosen by considering all directions of flow. The derivatives u_x, u_y, u_{xx} and u_{yy} are calculated at ith point using Eqs. (8)–(10) as follows:

$$u_x \approx \sum_{j=1}^{n_i} b_{ij}^x u(x_j, y_j), u_y \approx \sum_{j=1}^{n_i} b_{ij}^y u(x_j, y_j), u_{xx} \approx \sum_{j=1}^{n_i} b_{ij}^{xx} u(x_j, y_j)$$

and $u_{yy} \approx \sum_{j=1}^{n_i} b_{ij}^{yy} u(x_j, y_j)$
where b_{ij}^x, b_{ij}^{xx}, b_{ij}^y and b_{ij}^{yy} are similar to c_{ij} in Eq. (8). By using the above expressions in given PDE, we will get
$\sum_{j=1}^{n_i} (b_{ij}^{xx} + b_{ij}^{yy}) u(x_j, y_j) = f(x_i, y_i)$ for ith internal point.
Here, the solution is obtained for three values of shape parameter ε. The choice of shape parameter ε is finalized by comparing with the rate of convergence. The rate of convergence is calculated for uniform levels by using the formula:

$$Rate \text{ of convergence} = \frac{log(E^h / E^{h/2})}{log(2)}$$

where E^h and $E^{h/2}$ are RMS errors with step size h and $h/2$, respectively.

The rate of converges are calculated for three values of shape parameter ε tabulated in Tables 1 and 2 by using the v-cycle multilevel method and the bootstrap method,

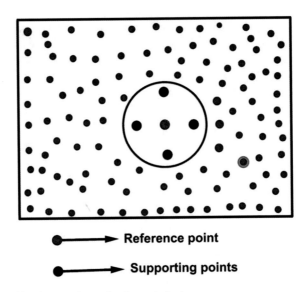

Fig. 1 Choice of local supporting nodes for central schemes

Table 1 Error and rate of convergence for Poisson equation by v-cycle multilevel RBF-FD method

Nodes	Error		
	$\varepsilon = 0.45$	$\varepsilon = 0.7$	$\varepsilon = 1.1$
10×10	3.3272097(−3)	1.5423918(−3)	2.7624057(−3)
Rate	−	−	−
19×19	8.4503595(−4)	3.9542242(−4)	7.8947091(−4)
Rate	1.98	1.96	1.81
37×37	4.5600595(−4)	**3.5928544(−5)**	1.7645840(−4)
Rate	0.89	**3.46**	2.16

Table 2 Error and rate of convergence for Poisson equation by bootstrap multilevel RBF-FD method

Nodes	Error		
	$\varepsilon = 0.5$	$\varepsilon = 0.7$	$\varepsilon = 1.1$
10×10	3.0271125(−3)	1.5424005(−3)	2.7623922(−3)
Rate	−	−	−
19×19	7.9376320(−4)	4.3595303(−4)	8.0183876(−4)
Rate	1.93	1.82	1.78
37×37	2.7665999(−4)	**1.0219010(−4)**	2.9636154(−4)
Rate	1.52	**2.09**	1.44

Table 3 Effect of v-cycle and bootstrap multilevel RBF-FD methods on CPU time for Poisson equation

No. of levels	Finest set of nodes	Coarsest set of nodes	v-cycle		Bootstrap	
			CPU time in Secs.	%	CPU time in Secs.	%
1	37 × 37	37 × 37	0.313	–	0.313	–
2	37 × 37	19 × 19	0.125	60.1	0.063	79.9
3	37 × 37	10 × 10	0.125	60.1	0.063	79.9

respectively. The rate of convergence for the v-cycle multilevel method is found to be 3.46 and 2.09 for the bootstrap method. By applying the v-cycle and bootstrap multilevel algorithm, 60.1% and 79.9% of the CPU time is saved when compared to the single finest level solution (Table 3).

3.2 Convection-Diffusion Problem

Consider the two-dimensional convection-diffusion equation

$$u_x + u_y - 0.1(u_{xx} + u_{yy}) = f(x, y), \ 0 \le x, y \le 1$$

with exact solution $u(x, y) = e^{5(x-1)(1-y)}$. The forcing term f and boundary conditions are calculated by using exact solutions. The discretized equation is

$$\sum_{j=1}^{n_i}(a_{ij}^x + a_{ij}^y - 0.1a_{ij}^{xx} - 0.1a_{ij}^{yy})u(x_j, y_j) = f(x_i, y_i)$$

for ith internal point.

The rate of convergence of the bootstrap local RBF method and v-cycle multilevel method are found to be 2.9 (Table 4) and 2.1 (Table 5), respectively. We are able to save 74.6% by applying both v-cycle and bootstrap multilevel RBF-FD methods (Table 6).

4 Convergence Analysis

After discretizing the governing PDE by the RBF-FD scheme, the system of linear equations will be in the form $AX = b$.

Theorem 1 ([11]) *A necessary and sufficient condition for convergence of an iterative method of the form* $\mathbf{x}^{(k+1)} = \mathbf{H}\mathbf{x}^{(k)} + \mathbf{c}, \ k = 0, 1, 2, \ldots$ *is that the eigenvalues of*

Table 4 Error and rate of convergence for convection-diffusion problem by bootstrap multilevel RBF-FD method

Nodes	Error		
	$\varepsilon = 0.5$	$\varepsilon = 1.0$	$\varepsilon = 1.6$
9×9	5.8692284(−3)	6.3706040(−3)	7.4418020(−3)
Rate	−	−	−
17×17	1.7575537(−3)	1.8084745(−3)	2.1762254(−3)
Rate	1.7	1.8	1.8
33×33	7.7410642(−4)	6.9674077(−4)	**3.003884(−4)**
Rate	1.2	1.4	**2.9**

Table 5 Error and rate of convergence for convection-diffusion equation by v-cycle multilevel RBF-FD method

Nodes	Error		
	$\varepsilon = 0.5$	$\varepsilon = 1.0$	$\varepsilon = 1.6$
9×9	5.8692782(−3)	6.3705789(−3)	7.4417945(−3)
Rate	−	−	−
17×17	1.6984533(−3)	1.7611138(−3)	2.1991490(−3)
Rate	1.8	1.8	1.8
33×33	4.4051811(−4)	6.3654530(−4)	**4.9909949(−4)**
Rate	1.9	1.5	**2.1**

Table 6 Effect of v-cycle and bootstrap multilevel RBF-FD methods on CPU time for convection-diffusion problem

No. of levels	Finest set of nodes	Coarsest set of nodes	v-cycle		Bootstrap	
			CPU time in Secs.	%	CPU time in Secs.	%
1	33×33	33×33	0.063	−	0.063	−
2	33×33	17×17	0.047	25.4	0.031	50.8
3	33×33	9×9	0.016	74.6	0.016	74.6

the iteration matrix satisfy $|\lambda_i(\mathbf{H})| < 1$, $i = 1(1)n$, where $\mathbf{H}$ is the iterative matrix of the system $AX = b$.

The spectral radius of the iterative matrix is calculated for Poisson equation and convection-diffusion equations and tabulated in Tables 7 and 8, respectively. It is found that the spectral radius is less than one, which satisfies the necessary and sufficient convergence condition for a stationary iterative method. Hence, our results through multilevel RBF-FD are convergent.

Table 7 Spectral radius for Poisson problem

Multilevel methods	10×10	19×19	37×37
v-cycle	0.8830222	0.9698463	0.9924038
Bootstrap	0.8830222	0.9698463	0.9924038

Table 8 Spectral radius for 2D convection-diffusion problem

Multilevel methods	9×9	17×17	33×33
v-cycle	0.8602205	0.9378465	0.9743825
Bootstrap	0.8602203	0.9387945	0.9756354

5 Conclusions

Two multilevel augmented local radial basis function (RBF-FD) meshless algorithms have been developed and implemented with the 2D Poisson equation and convection-diffusion equation. The developed schemes save 60.1% and 79.9% of the CPU time for the Poisson equation and 74.6% and 74.6% of the CPU time for the convection-diffusion equation with respect to the usual local RBF method by the v-cycle and boot-strap multilevel RBF-FD methods, respectively. The iteration matrix of the enhanced local RBF method satisfies the necessary and sufficient conditions for convergence.

References

1. Shu, C., Ding, H., Yeo, K.S.: Local radial basis function-based differential quadrature method and its application to solve two-dimensional incompressible Navier-Stokes equations. Comput. Methods Appl. Mech. Eng. **192**(7), 941–954 (2003)
2. Tolstykh, A.I., Lipavskii, M.V., Shirobokov, D.A.: High-accuracy discretization methods for solid mechanics. Arch. Mech. **55**(5–6), 531–553 (2003)
3. Wright, G.B., Fornberg, B.: Scattered node compact finite difference-type formulas generated from radial basis functions. J. Comput. Phys. **212**(1), 99–123 (2006)
4. Chandhini, G., Sanyasiraju, Y.V.S.S.: Local RBF-FD solutions for steady convection-diffusion problems. Int. J. Numer. Meth. Eng. **72**(3), 352–378 (2007)
5. Ding, H., Shu, C., Yeo, K.S., Xu, D.: Simulation of incompressible viscous flows past a circular cylinder by hybrid FD scheme and meshless least square-based finite difference method. Comput. Methods in Appl. Mech. Eng. **193**(9–11), 727–744 (2004)
6. Javed, A., Djidjeli, K., Xing, J.T., Cox, S.J.: A hybrid mesh free local RBF-cartesian FD scheme for incompressible flow around solid bodies. Int. J. Math. Comput. Phys. Electr. Comput. Eng. **7**, 957–966 (2013)
7. Hyman, J.M.: Mesh refinement and local inversion of elliptic partial differential equations. J. Comput. Phys. **23**(2), 124–134 (1977)
8. Barik, N.B., Sekhar, T.V.S.: An efficient local RBF meshless scheme for steady convection-diffusion problems. Int. J. Comput. Methods **14**(06), 1750064 (2017)

9. Barik, N.B., Sekhar, T.V.S.: A novel RBF-FD meshless scheme in curvilinear geometry for unbounded flows. Int. J. Comput. Methods Eng. Sci. Mech. **18**(4–5), 209–219 (2017)
10. Barik, N.B., Sekhar, T.V.S.: Accommodative FAS-FMG multilevel based meshfree augmented RBF-FD method for Navier-Stokes equations in spherical geometry, mathematics and computing. ICMC 2017. Commun. Comput. Inf. Sci. **655**, 141–151 (2017)
11. Jain, M.K., Iyengar, S.R.K., Jain, R.K.: Numerical Methods for Scientific and Engineering Computation, New Age International. (P) Limited, New Delhi, India (2003)

Some Fixed Point Theorems in Fuzzy Strong b-Metric Spaces

S. Chatterjee, Arunima Majumder, and T. Bag

Abstract In this paper, some fixed point theorems are established in strong fuzzy b-metric space for different types of generalized contraction mappings.

Keywords ψ-contractive mapping · Fixed point

1 Introduction

In 1975, Kramosil and Michalek [10] first introduced the concept of a fuzzy metric space which is a generalization of the statistical metric space, and this work provides an important basis for the construction of fixed point theory in fuzzy metric spaces. Again b-metric spaces, as a generalization of metric spaces, were introduced by Bhaktin [2] and Czerwik [3]. Several authors considered different types of contraction mappings and established many fixed point theorems in Kramosil and Michalek [10] type fuzzy metric spaces as well as in b-metric spaces (for references, please see [5–8, 11, 12, 14, 15, 20, 21]).

In the last few years, the concept of some generalized fuzzy metric spaces, viz. G-fuzzy metric space and D^*-fuzzy metric space have been introduced by different authors and some fixed point theorems have been developed in such spaces (for reference, please see [16, 18]).

A study on fuzzy strong b-metric space is a recent development. The idea of fuzzy b-metric space was introduced by Saddati [17]. It is the generalization of Kramosil and Michalek [10] as well as George and Veeramoni [4] type fuzzy metric. Following the definition of fuzzy b-metric space, the idea of fuzzy strong b-metric space has been introduced by Oner [13] and he developed some basic fundamental topological results.

Supported by organization UGC-SAP (DRS, Phase-III) with sanction order No. F.510/3/DRS-III/2015(SAPI).

S. Chatterjee · A. Majumder · T. Bag (✉)
Department of Mathematics, Visva-Bharati, Santiniketan 731235, West Bengal, India
e-mail: tarapadavb@gmail.com

The main motive of this paper is to develop some fixed point theorems for different types of contractive mappings in fuzzy strong b-metric spaces. Here, we consider one generalized contraction mapping and another is ψ-contractive mapping which is defined in the paper [12].

The organization of the paper is as follows.

Section 2 provides some preliminary results which are used to study the results in this paper. One fixed point theorem is established for generalized contraction mapping in Sect. 3. In Sect. 4, another fixed point theorem is proved for ψ-contractive mapping.

2 Preliminaries

Definition 1 ([9]) An ordered triplet (X, D, K) is called strong b-metric space and D is called strong b-metric on X if X is a non-empty set, $K \geq 1$ is a given real number and $D : X \times X \to [0, \infty)$ satisfies the following conditions for all $x, y, z \in X$:
(1) $D(x, y) = 0$ if and only if $x = y$,
(2) $D(x, y) = D(y, x)$,
(3) $D(x, z) \leq D(x, y) + K D(y, z)$.

Definition 2 ([13]) Let X be a non-empty set, $K \geq 1$, $*$ be a continuous t-norm and M be a fuzzy set on $X \times X \times (0, \infty)$ such that for all $x, y, z \in X$ and $t, s > 0$,

(1) $M(x, y, t) > 0$,
(2) $M(x, y, t) = 1$ iff $x = y$,
(3) $M(x, y, t) = M(y, x, t)$,
(4) $M(x, y, t) * M(y, z, s) \leq M(x, z, t + Ks)$,
(5) $M(x, y, .) : (0, \infty) \to [0, 1]$ is continuous.

Then M is called a fuzzy strong b-metric on X and $(X, M, *, K)$ is called a fuzzy strong b-metric space.

Note 1 By (3) of Definition 2, it can be easily checked that '$M(x, y, s) * M(y, z, t) \leq M(x, z, t + Ks)$'—this form of the inequality is also valid in a fuzzy strong b-metric space.

Example 1 ([13]) Let (X, D, K) be a strong b-metric space. Define $M_D(x, y, t) = \frac{t}{t+D(x,y)}$ for $t > 0$ and $x, y \in X$. Then $(X, M_D, *, K)$ where $* = product$ is a fuzzy strong b-metric space and is called standard fuzzy strong b-metric space induced by D.

Note 2 If we choose $* = min$, then $(X, M_D, *, K)$ is also a fuzzy strong b-metric space.

Definition 3 ([13]) Let $(X, M_D, *, K)$ be a fuzzy strong b-metric space, $x \in X$ and $\{x_n\}$ be a sequence in X. Then

(i) $\{x_n\}$ is said to converge to x if for any $t > 0$ and for any $r \in (0, 1)$, there exists a natural number n_0 such that $M(x_n, x, t) > 1 - r \; \forall \, n \geq n_0$.

(ii) $\{x_n\}$ is said to be a Cauchy sequence if for any $t > 0$ and for any $r \in (0, 1)$, there exists a natural number n_0 such that $M(x_n, x_m, t) > 1 - r \; \forall \, n, m \geq n_0$.

In the section "Fixed point theorem for contractive mapping", the fixed point theorem is motivated from the following theorem.

Theorem 1 ([1]) *Let (X, d) be a complete b-metric space. Let T be a mapping $T : X \to X$ such that*

$$d(Tx, Ty) \leq a \, max\{d(x, Tx), d(y, Ty), d(x, y)\} + b\{d(x, Ty) + d(y, Tx)\}$$

(1)

where $a, b > 0$ such that $a + 2bs \leq 1 \; \forall x, y \in X$ and $s \geq 1$ then T has a unique fixed point.

The motivation of the next fixed point theorem is ψ-contractive mapping in fuzzy metric space given by Wang [21]. Since fuzzy strong b-metric is a generalization of fuzzy metric space so, our result generalized the results of Wang [21].

Theorem 2 ([21]) *Let $(X, M, *)$ be an M-complete fuzzy metric space in the sense of Kramosil and Michalek with $*$ positive, and $f : X \to X$ be a fuzzy ψ-contractive mapping. If there is $x \in X$ such that $M(x, fx, t) > 0 \; \forall \, t > 0$, then f has a fixed point in X.*

3 Fixed Point Theorem for Contractive Type Mapping

Proposition 1 *Let (X, M, min, K) be a fuzzy strong b-metric space and $\{x_n\}$ be a sequence in X. Then $\bigwedge\{t > 0 : M(x_{n-1}, x_{n+1}, t) > 1 - \alpha\} \leq \bigwedge\{t > 0 : M(x_{n-1}, x_n, t) > 1 - \alpha\} + K \bigwedge\{t > 0 : M(x_n, x_{n+1}, t) > 1 - \alpha\}$ for each $\alpha \in (0, 1)$.*

Proof. Note that $M(x, z, s + kt) > M(x, y, s) * M(y, z, t)$ $\qquad (K > 1)$. Here $* = min$.

We have $\bigwedge\{t > 0 : M(x_{n-1}, x_n, \frac{t}{2}) > 1 - \alpha\} + \bigwedge\{t > 0 : M(x_n, x_{n+1}, \frac{t}{2K}) > 1 - \alpha\}$

$= 2 \bigwedge\{\frac{t}{2} > 0 : M(x_{n-1}, x_n, \frac{t}{2}) > 1 - \alpha\} + 2 \bigwedge\{\frac{t}{2} > 0 : M(x_n, x_{n+1}, \frac{t}{2K}) > 1 - \alpha\}$

$= 2[\bigwedge\{\frac{t}{2} > 0 : M(x_{n-1}, x_n, \frac{t}{2}) > 1 - \alpha\} + \bigwedge\{\frac{t}{2} > 0 : M(x_n, x_{n+1}, \frac{t}{2K}) > 1 - \alpha\}]$

$= 2 \bigwedge\{t > 0 : M(x_{n-1}, x_n, \frac{t}{2}) > 1 - \alpha, M(x_n, x_{n+1}, \frac{t}{2K}) > 1 - \alpha\}$

$\geq 2 \bigwedge\{t > 0 : M(x_{n-1}, x_{n+1}, t) > 1 - \alpha\}$

$\Rightarrow \bigwedge\{t > 0 : M(x_{n-1}, x_{n+1}, t) > 1 - \alpha\} \leq \frac{1}{2}[\bigwedge\{t > 0 : M(x_{n-1}, x_n, \frac{t}{2}) > 1 - \alpha\} + \bigwedge\{t > 0 : M(x_n, x_{n+1}, \frac{t}{2K}) > 1 - \alpha\}]$

$$\Rightarrow \bigwedge\{t > 0 : M(x_{n-1}, x_{n+1}, t) > 1 - \alpha\} \leq \bigwedge\{\tfrac{t}{2} > 0 : M(x_{n-1}, x_n, \tfrac{t}{2}) >$$
$$1 - \alpha\} + K \bigwedge\{\tfrac{t}{2K} > 0 : M(x_n, x_{n+1}, \tfrac{t}{2K}) > 1 - \alpha\}$$
$$\Rightarrow \bigwedge\{t > 0 : M(x_{n-1}, x_{n+1}, t) > 1 - \alpha\} \leq \bigwedge\{t > 0 : M(x_{n-1}, x_n, t) >$$
$$\alpha\} + K \bigwedge\{t > 0 : M(x_n, x_{n+1}, t) > 1 - \alpha\}.$$

Proposition 2 *Let* $(X, M, *, K)$ *be a fuzzy strong b-metric space and* $\{x_n\}$ *be a sequence in X. If* $x_n \to x^*$, *then* $\lim_{n\to\infty} \bigwedge\{t > 0 : M(x_n, x^*, t) > 1 - \alpha\} = 0 \ \forall \alpha \in (0, 1).$

Proof. Since $x_n \to x^*$, thus
$$\lim_{n\to\infty} M(x_n, x^*, t) = 1 \quad \forall t > 0$$
$$\Rightarrow \lim_{n\to\infty} M(x_n, x^*, t) > 1 - \alpha \quad \forall t > 0 \quad \forall \alpha \in (0, 1).$$
So for a given $\epsilon > 0$, for each $\alpha \in (0, 1)$, $\exists$ a natural number $n_0(\alpha)$ such that
$$M(x_n, x^*, \epsilon) > 1 - \alpha \quad \forall n > n_0(\alpha) \quad \forall \alpha \in (0, 1)$$
$$\Rightarrow \bigwedge\{t > 0 : M(x_n, x^*, t) > 1 - \alpha\} \leq \epsilon \quad \forall n > n_0(\alpha) \quad \forall \alpha \in (0, 1).$$
$$\Rightarrow \lim_{n\to\infty} \bigwedge\{t > 0 : M(x_n, x^*, t) > 1 - \alpha\} \leq \epsilon \quad \forall \alpha \in (0, 1).$$
Since $\epsilon > 0$ is arbitrary, we have
$$\Rightarrow \lim_{n\to\infty} \bigwedge\{t > 0 : M(x_n, x^*, t) > 1 - \alpha\} = 0 \quad \forall \alpha \in (0, 1).$$

Theorem 3 *Let* $(X, M, *, K)$ *be a complete strong fuzzy b-metric space where* $* = min$. *Let* $T : X \to X$ *be such that*

$$\bigwedge\{t > 0 : M(Tx, Ty, t) > 1 - \alpha\} \leq a \ \max\{\bigwedge\{t > 0 : M(x, Tx, t) > 1 - \alpha\},$$
$$\bigwedge\{t > 0 : M(y, Ty, t) > 1 - \alpha\}, \bigwedge\{t > 0 : M(x, y, t) > 1 - \alpha\}\}$$
$$+ b\{\bigwedge\{t > 0 : M(x, Ty, t) > 1 - \alpha\} + \bigwedge\{t > 0 : M(y, Tx, t) > 1 - \alpha\}\} \forall \alpha \in (0, 1), \forall x, y \in X$$
$$(2)$$

where $a, b > 0$ *is such that* $a + (1 + K)b < 1$ *and* $K > 1$.
        Then T has a unique fixed point.

Proof. Let $x_0 \in X$ and $\{x_n\}$ be a sequence in X defined by $x_n = Tx_{n-1} = T^n x_0$, $n = 1, 2, 3, \cdots$
    We have
$$\bigwedge\{t > 0 : M(x_n, x_{n+1}, t) > 1 - \alpha\} = \bigwedge\{t > 0 : M(Tx_{n-1}, Tx_n, t) > 1 - \alpha\}$$
$$\leq a \ \max\{\bigwedge\{t > 0 : M(x_{n-1}, Tx_{n-1}, t) > 1 - \alpha\}, \bigwedge\{t > 0 : M(x_n, Tx_n, t) > 1 - \alpha\},$$
$$\bigwedge\{t > 0 : M(x_{n-1}, x_n, t) > 1 - \alpha\}\} + b\{\bigwedge\{t > 0 : M(x_{n-1}, Tx_n, t) > 1 - \alpha\} + \bigwedge\{t > 0 : M(x_n, Tx_{n-1}, t) > 1 - \alpha\}\}$$
    (as T satisfies Inequality (2))
$$\Rightarrow \bigwedge\{t > 0 : M(x_n, x_{n+1}, t) > 1 - \alpha\} \leq a \ \max\{\bigwedge\{t > 0 : M(x_{n-1}, x_n, t) > 1 - \alpha\}, \bigwedge\{t > 0 : M(x_n, x_{n+1}, t) > 1 - \alpha\}, \bigwedge\{t > 0 : M(x_{n-1}, x_n, t) > 1 - \alpha\}\} + b\{\bigwedge\{t > 0 : M(x_{n-1}, x_{n+1}, t) > 1 - \alpha\} + \bigwedge\{t > 0 : M(x_n, x_n, t) > 1 - \alpha\}\}$$

$$\Rightarrow \bigwedge\{t > 0 : M(x_n, x_{n+1}, t) > 1 - \alpha\} \le a \quad max\{\bigwedge\{t > 0 : M(x_{n-1}, x_n, t) > 1 - \alpha\}, \bigwedge\{t > 0 : M(x_n, x_{n+1}, t) > 1 - \alpha\}\} + b \bigwedge\{t > 0 : M(x_{n-1}, x_{n+1}, t) > 1 - \alpha\}$$

$$\Rightarrow \bigwedge\{t > 0 : M(x_n, x_{n+1}, t) > 1 - \alpha\}$$

$$\le aM_1 + b \bigwedge\{t > 0 : M(x_{n-1}, x_n, t) > 1 - \alpha\}$$

$$+ bK \bigwedge\{t > 0 : M(x_n, x_{n+1}, t) > 1 - \alpha\} \quad \forall \alpha \in (0, 1) \tag{3}$$

where $M_1 = max\{\bigwedge\{t > 0 : M(x_{n-1}, x_n, t) > 1 - \alpha\}, \bigwedge\{t > 0 : M(x_n, x_{n+1}, t) > 1 - \alpha\}\}$.

Now two cases may arise

Case I Suppose $M_1 = \bigwedge\{t > 0 : M(x_n, x_{n+1}, t) > 1 - \alpha\}$. Then we have

$$\bigwedge\{t > 0 : M(x_n, x_{n+1}, t) > 1 - \alpha\} \le a \bigwedge\{t > 0 : M(x_n, x_{n+1}, t) > 1 - \alpha\} + b \bigwedge\{t > 0 : M(x_{n-1}, x_n, t) > 1 - \alpha\} + bK \bigwedge\{t > 0 : M(x_n, x_{n+1}, t) > 1 - \alpha\}$$

$$\Rightarrow (1 - a - bK) \bigwedge\{t > 0 : M(x_n, x_{n+1}, t) > 1 - \alpha\} \le b \bigwedge\{t > 0 : M(x_{n-1}, x_n, t) > 1 - \alpha\}$$

$$\Rightarrow \bigwedge\{t > 0 : M(x_n, x_{n+1}, t) > 1 - \alpha\} \le \frac{b}{1-a-bK} \bigwedge\{t > 0 : M(x_{n-1}, x_n, t) > 1 - \alpha\}$$

$$\Rightarrow \bigwedge\{t > 0 : M(x_n, x_{n+1}, t) > 1 - \alpha\} \le r \bigwedge\{t > 0 : M(x_{n-1}, x_n, t) > 1 - \alpha\}$$

where $r = \frac{b}{1-a-bK} < 1$.

Continuing this process we get

$$\bigwedge\{t > 0 : M(x_n, x_{n+1}, t) > 1 - \alpha\} \le r^n \bigwedge\{t > 0 : M(x_0, x_1, t) > 1 - \alpha\}. \tag{4}$$

Case II Suppose $M_1 = \bigwedge\{t > 0 : M(x_{n-1}, x_n, t) > 1 - \alpha\}$. Then we have

$$\bigwedge\{t > 0 : M(x_n, x_{n+1}, t) > 1 - \alpha\} \le a \bigwedge\{t > 0 : M(x_{n-1}, x_n, t) > 1 - \alpha\} + b \bigwedge\{t > 0 : M(x_{n-1}, x_n, t) > 1 - \alpha\} + bK \bigwedge\{t > 0 : M(x_n, x_{n+1}, t) > 1 - \alpha\}$$

$$\Rightarrow (1 - bK) \bigwedge\{t > 0 : M(x_n, x_{n+1}, t) > 1 - \alpha\} \le (a + b) \bigwedge\{t > 0 : M(x_{n-1}, x_n, t) > 1 - \alpha\}$$

$$\Rightarrow \bigwedge\{t > 0 : M(x_n, x_{n+1}, t) > 1 - \alpha\} \le \frac{a+b}{1-bK} \bigwedge\{t > 0 : M(x_{n-1}, x_n, t) > 1 - \alpha\}$$

$$\Rightarrow \bigwedge\{t > 0 : M(x_n, x_{n+1}, t) > 1 - \alpha\} \le s \bigwedge\{t > 0 : M(x_{n-1}, x_n, t) > 1 - \alpha\}$$

where $s = \frac{a+b}{1-bK} < 1$.

Continuing this process we have

$$\bigwedge\{t > 0 : M(x_n, x_{n+1}, t) > 1 - \alpha\} \le s^n \bigwedge\{t > 0 : M(x_0, x_1, t) > 1 - \alpha\}. \tag{5}$$

Now we show that $\{x_n\}$ is a Cauchy sequence in X.

For $m > n$, we have

$$\bigwedge\{t > 0 : M(x_n, x_m, t) > 1 - \alpha\} \le \bigwedge\{t > 0 : M(x_n, x_{n+1}, t) > 1 - \alpha\} + K \bigwedge\{t > 0 : M(x_{n+1}, x_m, t) > 1 - \alpha\}$$

$$\leq \bigwedge\{t > 0 : M(x_n, x_{n+1}, t) > 1 - \alpha\} + K \bigwedge\{t > 0 : M(x_{n+1}, x_{n+2}, t) > 1 - \alpha\} + K^2 \bigwedge\{t > 0 : M(x_{n+2}, x_m, t) > 1 - \alpha\}$$

$$\leq \bigwedge\{t > 0 : M(x_n, x_{n+1}, t) > 1 - \alpha\} + K \bigwedge\{t > 0 : M(x_{n+1}, x_{n+2}, t) > 1 - \alpha\} + K^2 \bigwedge\{t > 0 : M(x_{n+2}, x_{n+3}, t) > 1 - \alpha\} + K^3 \bigwedge\{t > 0 : M(x_{n+3}, x_m, t) > 1 - \alpha\}$$

$$\leq (r^n + K r^{n+1} + K^2 r^{n+2} + K^3 r^{n+3} + ...) \bigwedge\{t > 0 : M(x_0, x_1, t) > 1 - \alpha\}$$

$$= r^n\{1 + (Kr) + (Kr)^2 + (Kr)^3 + ...\} \bigwedge\{t > 0 : M(x_0, x_1, t) > 1 - \alpha\}$$

$$\Rightarrow \lim_{m,n \to \infty} \bigwedge\{t > 0 : M(x_n, x_m, t) > 1 - \alpha\} = 0 \ (\because r < 1) \ \forall \alpha \in (0, 1) \qquad (6)$$

Choose $\epsilon > 0$. Then for each $\alpha \in (0, 1)$, $\exists (+ve)$ integer $n_0(\alpha)$ such that

$$\bigwedge\{t > 0 : M(x_n, x_m, t) > 1 - \alpha\} < \epsilon \ \forall m, n > n_0(\alpha)$$

$$\Rightarrow M(x_n, x_m, \epsilon) > 1 - \alpha \quad \forall m, n > n_0(\alpha)$$

$$\Rightarrow \lim_{m,n \to \infty} M(x_n, x_m, \epsilon) > 1 - \alpha \quad \forall \alpha \in (0, 1)$$

$$\Rightarrow \lim_{m,n \to \infty} M(x_n, x_m, \epsilon) = 1.$$

Since $\epsilon > 0$ is arbitrary,

$$\lim_{m,n \to \infty} M(x_n, x_m, \epsilon) = 1 \quad \forall t > 0.$$

Thus $\{x_n\}$ is a Cauchy sequence in X.

Since X is complete, $\{x_n\}$ converges to some $x^* \in X$.

Now we show that x^* is the fixed point of T.

We have

$$\bigwedge\{t > 0 : M(x^*, Tx^*, t) > 1 - \alpha\} \leq \bigwedge\{t > 0 : M(x^*, x_{n+1}, t) > 1 - \alpha\} + K \bigwedge\{t > 0 : M(x_{n+1}, Tx^*, t) > 1 - \alpha\}$$

$$= \bigwedge\{t > 0 : M(x^*, x_{n+1}, t) > 1 - \alpha\} + K \bigwedge\{t > 0 : M(Tx_n, Tx^*, t) > 1 - \alpha\}$$

$$\leq \bigwedge\{t > 0 : M(x^*, x_{n+1}, t) > 1 - \alpha\} + Ka \ max[\bigwedge\{t > 0 : M(x_n, Tx_n, t) > 1 - \alpha\}, \bigwedge\{t > 0 : M(x^*, Tx^*, t) > 1 - \alpha\}, \bigwedge\{t > 0 : M(x_n, x^*, t) > 1 - \alpha\}] + Kb[\bigwedge\{t > 0 : M(x_n, Tx^*, t) > 1 - \alpha\} + \bigwedge\{t > 0 : M(x^*, Tx_n, t) > 1 - \alpha\}]$$

or $\bigwedge\{t > 0 : M(x^*, Tx^*, t) > 1 - \alpha\}$

$$\leq \bigwedge\{t > 0 : M(x^*, x_{n+1}, t) > 1 - \alpha\} + Ka \ max[\bigwedge\{t > 0 : M(x_n, x_{n+1}, t) > 1 - \alpha\}, \bigwedge\{t > 0 : M(x^*, Tx^*, t) > 1 - \alpha\}, \bigwedge\{t > 0 : M(x_n, x^*, t) > 1 - \alpha\}] + Kb[\bigwedge\{t > 0 : M(x_n, Tx^*, t) > 1 - \alpha\} + \bigwedge\{t > 0 : M(x^*, x_{n+1}, t) > 1 - \alpha\}]$$

$$\leq \bigwedge\{t > 0 : M(x^*, x_{n+1}, t) > 1 - \alpha\} + Ka \ max[\bigwedge\{t > 0 : M(x_n, x_{n+1}, t) > 1 - \alpha\}, \bigwedge\{t > 0 : M(x^*, Tx^*, t) > 1 - \alpha\}, \bigwedge\{t > 0 : M(x_n, x^*, t) > 1 - \alpha\}] + Kb \bigwedge\{t > 0 : M(x_n, x^*, t) > 1 - \alpha\} + K^2 b \bigwedge\{t > 0 : M(x^*, Tx^*, t) > 1 - \alpha\} + Kb \bigwedge\{t > 0 : M(x^*, x_{n+1}, t) > 1 - \alpha\}$$

or $\quad (1 - K^2 b) \bigwedge\{t > 0 : M(x^*, Tx^*, t) > 1 - \alpha\} \leq (1 + Kb) \bigwedge\{t > 0 : M(x^*, x_{n+1}, t) > 1 - \alpha\} + Ka \ max[\bigwedge\{t > 0 : M(x_n, x_{n+1}, t) > 1 - \alpha\}, \bigwedge\{t > 0 : M(x^*, Tx^*, t) > 1 - \alpha\}, \bigwedge\{t > 0 : M(x_n, x^*, t) > 1 - \alpha\}] + Kb \bigwedge\{t > 0 : M(x_n, x^*, t) > 1 - \alpha\}.$

That is, $(1 - K^2 b) \bigwedge\{t > 0 : M(x^*, Tx^*, t) > 1 - \alpha\} \leq (1 + Kb) \bigwedge\{t > 0 : M(x^*, x_{n+1}, t) > 1 - \alpha\} + Ka M_2 + Kb \bigwedge\{t > 0 : M(x_n, x^*, t) > 1 - \alpha\}$

where $M_2 = max[\bigwedge\{t > 0 : M(x_n, x_{n+1}, t) > 1 - \alpha\}, \bigwedge\{t > 0 : M(x^*, Tx^*, t) > 1 - \alpha\}, \bigwedge\{t > 0 : M(x_n, x^*, t) > 1 - \alpha\}].$

Case I Let $M_2 = \bigwedge\{t > 0 : M(x_n, x_{n+1}, t) > 1 - \alpha\}$. Then we have

$(1 - K^2b) \bigwedge\{t > 0 : M(x^*, Tx^*, t) > 1 - \alpha\} \leq (1 + Kb) \bigwedge\{t > 0 : M(x^*, x_{n+1}, t) > 1 - \alpha\} + Ka \bigwedge\{t > 0 : M(x_n, x_{n+1}, t) > 1 - \alpha\} + Kb \bigwedge\{t > 0 : M(x_n, x^*, t) > 1 - \alpha\}$

$\leq (1 + Kb) \bigwedge\{t > 0 : M(x^*, x_{n+1}, t) > 1 - \alpha\} + Ka \bigwedge\{t > 0 : M(x_n, x^*, t) > 1 - \alpha\} + K^2a \bigwedge\{t > 0 : M(x^*, x_{n+1}, t) > 1 - \alpha\} + Kb \bigwedge\{t > 0 : M(x_n, x^*, t) > 1 - \alpha\}$

$= (1 + Kb + K^2a) \bigwedge\{t > 0 : M(x^*, x_{n+1}, t) > 1 - \alpha\} + K(a + b) \bigwedge\{t > 0 : M(x_n, x^*, t) > 1 - \alpha\}$

$$\Rightarrow \bigwedge\{t > 0 : M(x^*, Tx^*, t) > 1 - \alpha\}$$
$$\leq \frac{1 + Kb + K^2a}{1 - K^2b} \bigwedge\{t > 0 : M(x^*, x_{n+1}, t) > 1 - \alpha\}$$
$$+ \frac{K(a + b)}{1 - K^2b} \bigwedge\{t > 0 : M(x_n, x^*, t) > 1 - \alpha\} \quad \forall \alpha \in (0, 1). \tag{7}$$

By using Proposition 3.2, from (7) by taking $n \to \infty$ it follows that

$\bigwedge\{t > 0 : M(x^*, Tx^*, t) > 1 - \alpha\} = 0 \quad \forall \alpha \in (0, 1).$
$\Rightarrow M(x^*, Tx^*, t) > 1 - \alpha \quad \forall t > 0 \quad \forall \alpha \in (0, 1).$
$\Rightarrow M(x^*, Tx^*, t) = 1 \quad \forall t > 0$
$\Rightarrow Tx^* = x^*.$

Thus x^* is a fixed point of T.

Case II Suppose $M_2 = \bigwedge\{t > 0 : M(x_n, x^*, t) > 1 - \alpha\}$. Then

$(1 - K^2b) \bigwedge\{t > 0 : M(x^*, Tx^*, t) > 1 - \alpha\} \leq (1 + Kb) \bigwedge\{t > 0 : M(x^*, x_{n+1}, t) > 1 - \alpha\} + Ka \bigwedge\{t > 0 : M(x_n, x^*, t) > 1 - \alpha\} + Kb \bigwedge\{t > 0 : M(x_n, x^*, t) > 1 - \alpha\}$

$\Rightarrow \bigwedge\{t > 0 : M(x^*, Tx^*, t) > 1 - \alpha\} \leq \frac{1+Kb}{1-K^2b} \bigwedge\{t > 0 : M(x^*, x_{n+1}, t) > 1 - \alpha\} + \frac{K(a+b)}{1-K^2b} \bigwedge\{t > 0 : M(x_n, x^*, t) > 1 - \alpha\} \quad \forall \alpha \in (0, 1).$

$\Rightarrow \bigwedge\{t > 0 : M(x^*, Tx^*, t) > 1 - \alpha\} = 0 \ \forall \alpha \in (0, 1)$ (by Proposition 3.2)
$\Rightarrow Tx^* = x^*$ (as above).

Case III If $M_2 = \bigwedge\{t > 0 : M(x^*, Tx^*, t) > 1 - \alpha\}$, then in a similar way as above it can be shown that $Tx^* = x^*$.

Uniqueness of fixed point

If possible suppose that x' is another fixed point of T.

Then we have $Tx' = x'$. Now

$\bigwedge\{t > 0 : M(x^*, x', t) > 1 - \alpha\} = \bigwedge\{t > 0 : M(Tx^*, Tx', t) > 1 - \alpha\}$
$\leq a \quad max[\bigwedge\{t > 0 : M(x^*, Tx^*, t) > 1 - \alpha\}, \bigwedge\{t > 0 : M(x', Tx', t) > 1 - \alpha\}, \bigwedge\{t > 0 : M(x^*, x', t) > 1 - \alpha\}] + b[\bigwedge\{t > 0 : M(x^*, Tx', t) > 1 - \alpha\} + \bigwedge\{t > 0 : M(x', Tx^*, t) > 1 - \alpha\}]$

$= a \qquad max[\bigwedge\{t > 0 : M(x^*, x^*, t) > 1 - \alpha\}, \bigwedge\{t > 0 : M(x', x', t) > 1 - \alpha\}, \bigwedge\{t > 0 : M(x^*, x', t) > 1 - \alpha\}] + b[\bigwedge\{t > 0 : M(x^*, x', t) > 1 - \alpha\} + \bigwedge\{t > 0 : M(x', x^*, t) > 1 - \alpha\}]$

$= a \bigwedge\{t > 0 : M(x^*, x', t) > 1 - \alpha\} + 2b \bigwedge\{t > 0 : M(x^*, x', t) > 1 - \alpha\}]$

$\Rightarrow \bigwedge\{t > 0 : M(x^*, x', t) > 1 - \alpha\} \le (a + 2b) \bigwedge\{t > 0 : M(x^*, x', t) > 1 - \alpha\}]$

$\forall \alpha \in (0, 1)$.

$\Rightarrow a + 2b > 1$ which is a contradiction ($\because a + 2b < 1$).

Hence $x^* = x'$.

To justify the above theorem, we consider the following example.

Example 2 Let $X = \{1, 2, 3\}$, $D : X \times X \to [0, \infty)$ be defined by

$D(1, 1) = D(2, 2) = D(3, 3) = 0$

$D(1, 2) = D(2, 1) = 2, D(2, 3) = D(3, 2) = 1, D(1, 3) = D(3, 1) = 6$.

Then D is a complete strong b-metric on X with K=4 (please see [19]).

Define $M(x, y, t) = \frac{t}{t+D(x,y)}$; then $(X, M, *)$ is a complete fuzzy strong b-metric space with $* = min$.

Define a map $T : X \to X$ by $T1 = 1, T2 = 1, T3 = 1$.

For $x, y \in X$, we have $M(Tx, Ty, t) = \frac{t}{t+D(Tx,Ty)} = \frac{t}{t+D(1,1)} = \frac{t}{t+0} = 1$ $\forall t > 0$.

Thus, $\bigwedge\{t > 0 : M(Tx, Ty, t) > 1 - \alpha\} = 0$ $\forall \alpha \in (0, 1)$.

Thus, the condition (2) is satisfied for suitable $a, b > 0$ such that $a + (1 + K) b < 1$.

Consequently, $x = 1$ is the unique fixed point of T.

4 Fixed Point Theorem for a Pair of ψ-Contractive Type Mapping

In this section, we prove a common fixed point theorem for a pair of ψ-contractive mappings.

Let $\psi : [0, 1] \to [0, 1]$ such that

(i) ψ is non-decreasing and left-continuous.

(ii) $\psi(t) > t$ $\forall t \in (0, 1)$.

We denote

$\Psi = \{\psi : [0, 1] \to [0, 1]; \psi$ satisfies $(i) - (ii)\}$.

Lemma 1 ([20]) If $\psi \in \Psi$, then $\psi(1) = 1$.

Proof. If $\psi(1) \in (0, 1)$ then by the left-continuity property of ψ, $\exists\, t \in (0, 1)$ such that $\psi(1) \le \psi(t)$. This contradicts the property that ψ is non-decreasing. Thus $\psi(1) = 1$.

Lemma 2 ([20]) *If $\psi \in \Psi$ then $\lim_{n\to\infty} \psi^n(t) = 1 \quad \forall t \in (0, 1)$.*

Proof. We have by the property (2) of ψ, $\psi^n(t) > t^n \, \forall t \in (0, 1)$. Now $\lim_{n\to\infty} \psi^n(t) \geq \lim_{n\to\infty} t^n$. Thus, we get $\lim_{n\to\infty} \psi^n(t) = 1$.

Definition 4 Let X be a non-empty set and M be a fuzzy strong b-metric on $X^2 \times [0, \infty)$. Let $f, g : X \to X$ and (f, g) be a pair of ψ-contractive mappings if there exists $\psi \in \Psi$ such that for every $x, y \in X$ the following condition holds:
$M(f(x), g(y), t) > \psi(m(x, y, t))$
where $m(x, y, t) = min\{M(x, y, t), M(f(x), x, t), M(y, g(y), t)\}$.

Fix $x_0 \in X$, and define the sequence $\{x_n\}$ by
$x_1 = f(x_0), x_2 = g(x_1), \dots\dots x_{2n+1} = f(x_{2n}), x_{2n+2} = g(x_{2n+1}), \cdots$
Then $\{x_n\}$ is called an (f, g)-sequence of initial point x_0.

Lemma 3 *If X be a non-empty set and M is a fuzzy strong b-metric on $X^2 \times (0, \infty)$, let $f, g : X \to X$ be two self-mappings. Then $\lim_{n\to\infty} M(x_{n+1}, x_n, t) = 1 \forall t > 0$, where $\{x_n\}$ is an (f, g)-sequence of initial point x_0.*

Proof. If $M(x_{n+1}, x_n, t_0) = 1$ for some $n \in N$, $t_0 \in (0, \infty)$ then $M(x_{m+1}, x_m, t_0) = 1 \, \forall m > n$.

From $M(x_2, x_1, t_0) = M(x_1, x_2, t_0)$
$\qquad = M(f(x_0), g(x_1), t_0)$
$\qquad > \psi(m(x_0, x_1, t_0))$
$\qquad > M(x_0, x_1, t_0) > 0$.
Again $M(x_2, x_3, t) > \psi M(x_1, x_2, t_0)$
$\qquad > \psi^2(M(x_0, x_1, t_0)) > 0$.
By induction, $M(x_{n+1}, x_n, t_0) > M(x_n, x_{n-1}, t_0) > \psi^n(M(x_0, x_1, t_0)) > 0$.
Taking limit as $n \to \infty$ on both sides, we have
$\lim_{n\to\infty} M(x_{n+1}, x_n, t_0) = 1$.
Thus, $\lim_{n\to\infty} M(x_{n+1}, x_n, t) = 1, \, \forall t > 0$.

Lemma 4 *Let $(X, M, *)$ be a fuzzy strong b-metric space and let $f, g : X \to X$ be two self-mappings. Assume that (f, g)-is a pair of ψ-contractive mappings. Then (f, g)-sequence $\{x_n\}$ of initial point x_0 is Cauchy.*

Proof. If possible suppose $\{x_n\}$ is not Cauchy. Then there exists an $\alpha_0 \in (0, 1)$ and $t' > 0$ such that for each $k \in \mathbb{N}$, there exists
$m(k), n(k) \in \mathbb{N}$ with $m(k) > n(k) \geq k$ and
$M(x_{m(k)}, x_{n(k)}, t') \leq 1 - \alpha_0$.

Let, for each $k, m(k)$, there be the least positive integer exceeding $n(k)$ satisfying this property,
i.e., $M(x_{m(k)-1}, x_{n(k)}, t') > 1 - \alpha_0$ and $M(x_{m(k)}, x_{n(k)}, t') \leq 1 - \alpha_0$.

Since $M(x, y, \cdot)$ is continuous, $\exists t_0 \in (0, t')$ such that
$M(x_{m(k)-1}, x_{n(k)}, t_0) > 1 - \alpha_0$.

Thus, for a positive integer k,

$$1 - \alpha_0 > M(x_{m(k)}, x_{n(k)}, t')$$
$$> M(x_{m(k)-1}, x_{n(k)}, t_0') * M(x_{m(k)-1}, x_{m(k)}, \tfrac{t'-t_0}{K}).$$

Taking limit as $k \to \infty$ on both sides, we get

$$1 - \alpha_0 \geq \lim_{k \to \infty} M(x_{m(k)}, x_{n(k)}, t') > (1 - \alpha_0) * 1$$
$$= 1 - \alpha_0.$$

Thus, $\lim_{k \to \infty} M(x_{m(k)}, x_{n(k)}, t') = 1 - \alpha_0$ ($\because$ of Lemma 4.1).

Now $M(x_{m(k)}, x_{n(k)}, t') \geq M(x_{m(k)}, x_{n(k)+1}, t_0) * M(x_{n(k)+1}, x_{n(k)}, \tfrac{t'-t_0}{K})$.

Now again for $t_0 > 0$, there exists $t'' \in (0, t_0)$ such that
$M(x_{m(k)-1}, x_{n(k)}, t'') > 1 - \alpha_0.$

Now, $M(x_{m(k)}, x_{n(k)}, t') > M(x_{m(k)+1}, x_{n(k)+1}, t'') * M(x_{m(k)}, x_{m(k)+1}, \tfrac{t'-t''}{K}) *$
$M(x_{n(k)+1}, x_{n(k)}, \tfrac{t'-t_0}{K})$

$> \psi[M(x_{m(k)}, x_{n(k)}, t'') * M(x_{m(k)}, x_{m(k)+1}, \tfrac{t-t''}{K}) * M(x_{n(k)+1}, x_{n(k)}, \tfrac{t-t_0}{K})].$

Letting $k \to +\infty$ on both sides, we have

$$1 - \alpha_0 > \psi(1 - \alpha_0) > 1 - \alpha_0 \qquad (\because t'' < t' \therefore 1 - \alpha_0 \leq$$
$$\lim_{k \to \infty} M(x_{m(k)}, x_{n(k)}, t'') \leq \lim_{k \to \infty} M(x_{m(k)}, x_{n(k)}, t') = 1 - \alpha_0)$$

which is a contradiction.

Thus $\{x_n\}$ is a Cauchy sequence.

Theorem 4 *Let $(X, M, *)$ be a complete fuzzy strong b-metric space and M be continuous with respect to each of its components, and let $f, g : X \to X$ be two self-mappings. Assume that (f,g) is a pair of ψ-contractive mappings. Then f and g have a unique common fixed point.*

Proof. By Lemma 4, it follows that the (f,g)-sequence $\{x_n\}$ of initial point x_0 is Cauchy.

Since X is complete, there exists $x \in X$ such that $\lim_{n \to \infty} x_n = x$.

If $f(x) = x$ and $g(x) = x$, then the theorem is proved.

If possible suppose that $f(x) \neq x$, then there exists $t > 0$ such that
$0 < M(x, f(x), t) < 1$.

(By Definition 2 of (1) and (2), $0 < M(x, f(x), t) < 1$ is valid.)

Now, $M(f(x), x_{2n}, t) = M(f(x), g(x_{2n-1}), t) > \psi(m(x, x_{2n-1}, t))$.

Now, $m(x, x_{2n-1}, t)$
$= min\{M(x, x_{2n-1}, t), M(x, f(x), t), M(x_{2n-1}, g(x_{2n-1}), t)\}$
$= min\{M(x, x_{2n-1}, t), M(x, f(x), t), M(x_{2n-1}, x_{2n}, t)\}$.

Taking $\lim$ $n \to \infty$, we have $M(x, f(x), t) > \psi(M(x, f(x), t)) > M(x, f(x), t)$—a contradiction.

$\therefore f(x) = x$ and similarly $g(x) = x$.

Thus x is a common fixed point of f and g.

Now for uniqueness suppose x and y are two fixed points of f and g. If $x \neq y$, there exists $t > 0$ such that $0 < M(x, y, t) < 1$ and hence $M(x, y, t) = M(f(x), g(y), t) > \psi(m(x, y, t)) > M(x, y, t)$—a contradiction.

$\therefore x = y$.

Corollary 1 *Let $(X, M, *)$ be a complete fuzzy strong b-metric space and let $f :$ $X \rightarrow X$ be a self-mapping. Assume that f is a ψ-contractive mapping. Then f has a unique fixed point.*

Proof. Proof follows immediately by Theorem 4 by putting $f = g$.

Example 3 To justify the above theorem, we consider the following example.
Let $X = \{1, 2, 3\}$, $\psi(t) = \sqrt{t}$, $\forall t \in [0, 1]$, $M(x, y, t) = (\frac{t}{t+1})^{|x-y|}$ $\forall x, y \in X$.
Then X is a fuzzy strong b-metric space; let $f : X \rightarrow X$ be a self-mapping defined by $f(1) = 2, f(2) = 2, f(3) = 2$.
Now for any $x, y \in X$, we have
$M(fx, fy, t) = (\frac{t}{t+1})^{|fx-fy|}$
$= (\frac{t}{t+1})^{|2-2|}$
$= 1 \geq \psi(m(x, y, t))$. Consequently, f satisfies the ψ-contractive condition of Theorem 4. Then $x = 2$ is the unique fixed point of f.

5 Conclusion

Some basic results on completeness and compactness in fuzzy cone metric spaces have been studied. Cantor's intersection theorem has been established in a fuzzy setting. The notion of a totally fuzzy bounded set is introduced, and we have studied some relation with compact fuzzy cone metric space. We think that researchers in the field of fuzzy cone metric space will be benefited by using the results of this paper.

References

1. Agarwal, S., Qureshi, K., Nema, J.: A fixed point theorem for b-metric space. Int. J. Pure Appl. Math. Sci. (2016). ISSN 0972-9828
2. Bakhtin, I.: The contraction mapping principle in quasimetric spaces. Func. An. Gos. Ped. Inst. Unianowsk **30**, 26–37 (1989)
3. Czerwik, S.: Contraction mappings in b-metric spaces. Acta Math. Inform. Univ. Ostrav. **1**(1), 5–11 (1993)
4. George, A., Veeramani, P.: On some results in fuzzy metric spaces. Fuzzy Sets Syst. **64**(3), 395–399 (1994)
5. Grabiec, M.: Fixed points in fuzzy metric spaces. Fuzzy Sets Syst. **27**(3), 385–389 (1988)
6. Gupta, V., Ege, O., Saini, R., sen, M.D.L.: Various fixed point results in complete gb-metric spaces. Dyn. Syst. Appl. **30**(2), 277–293 (2021)
7. Kadelburg, Z., Radenovic, S.: Notes on some recent papers concerning f-contractions in b-metric spaces. Constr. Math. Anal **1**(2), 108–112 (2018)
8. Karapinar, E.: A short survey on the recent fixed point results on b-metric spaces. Constr. Math. Anal. **1**(1) (2018)
9. Kirk, W., Shahzad, N.: Fixed Point Theory in Distance Spaces. Springer (2014)
10. Kramosil, I., Michálek, J.: Fuzzy metrics and statistical metric spaces. Kybernetika **11**(5), 336–344 (1975)

11. Miheţ, D.: On fuzzy contractive mappings in fuzzy metric spaces. Fuzzy Sets Syst. **158**(8), 915–921 (2007)
12. Miheţ, D.: Fuzzy ψ-contractive mappings in non-Archimedean fuzzy metric spaces. Fuzzy Sets Syst. **159**(6), 739–744 (2008)
13. Oner, T.: On topology of fuzzy strong b-metric spaces. J. New Theory **21**, 59–67 (2018)
14. Rakić, D., Došenović, T., Mitrović, Z.D., de la Sen, M., Radenović, S.: Some fixed point theorems of ćirić type in fuzzy metric spaces. Mathematics **8**(2), 297 (2020)
15. Rakić, D., Mukheimer, A., Došenović, T., Mitrović, Z.D., Radenović, S.: On some new fixed point results in fuzzy b-metric spaces. J. Inequalities Appl. **2020**(99), (2020)
16. Rao, K., Altun, I., Bindu, S.H.: Common coupled fixed-point theorems in generalized fuzzy metric spaces. Adv. Fuzzy Syst. **2011** (2011)
17. Saadati, R.: On the topology of fuzzy metric type spaces. Filomat **29**(1), 133–141 (2015)
18. Sun, G.P., Yang, K.: Generalized fuzzy metric spaces with properties. Res. J. Appl. Sci. Eng. Technol. **2**(7), 673–678 (2010)
19. Van An, T., Van Dung, N., et al.: Answers to Kirk-Shahzad's questions on strong b-metric spaces. Taiwan. J. Math. **20**(5), 1175–1184 (2016)
20. Vetro, C.: Fixed points in weak non-Archimedean fuzzy metric spaces. Fuzzy Sets Syst. **162**(1), 84–90 (2011)
21. Wang, S.: Answers to some open questions on fuzzy ψ-contractions in fuzzy metric spaces. Fuzzy Sets Syst. **222**, 115–119 (2013)

Fuzzy Random Continuous Review Inventory Model with Controllable Lead-Time and Exponential Crashing Cost

Wasim Firoz Khan and Oshmita Dey

Abstract In this paper, a continuous review inventory model is studied in a mixed imprecise and uncertain environment. A methodology is proposed to quantify both imprecise (fuzzy) and stochastic (random) information simultaneously. This enables the simultaneous inclusion of both the inherent randomness present in any real-life inventory situation and the subjective evaluation of the decision maker into the model. A methodology is developed by assuming the annual customer demand to be a continuous fuzzy random variable following normal distribution with associated fuzzy probability density function. Also the lead-time is assumed to be a control parameter and a lead-time crashing cost, in the form of an exponential function, is incorporated into the total inventory cost. The optimal values of the decision variables are determined so that the crisp equivalent of total expected annual inventory cost is a minimum. Numerical analysis is presented to illustrate the proposed model and provide some insights.

Keywords Inventory · Continuous review · Fuzzy random variable · Normal distribution · Lead-time crashing cost

1 Introduction

Over the years, the (Q, r) continuous review system (Hadley and Whitin [1]) has played a very significant role in the analysis of inventory models under various assumptions. A significant amount of literature is available in this regard. Most of these works, however, are in the deterministic and/or stochastic framework. With the development of the fuzzy set theory by Zadeh [2], several researchers have studied the continuous review system under the fuzzy framework and a substantial amount of literature exists in this regard as well. However, neither the stochastic models nor the fuzzy models take into account the simultaneous occurrence of fuzziness and

W. F. Khan · O. Dey (✉)
Department of Mathematics, Techno India University, Kolkata, West Bengal 700091, India
e-mail: oshmi_kgp@yahoo.co.uk

© The Author(s), under exclusive license to Springer Nature Singapore Pte Ltd. 2023
P. Gyei-Kark et al. (eds.), *Engineering Mathematics and Computing*,
Studies in Computational Intelligence 1042,
https://doi.org/10.1007/978-981-19-2300-5_3

randomness. But, it is observed that fuzziness and randomness do not appear in a mutually exclusive way in a real-life inventory situation. Various reasons, viz., lack of historical data, unreliable information, unknown reproduction conditions, varying linguistic expressions from expert to expert, etc., may all lead to the occurrence of fuzziness and randomness simultaneously in the inventory situations (Dey and Chakraborty [3]). In such cases, the purely probabilistic models and/or the purely fuzzy models fail to capture the simultaneous existence of imprecision and uncertainty in the collected data. A more suitable approach in this regard is to quantify both these types of uncertainties (i.e., fuzziness and randomness) simultaneously. One of the effective ways of doing so is the use of fuzzy random variables. In this vein, Dutta et al. [4] developed a single period inventory model where the annual customer demand was considered to be a discrete fuzzy random variable of the form $(\tilde{D}_i, p_i)$, where $\tilde{D}_i$ is fuzzy random variable customer demand with associated probabilities p_i. Dey and Chakraborty [3] then developed the periodic review system under the fuzzy random framework with a similar assumption. Since then, a number of researchers have developed inventory models under the fuzzy random framework.

Now in most of the existing literature on fuzzy random inventory models, it is observed that the discrete fuzzy random variable is taken to quantify the fuzzy random information. But, in most real inventory problems, especially when the information about customer demand contains abundant or sufficient statistical data, then the discrete $(\tilde{D}_i, p_i)$ form of the demand information makes the mathematical calculations very difficult and time consuming. But ignoring fuzziness also causes a loss of information. So, a methodology needs to be developed so as to quantify the available fuzzy random information while ensuring computational flexibility (Dey and Chakraborty [5]). With this point of view, Dey and Chakraborty [5] proposed a methodology to simultaneously quantify fuzzy and random information by way of a uniformly distributed continuous fuzzy random variable customer demand. But it is quite well established in stochastic inventory models that the normal distribution is more suitable to describe stochastic data compared to the uniform distribution. In this vein, Khan and Dey [6, 7] developed the periodic review and continuous review inventory models, respectively, considering the annual customer demand to be a normally distributed fuzzy random variable of the form $\tilde{D} = (D - \Delta_1, D, D + \Delta_2)$ (Liu and Liu [8]) with associated fuzzy probability density function, where $\tilde{D}$ is annual customer demand and Δ_1, Δ_2 are the left and right spreads of $\tilde{D}$ respectively. Here, $\Delta_1, \Delta_2(> 0)$ are set by the decision maker. This therefore allows the decision makers to incorporate their own experience, intuition, and understanding of the real-life inventory situation into the model. However, one of the restrictions of these models was that, in both the models, the lead-time was taken to be a constant. This is very rarely the case in a real-life inventory situation. The present model, therefore, makes an attempt to further extend literature by including the problem of lead-time control.

In inventory management, lead-time is defined as the duration of time between the placing of an order and its actual arrival in the inventory. Over the years, lead-time control has been one of the focal areas of research in inventory modeling. This is due to the fact that lead-time control ensures the smooth and efficient running of any business since a shorter lead-time provides many advantages, viz., lowering the safety stock level, reducing the loss during stock-out period, providing adequate service level to the customer, increasing the competitive advantages in business, etc. (Liao and Shyu [9]). In this regard, Liao and Shyu [9] introduced an inventory model where the lead-time was considered as a decision variable. Since then a number of researchers have focused their attention on developing various inventory models with lead-time control and a significant amount of literature is available in this context. For instance, Pan and Hsiao [10] studied the integrated inventory models with controllable lead-time and backorder discount considerations. Chang et al. [11] developed an integrated vendor-buyer cooperative inventory models with controllable lead-time and ordering cost reduction. Wei and Bing [12] developed a distribution-free continuous review inventory model with controllable lead-time and setup cost. Uthayakumar and Priyan [13] studied permissible delay in payments in a two-echelon inventory system with controllable setup cost and lead-time. Priyan and Uthayakumar [14] proposed a continuous review inventory model with lost sales rate, controllable lead-time and order processing cost. Several other researchers, viz., Chang et al. [15], Rong [16], Wang [17], Dey and Chakraborty [18], Shah and Soni [19], Bhuiya and Chakraborty [20], Soni and Patel [21], Rong [22], Kurdhi [23], Shin et al. [24], etc., to name a few, have also made a significant contribution in this regard. Recently, Malik [25] developed a multi-product continuous-review inventory model with uncertain demand and variation control in lead-time. Karthick [26] described an imperfect production model with controllable lead-time where the customer demand was taken as fuzzy in nature. Shee and Chakrabarti [27] studied a fuzzy two-echelon supply chain model for deteriorating items with time-varying holding cost with lead-time as a decision variable. Tharani and Uthaykumar [28] proposed a novel approach to safety stock management in an integrated supply chain with controllable lead-time and ordering cost reduction. Kurdhi et al. [29] developed a periodic review inventory system with controllable safety stock, lead-time, and price discount under partially backlogged shortages. Krishnaraj [30] studied an inventory model involving lead-time crashing cost as a Weibull distribution. Castellano et al. [31] developed a periodic review policy for a coordinated single vendor-multiple buyers supply chain with controllable lead-time under a distribution-free approach. Very recently, Dey et al. [32] analyzed a supply chain model with controllable lead-time and variable demand for a smart manufacturing system.

Most of these works which consider lead-time control, however, do so by way of a component-wise crashing cost. This is due to the fact that lead-time often consists of a number of components, like, arrangement of order, transportation of order, supplier lead-time, delivery time, etc. Also, each of the above-mentioned components may be crashed from their usual duration to a shorter duration by various means, viz., hiring extra labor, using special transport for faster delivery, workers working over time, etc. (Tersine [33]) This obviously involves investing an extra amount of money which is known as the crashing cost. A significant volume of research in lead-time control follows this approach. However, as observed by some researchers, viz., Dey and Chakraborty [18], Priyan and Uthayakumar [14], it may not always be possible or feasible to calculate the component-wise crashing cost. This is due to the fact that,

2.2 *Fuzzy Random Variable and Its Expectation*

A fuzzy random variable $\tilde{D}(\omega)$ is described as $\tilde{D}(\omega) = (D(\omega) - \Delta_1, D(\omega), D(\omega) + \Delta_2)$, where (Ω, B, P) is a probability space and $\omega \in \Omega$. Here Δ_1 and Δ_2 are the left and right spreads of $\tilde{D}$, respectively, and $D(\omega)$ follows some continuous distribution, for all $\omega \in \Omega$, and $0 < \Delta_1 < D(\omega)$ and $\Delta_2 > 0$ (Liu and Liu [8]).

If $\tilde{X}$ is a fuzzy random variable with associated probability density function $\tilde{f}(\tilde{x})$, then the expectation of $\tilde{X}$ is given by $E\tilde{X} = \int_{-\infty}^{\infty} \tilde{x} \tilde{f}(\tilde{x}) dx$ and its α-cut is given by $E[\tilde{x}] = [E(x_\alpha^-), E(x_\alpha^+)]$.

2.3 *Defuzzification of a Fuzzy Number*

The possibilistic mean value of $\tilde{A}$ (Carlsson and Fuller [34]) is defined as $\overline{M}(\tilde{A}) = \frac{M_*(\tilde{A}) + M^*(\tilde{A})}{2}$, where $M_*(\tilde{A})$ and $M^*(\tilde{A})$ are the lower and upper possibilistic mean values of $\tilde{A}$, respectively, and given by $M_*(\tilde{A}) = \frac{\int_0^1 A_\alpha^- d\alpha}{\int_0^1 \alpha d\alpha}$ and $M^*(\tilde{A}) = \frac{\int_0^1 A_\alpha^+ d\alpha}{\int_0^1 \alpha d\alpha}$. It can also be written as

$$\overline{M}(\tilde{A}) = \int_0^1 (A_\alpha^- + A_\alpha^+) d\alpha = \frac{a_1 + a_3}{6} + \frac{2a_2}{3} \tag{1}$$

where $\alpha \in [0, 1]$ and, $A_\alpha^- = a_1 + \alpha(a_2 - a_1)$, $A_\alpha^+ = a_3 - \alpha(a_3 - a_2)$.

3 Mathematical Model

3.1 *Notations of the Model*

Q	Optimal order quantity (decision variable)
r	Reorder point (decision variable)
L	Lead-time (decision variable)
A	Ordering cost per item
s	Fixed shortage cost per unit demand during stockout
h	Unit holding cost per item
$R(L)$	Lead-time crashing cost
m	Mean of the normal distribution
σ	Standard deviation (sd.) of the normal distribution
$\tilde{D}$	Annual customer demand (continuous fuzzy random variable)
$\tilde{D}_L$	Lead-time demand (continuous fuzzy random variable)
LR, RR	Left and right shape function of $\tilde{D}_L$ respectively
$\tilde{f}(\tilde{D})$	Fuzzy probability density function of $\tilde{D}$
Δ_1, Δ_2	Left and right spread of $\tilde{D}$, respectively
SS	Safety stock
$\overline{M}(\tilde{D}_L - R)^+$	Expected shortage
TC	Total annual cost

3.2 Assumptions of the Model

- a (Q, r) continuous review inventory system is considered with Q, r as control parameters
- the annual customer demand is assumed to be a normally distributed continuous fuzzy random variable of the form $\tilde{D} = (D - \Delta_1, D, D + \Delta_2)$ with its associated fuzzy probability density function $\tilde{f}(\tilde{D})$
- the lead-time demand is considered to be (the annual demand) × (duration of the lead-time) i.e., $\tilde{D}_L = \tilde{D}L$ (Dey and Chakraborty [3]). Therefore, the lead-time demand is also normally distributed fuzzy random variable.
- the lead-time is considered to be a control parameter and the lead-time crashing cost per order is $R(L)$, which is assumed to be an exponential function involving L and is defined as below (Priyan and Uthayakumar [14]):

$$R(L) = \begin{cases} 0 & L = L_0 \\ e^{C/L} & L_b \le L < L_0 \end{cases}$$

where $C(> 0)$ is a constant and L_0 and L_b are the existing and the shortest lead-times, respectively.
- Shortages are allowed and fully backordered.

3.3 Normally Distributed Fuzzy Random Variable Demand

The normally distributed fuzzy random variable annual customer demand, as proposed by Khan and Dey [6], is presented briefly below.

The membership function of the fuzzy random variable demand $\tilde{D}$ is

$$\mu_{\tilde{D}}(x) = \begin{cases} \frac{x - (D - \Delta_1)}{\Delta_1}, & D - \Delta_1 \le x \le D \\ \frac{(D + \Delta_2) - x}{\Delta_2}, & D \le x \le D + \Delta_2 \end{cases} \tag{2}$$

Based on the probability density function (pdf) of the normal distribution, i.e., $y = f(x) = \frac{e^{-(x-m)^2/2\sigma^2}}{\sigma\sqrt{2\pi}}$ and using extension principle, the corresponding fuzzy pdf $\tilde{f}(\tilde{D})$ is obtained as $(f(\underline{D}), f(D), f(\overline{D}))$. Khan and Dey [6] where,

$$f(\underline{D}) = \frac{e^{-(D-m+\Delta_2)^2/2\sigma^2}}{\sigma\sqrt{2\pi}}, \ f(D) = \frac{e^{-(D-m)^2/2\sigma^2}}{\sigma\sqrt{2\pi}}, \ f(\overline{D}) = \frac{e^{-(D-m-\Delta_1)^2/2\sigma^2}}{\sigma\sqrt{2\pi}} \tag{3}$$

Now, by setting $\Delta_1 = \Delta_2 = 0$, then the fuzzy pdf, $\tilde{f}(\tilde{D})$ reduces crisp equivalent, i.e., $f(D) = \frac{e^{-(D-m)^2/2\sigma^2}}{\sigma\sqrt{2\pi}}$, which is the pdf of the normal distribution.

Since, $\tilde{D} = (D - \Delta_1, D, D + \Delta_2)$ and $\tilde{D}_L = \tilde{D} \times L$, then the expected lead-time demand is obtained of the form $(\underline{D}_L, D_L, \overline{D}_L)$. Khan and Dey [6] where,

$$DL(D_{Lower}) = (m - \Delta_1 - \Delta_2)L, \, DM(D_{Middle}) = mL, \, DU(D_{Upper}) = (m + \Delta_1 + \Delta_2)L \tag{4}$$

3.4 Determination of the Expected Shortage

When the lead-time demand exceeds the reorder point r, then shortages arise. To investigate the expected shortage, two separate cases arise according to the position of $r \in [DL, DU]$ subject to the criterion that $SS \geq 0$ holds (Dey and Chakraborty [3]). Therefore using Eq. (4), LR and RR are obtained as follows (Khan and Dey [6]):

$$LR = \frac{(r - DL)}{(DM - DL)} = \frac{r - (m - \Delta_1 - \Delta_2)L}{mL - (m - \Delta_1 - \Delta_2)L}$$
$$RR = \frac{(DU - r)}{(DU - DM)} = \frac{-r + (m + \Delta_1 + \Delta_2)L}{-mL + (m + \Delta_1 + \Delta_2)L} \tag{5}$$

Therefore, using Eq. (5), shortage is obtained from the following two cases (Dey and Chakraborty [3]):

Case 1: when r lies between DL and DM

$$\overline{b}(r) = \overline{M}(\tilde{D}_L - r)^+$$
$$= \frac{DU}{2} - \frac{(DU - DM)}{3} + \frac{DL}{2} - \frac{(LR)^2 DL}{2} + \frac{(DM - DL)}{3}$$
$$- \frac{(LR)^3(DM - DL)}{3} - r(1 - \frac{(LR)^2}{2}) \tag{6}$$

Case 2: when, r lies between DM and DU

$$\overline{b}(r) = \overline{M}(\tilde{D}_L - R)^+ = \frac{(RR)^2 DU}{2} - \frac{(RR)^3(DU - DM)}{3} - \frac{(RR)^2 r}{2} \tag{7}$$

3.5 Mathematical Model

The total annual inventory cost for the (Q, r) continuous review system is given as follows (Hadley and Whitin [1]):

$$TC(Q, r) = \frac{AD}{Q} + h(Q/2 + r - D_L) + \frac{Ds}{Q}\bar{b}(r) \tag{8}$$

where the control parameters are the order quantity Q and the reorder point r. Here, to reduce the length of the lead-time, the lead-time crashing cost $R(L)$ is added to the total annual inventory cost, thereby including the lead-time L as an additional control parameter.

Therefore, the total annual inventory cost is rewritten as

$$TC(Q, r, L) = \frac{AD}{Q} + h(Q/2 + r - D_L) + \frac{Ds}{Q}\bar{b}(r) + \frac{D}{Q}R(L) \tag{9}$$

Therefore, the fuzzy random total annual cost for continuous review system with crashing cost is obtained as

$$\tilde{T}C(Q, r, L) = \frac{A\tilde{D}}{Q} + h(Q/2 + r - \tilde{D}_L) + \frac{\tilde{D}s}{Q}\bar{b}(r) + \frac{\tilde{D}}{Q}R(L) \tag{10}$$

where $\tilde{D}$ is a normally distributed fuzzy random variable.

This is reduced to its fuzzy equivalent (fuzzy number) by taking its expectation (probabilistic) and the expected total annual inventory cost is obtained as

$$E\tilde{T}C(Q, r, L) = \frac{A}{Q}\int_{-\infty}^{\infty}\tilde{D}f(\tilde{D})dD + h(Q/2 + r - L\int_{-\infty}^{\infty}\tilde{D}f(\tilde{D})dD)$$
$$+ \frac{s}{Q}\int_{-\infty}^{\infty}\tilde{D}f(\tilde{D})dD\bar{b}(r) + \frac{R(L)}{Q}\int_{-\infty}^{\infty}\tilde{D}f(\tilde{D})dD \tag{11}$$

The possibilistic mean (Carlsson and Fuller [34]) of the fuzzy number $\int_{-\infty}^{\infty}\tilde{D}\tilde{f}(\tilde{D})dD$ is obtained as in Khan and Dey [6]:

$$\bar{M}\left(\int_{-\infty}^{\infty}\tilde{D}\tilde{f}(\tilde{D})dD\right) = \int_{-\infty}^{\infty}\bar{M}(\tilde{D}\tilde{f}(\tilde{D})dD) = \int_{-\infty}^{\infty}\left(\frac{\underline{Df} + \overline{Df}}{6} + \frac{2Df}{3}\right)dD = m$$

Using the above result, the crisp equivalent of the fuzzy random total annual cost is obtained as

$$ETC(Q, r, L) = \frac{mA}{Q} + h(Q/2 + r - mL) + \frac{ms}{Q}\bar{b}(r) + \frac{m}{Q}R(L) \tag{12}$$

3.6   Derivation of the Optimal Solution

The problem is to obtain the optimal values of Q, r and the minimum value of $L \in [L_b, L_0]$ so that ETC is minimized.

Case 1: when $r \in [DL, DM]$

The total annual cost function ETC is obtained as

$$ETC(Q, r, L) = \frac{mA}{Q} + h[Q/2 + r - mL] + \frac{m}{Q}R(L)$$

$$+ \frac{ms}{Q}\left(\frac{DU}{2} - \frac{(DU - DM)}{3} + \frac{DL}{2} - \frac{(LR)^2DL}{2} + \frac{(DM - DL)}{3}\right.$$

$$\left. - \frac{(LR)^3(DM - DL)}{3} - r(1 - \frac{(LR)^2}{2})\right) \tag{13}$$

It is to be noted that the cost expression obtained above need not necessarily be a convex function in all three control parameters Q, r, L. However, it can be shown, as follows, that for fixed values of Q, r, ETC is a convex function in L. Taking, twice partial derivatives of $ETC(Q, r, L)$ with respect to L, we get

$$\frac{\partial^2 ETC(Q, r, L)}{\partial L^2} = \frac{m}{L^4 Q}(ce^{c/L}(c + L) - \frac{3r^2s(r + L(-m + H))}{H^2}) > 0$$

Therefore $ETC(Q, r, L)$ is a convex function of $L \in [L_b, L_{b-1}]$ for fixed Q and r. That is, $ETC(Q, r, L)$ has a minimum value at any point in the above interval. So, to determine the optimal values of Q, r, L which minimizes the total annual cost, the value of $L \in [L_b, L_{b-1}]$ is set. For this fixed value, the first derivatives of ETC w.r.t Q, r are derived and set equal to zero. Solving these equations simultaneously, the values of Q, r are determined and the total annual cost is derived. This process is repeated for values of $L \in [L_b, L_{b-1}]$. It is observed that the total annual cost first decreases and then starts increasing after a certain value. This value of L, therefore, yields the minimum total annual inventory cost and the corresponding optimal values of Q, r.

Therefore, deriving the first partial derivatives of ETC (Eq. 13) w.r.t. Q, r we get

$$\frac{\partial ETC(Q, r, L)}{\partial r} = \frac{1}{2L^2QH^2}$$

$$\left[-3mr^2H + 6Lmrs(m - H) + L^2(-3m^3s + 6m^2sH + 2hQH^2 - 5msH^2)\right]$$

and

$$\frac{\partial ETC(Q, r, L)}{\partial Q} = \frac{1}{2L^2Q^2H^2}$$

$$[mr^3s + 3Lmr^2s(-m + H) + L^2(3m^3rs - 6m^2rsH) + hQ^2H^2$$

$$-m(2A + 2R(L) - 5rs)H^2 + L^3ms(-m^3 + 3m^2H - 5mH^2 + H^3)]$$

where H = sum of the spreads of $\tilde{D}$, i.e., $H = (\Delta_1 + \Delta_2)$ and,

$$R(L) = \begin{cases} 0 & L = L_0 \\ e^{C/L} & L_b \le L < L_0 \end{cases}$$

Equating the first partial derivatives to zero, we get

$$\frac{\partial ETC(Q, r, L)}{\partial r} = 0, \quad \frac{\partial ETC(Q, r, L)}{\partial Q} = 0. \tag{14}$$

The two equations are solved simultaneously to obtain (Q, r) and determine the corresponding ETC for $L \in [L_b, L_0]$.

Case 2: when $r \in [DM, DU]$

The total annual cost function ETC is obtained as:

$$\begin{aligned} ETC(Q, r, L) &= \frac{mA}{Q} + h[Q/2 + r - mL] + \frac{m}{Q}R(L) \\ &+ \frac{ms}{Q}\left(\frac{(RR)^2DU}{2} - \frac{(RR)^3(DU - DM)}{3} - \frac{(RR)^2r}{2}\right) \end{aligned} \tag{15}$$

Similar to the results obtained in case 1, it can be shown here as well that ETC is convex in L for fixed values of Q, r as follows:

$$\frac{\partial^2 ETC(Q, r, L)}{\partial L^2} = \frac{1}{L^4Q(\Delta_1 + \Delta_2)^2}[ce^{c/L}(c + 2L)m(\Delta_1 + \Delta_2)^2 + mr^2s(L(m + \Delta_1 + \Delta_2) - r)] > 0$$

Therefore, as in case 1, for fixed values of $L \in [L_b, L_0]$, setting the first partial derivatives of ETC w.r.t. Q, r to zero we get

$$\frac{\partial TC(Q, r, L)}{\partial r} = \frac{1}{2L^2QH^2}[-mr^2s - 2Lmrs(m + H) + L^2(m^3s + 2m^2sH) - 2hQH^2 + msH^2]$$

and $\dfrac{\partial ETC(Q, r, L)}{\partial Q} = \dfrac{h}{2} - \dfrac{Am}{Q^2} - \dfrac{mR(L)}{Q^2} - \dfrac{ms(-r + L(m + H))^3}{6L^2Q^2H^2}$

Equating the first partial derivatives to zero, we get

Table 1 Derivation of the optimal solution

L	Q	r	R(L)	TC
0.01	120.93	6.72	403.43	6083.04
0.02	73.60	14.23	20.08	3791.88
0.03	**71.93**	**21.39**	**7.39**	**3766.31**
0.04	71.90	28.52	4.48	3821.44
0.05	72.17	35.64	3.32	3891.11
0.06	72.07	42.77	0	3942.80

Table 2 Effect of the spreads

Parameters	Values	r	Q	L	TC
$\Delta_1(\Delta_2 = 150)$	75	21.05	71.79	0.03	3742.49
	100	21.39	71.93	0.03	3766.31
	125	21.73	72.07	0.03	3790.12
$\Delta_2(\Delta_1 = 100)$	125	21.05	71.79	0.03	3742.49
	150	21.39	71.93	0.03	3766.31
	175	21.73	72.07	0.03	3790.12

$$\frac{\partial ETC(Q, r, L)}{\partial r} = 0, \ \frac{\partial ETC(Q, r, L)}{\partial Q} = 0. \tag{16}$$

As in case 1, the two equations are solved simultaneously to obtain (Q, r) and determine the corresponding ETC for $L \in [L_b, L_0]$. The proposed methodology is illustrated with numerical examples in the following section.

4 Numerical Example and Analysis of the Numerical Result

For numerical example, the following data set is used:

$$A = 200, h = 50, s = 40, m = 600, \Delta_1 = 100, \Delta_2 = 150, c = 0.06, \text{ and}$$

$$R(L) = \begin{cases} 0 & L = 0.06 \\ e^{C/L} & 0.01 \leq L < 0.06 \end{cases}$$

Table 3 Effect of the mean demand m

m	r	Q	L	$R(L)$	TC
450	16.57	62.59	0.03	7.39	3283.38
550	19.80	68.96	0.03	7.39	3613.09
600	21.39	71.93	0.03	7.39	3766.31
650	22.97	74.78	0.03	7.39	3913.04
750	26.12	80.18	0.03	7.39	4190.16

Table 4 Effect of the lead-time L_0

L_0	r	Q	L^*	$R(L^*)$	TC
0.04	14.27	71.46	0.02	7.38	3686.82
0.05	14.25	72.48	0.02	14.25	3726.87
0.06	21.39	71.93	0.03	7.38	3766.31
0.07	21.37	72.43	0.03	10.31	3790.61
0.08	21..36	73.12	0.03	14.39	3824.24

The optimal values of the control parameters are obtained as $Q^* = 71.93$, $r^* = 21.39$, $L^* = 0.03$ years and the minimum total annual inventory cost is $ETC =$ Rs.3766.31.

Table 1 shows that, with increasing value of L, the total annual cost ETC first decreases and then starts increasing thereby yielding the optimal values of the control parameters Q, r, L for which the total annual cost is a minimum. This is as expected, since, for fixed values of Q, r, ETC is convex in L. It is also observed that the reorder point r increases as the lead-time L increases. This is expected because, an increase in lead-time, signifies an increase in the lead-time demand D_L. To preserve the non-negativity criterion of the safety stock, this leads to a higher reorder point.

From Table 2, it is observed that increasing Δ_1 (with Δ_2 unchanged) results in an increase in the total cost incurred. Similar observation is made for Δ_2 (with Δ_1 unchanged).

From Table 3, it is observed that when the mean demand m increases, the optimal lot-size Q^*, the optimal reorder point r^* and total annual cost $ETC^*(Q, r, L)$ increase. This result is expected because higher customer demand may amplify the order quantity and the reorder level leading to a higher inventory cost. The effect of the existing lead-time, L_0 has also been shown in Table 4. It is seen that, as L_0 increases, the annual inventory cost increases as well. This is intuitively correct, since a greater lead-time would imply a greater crashing cost required to reduce it, thereby increasing the total annual cost incurred.

5 Concluding Remarks and Scope for Future Research

This paper develops a (Q, r) continuous review inventory model with controllable lead-time under a generalized fuzzy random framework. The annual customer demand is taken to be a normally distributed fuzzy random variable with an associated fuzzy probability density function. The lead-time is controlled by means of an exponential crashing cost. A methodology is developed to determine the optimal values of the lot-size, reorder point, and the lead-time to minimize the crisp equivalent of the total annual inventory cost. A restriction of the present model is that the process quality is assumed to be perfect. As a scope of future research, this model can be further extended to include quality control. The problem of setup cost reduction together with process quality and lead-time control may also be investigated. Other models viz., the periodic review system, integrated inventory models, etc., may also be developed in this fuzzy random framework using normally distributed fuzzy random variable. A restriction of the proposed model is also that the cost parameters are assumed to be deterministic here. This restriction may also be relaxed in future research.

References

1. Hadley, G., Whitin, T.M.: Analysis of Inventory Systems. Prentice Hall Inc., Englewood Cliffs: N.J. (1963)
2. Zadeh, L.A.: Fuzzy sets. Inf. Control **8**, 338–353 (1965)
3. Dey, O., Chakraborty, D.: Fuzzy periodic review system with fuzzy random variable demand. Eur. J. Oper. Res. **198**, 113–120 (2009)
4. Dutta, P., Chakraborty, D., Roy, A.R.: A single-period inventory model with fuzzy random variable demand. Math. Comput. Model. **41**, 915–922 (2005)
5. Dey, O., Chakraborty, D.: A fuzzy random continuous review inventory system. Int. J. Prod. Econ. **132**, 101–106 (2011)
6. Khan, W.F., Dey, O.: Periodic review inventory model with normally distributed fuzzy random variable demand. Int. J. Syst. Sci. Oper. Logist. (2017). https://doi.org/10.1080/23302674.2017.1361481
7. Khan, W.F., Dey, O.: Continuous review inventory model with normally distributed fuzzy random variable demand. Int. J. Appl. Comput. Math. (2018). https://doi.org/10.1007/s40819-018-0564-0
8. Liu, Y.K., Liu, B.: A class of fuzzy random optimization: expected value model. Inf. Sci. **155**, 89–102 (2003)
9. Liao, C.J., Shyu, C.H.: An analytical determination of lead time with normal demand. Int. J. Oper. Prod. Manag. **11**(9), 72–78 (1991)
10. Pan, J.C., Hsiao, Y.C.: Integrated inventory models with controllable lead time and backorder discount considerations. Int. J. Prod. Econ. **93–94**, 387–397 (2005)
11. Chang H.C., Ouyang L.Y., Wu, K.S., Ho, C.H.: Integrated vendor-buyer cooperative inventory models with controllable lead time and ordering cost reduction. Eur. J. Oper. Res. **170**(2), 481–495 (2006). https://doi.org/10.1016/j.ejor.2004.06.029
12. Ma, W-M., Qiu, B-B.: Distribution-free continuous review inventory model with controllable lead time and setup cost in the presence of a service level constraint. Hindawi Publ. Corp. Math. Probl. Eng. (2012). https://doi.org/10.1155/2012/867847

13. Uthayakumar, R., Priyan, S.: Permissible delay in payments in the two-echelon inventory system with controllable setup cost and lead time under service level constraint. Int. J. Inf. Manag. Sci. **24**, 193–211 (2013)
14. Priyan, S., Uthayakumar, R.: Continuous review inventory model with controllable lead time, lost sales rate and order processing cost when the received quantity is uncertain. J. Manuf. Syst. **34**, 23–33 (2015)
15. Chang, H.C., Yao, J.S., Ouyang, L.Y.: Fuzzy mixture inventory model with variable lead-time based on probabilistic fuzzy set and triangular fuzzy number. Math. Comput. Model. **39**, 287–304 (2004)
16. Rong, M., Mahapatra, N.K., Maiti, M.: A multi-objective wholesaler-retailers inventory-distribution model with controllable lead-time based on probabilistic fuzzy set and triangular fuzzy number. Appl. Math. Model. **32**(12), 2670–2685 (2008)
17. Wang, X.: Continuous review inventory model with variable lead time in a fuzzy random environment. Expert Syst. Appl. **38**, 11715–11721 (2011)
18. Dey, O., Chakraborty, D.: A fuzzy random periodic review system with variable lead-time and negative exponential crashing cost. Appl. Math. Model. **36**, 6312–6322 (2012)
19. Shah, N.H., Soni, H.N.: Continuous review inventory model with fuzzy stochastic demand and variable lead time. Int. J. Appl. Ind. Eng. **1**(2), 7–24 (2012)
20. Bhuiya, S.K., Chakraborty, D.: A fuzzy random periodic review inventory model involving controllable back-order rate and variable lead-time. Math. Comput. (2015). https://doi.org/10.1007/978-81-322-2452-5-21
21. Soni, H.N., Patel, K.A.: Continuous review inventory model with reducing lost sales rate under fuzzy stochastic demand and variable lead time. Int. J. Procure. Manag. **8**, 546–569 (2015)
22. Rong, M., Maiti, M.: On an EOQ model with service level constraint under fuzzy-stochastic demand and variable lead-time. Appl. Math. Model. **39**(17), 5230–5240 (2015)
23. Kurdhi, N., Sutanto N.A., Kristanti N.A, Lestari, M.P.: Continuous review inventory models under service level constraint with probabilistic fuzzy number during uncertain received quantity. Int. J. Serv. Oper. Manag. **23**(4), 443 (2016). https://doi.org/10.1504/IJSOM.2016.075247
24. Shin, D., Guchhait, R., Sarkar, B., Mittal, M.: Controllable lead time, service level constraint, and transportation discounts in a continuous review inventory model. RAIRO-Oper. Res. **50**(4–5), 921–934 (2016)
25. Malik, A.I., Sarkar, B.: Optimizing a multi-product continuous-review inventory model with uncertain demand, quality improvement, setup cost reduction, and variation control in lead time. IEEE Access **6**, 36176-36187 (2018). https://doi.org/10.1109/ACCESS.2018.2849694
26. Karthick, B., Uthayakumar, R.: Optimizing an imperfect production model with varying setup cost, price discount, and lead time under fuzzy demand. Process. Integr Optim Sustain. **5**, 13–29 (2021). https://doi.org/10.1007/s41660-020-00133-8
27. Shee, S., Chakrabarti, T.: A fuzzy two-echelon supply chain model for deteriorating items with time varying holding cost involving lead time as a decision variable. In: Optimization and Inventory Management. Asset Analytics (Performance and Safety Management). Springer, Singapore (2020). https://doi.org/10.1007/978-981-13-9698-4_21
28. Tharani, S., Uthayakumar, R.: A novel approach to safety stock management in an integrated supply chain with controllable lead time and ordering cost reduction using present value. RAIRO-Oper. Res. **54**, 1327–1346 (2020)
29. Kurdhi, N.A., Doewes, R.I., Yuliana.: A periodic review inventory system under stochastic demand with controllable safety stock, lead time, and price discount under partially backlogged shortages. Int. J. Value Chain. Manag. **11**, 328 – 345 (2020)
30. Krishnaraj, R.B.: Modeling for inventory involving lead time crashing cost as a Weibull distribution with backorder price discount. J. Interdiscip. Cycle Res. **12**, 379–383 (2020)
31. Castellano, D., Gallo, M., Santillo, L.C.: A periodic review policy for a coordinated single vendor-multiple buyers supply chain with controllable lead time and distribution-free approach. 4OR-Q. J. Oper. Res. (2020). https://doi.org/10.1007/s10288-020-00448-9

32. Dey, B.K., Bhuniya, S., Sarkar, B.: Involvement of controllable lead time and variable demand for a smart manufacturing system under a supply chain management. Expert. Syst. Appl. **184**, (2021). https://doi.org/10.1016/j.eswa.2021.115464
33. Tersine, R.J.: Principles of Inventory and Materials Management. North Holland, New York (1982)
34. Carlsson, C., Fuller, R.: On possiblistic mean value and variance of fuzzy numbers. Fuzzy Sets Syst. **122**, 315–326 (2001)

Existence of Quadruple Fixed Point Results in Ordered K-Metric Space Through C-Distance with Application in Integral Equation

Sudipta Kumar Ghosh and C. Nahak

Abstract In 2011, Berinde and Borcut [9] proved tripled coincidence point theorems in partially ordered metric spaces. In this article, we extend the result of Berinde et al. from tripled to quadruple in a more generalized way, i.e., we extend the result using c-distance under partially ordered cone metric space. In the end, one example is given to justify our main result. To validate our result, we also provide one application in integral equation.

Keywords Quadruple fixed point · c-distance · w^*-compatibility · Ordered sets · g-mixed monotone property

1 Introduction

The concept of metric space was first proposed by $\acute{F}$rechet [14] in 1905. The Banach−Cacciopoli theorem (1922) [8], which was established in the context of complete metric space, is one of the fundamental theorems in the literature of fixed point theory. This theorem became very famous for its versatile application in the non-linear analysis (like integral equations, fractional derivatives, matrix equations, dynamic programming, differential equations, and many more). Ran and Reurings [34] established existing and uniqueness results of a fixed point for operators satisfying a particular type of contraction in an ordered metric space structure, where they applied their work in solving the non-linear matrix equations. In recent times many interesting results related to fixed point theory in complete metric space have come out (for details we refer to the reader [1, 7, 10, 15, 24, 26, 29, 35–39, 43]). On the other site, the concept of cone metric space or K-metric space was first studied by Zabrejko [50]. In the year 2007, Huang and Zhang [16] redefined the definition

S. K. Ghosh (✉) · C. Nahak
Department of Mathematics, Indian Institute of Technology Kharagpur, Kharagpur 721302, India
e-mail: ghosh.sudipta516@gmail.com

C. Nahak
e-mail: cnahak@maths.iitkgp.ac.in

of K-metric space as cone metric space (in brief CMS). Actually one can view a metric space as a particular type of CMS. Later many author studied this space and many results related to fixed point theory have been generalized to CMS (see [2–5, 17, 20, 21, 30–33, 41, 45, 46, 48]). The idea of w-distance was first proposed by Kada et al. [19] in 1996. Following these initial paper, the existence and uniqueness of many well known fixed point results have been investigated largely in the setting of w-distance (for details see [6, 18, 25, 40, 44]). Later in 2011 Cho et al. [12] (see also [49]) extended the concept of w-distance in CMS and named as c-distance. Sintunavarat et al. [47] studied the result of Cho et al. [12]. In the context of partially ordered metric spaces, Berinde et al. [9] proved tripled coincidence fixed point results, which generalized the result of Sabetghadam et al. [42]. Motivated by the work of Sintunavarat et al. [47] we extend the result of Berinde et al. [9] from tripled fixed point theorem to quadrupled fixed point by applying c-distance. We present an example to justify our new finding. Finally, we give an application in system of nonlinear integral equations which have been studied by using mixed monotone operator theory.

2 Preliminaries

We start this section by recalling some basic definitions, lemma, corollary, and remarks. From now we write MMP, NDS, NIS to mean mixed monotone property, non-decreasing sequence, non-increasing sequence, respectively.

Definition 1 ([41]) Let E be a real Banach space and $P \subseteq E$. P is called a cone if
(i) $P \neq \emptyset$, closed and $P \neq \{\theta_E\}$, (θ_E is the zero element);
(ii) $cy_1 + dy_2 \in P$ for all $y_1, y_2 \in P$ and $c, d \in [0, +\infty)$;
(iii) $P \cap (-P) = \{\theta_E\}$.

Suppose $P \subseteq E$ be a given cone, then we can endow the real Banach space E by a partial relation w.r.t P define by $y_1 \preceq_E y_2$ if and only if $y_2 - y_1 \in P$. $y_1 \prec_E y_2$ implies $y_1 \preceq_E y_2$ and $y_1 \neq y_2$, while $y_1 \ll y_2$ implies $y_2 - y_1 \in$ int P, where int P stand for interior of P. The cone P is called normal if there exists a real number $M > 0$ such that for all $y_1, y_2 \in E$,

$$\theta_E \preceq_E y_1 \preceq_E y_2 \text{ implies } \| y_1 \| \leq M \| y_2 \| .$$

Here, M is said to be a normal constant of P if it is the smallest positive number that satisfies the above inequality.

Definition 2 ([41]) Let $X \neq \emptyset$. Assume the function $d : X \times X \mapsto E$ satisfies
(i) $\theta_E \preceq_E d(y_1, y_2)$ for all $y_1, y_2 \in X$ and $d(y_1, y_2) = \theta_E$ if and only if $y_1 = y_2$;
(ii) $d(y_1, y_2) = d(y_2, y_1)$ for all $y_1, y_2 \in X$;
(iii) $d(y_1, y_2) \preceq_E d(y_1, y_3) + d(y_3, y_2)$ for all $y_1, y_2, y_3 \in X$.

Then, d is called a cone metric on X and (X,d) is called a cone metric space (in brief CMS).

Definition 3 ([41]) Let (X,d) be a CMS, $x \in X$ and $\{x_n\}_{n=1}^{+\infty}$ a sequence in X. Then
(i) $\{x_n\}_{n=1}^{+\infty}$ is said to be convergent to x if for every $c \in E$ with $\theta_E \ll c$ there is a N $\in \mathbb{N}$ such that $d(x_n, x) \ll c$ for all $n \geq N$. We denote this by $\lim_{n \to +\infty} x_n = x$ or $x_n \to x$.
(ii) $\{x_n\}_{n=1}^{+\infty}$ is said to be a Cauchy sequence if for every $c \in E$ with $\theta_E \ll c$ there is a $N \in \mathbb{N}$ such that $d(x_n, x_m) \ll c$ for all $n, m \geq N$.
(iii) (X,d) is said to be complete CMS if every Cauchy sequence is convergent.

Remark 1 ([16])
(i) Let $P \subseteq E$ be a given cone, where E stand for real Banach space and $x \preceq_E vx$, where $x \in E$ and $0 < v < 1$, then $x = \theta_E$.
(ii) Let $c \in int P$ with $\theta_E \preceq_E x_n$ and $x_n \to \theta_E$, then there exists $N \in \mathbb{N}$ such that $x_n \ll c$ for all $n \geq N$.

Definition 4 ([12]) Let (X, d) be a CMS. Then a function $q : X \times X \to E$ is called a c-distance on X if the following are satisfied:
(QC_1) $\theta_E \preceq_E q(y_1, y_2)$ for all $y_1, y_2 \in X$;
(QC_2) $q(y_1, y_2) \preceq_E q(y_1, y_3) + q(y_3, y_2)$ for all $y_1, y_2, y_3 \in X$;
(QC_3) for all $y \in X$ and $n \geq 1$, if $q(y, z_n) \preceq_E u$ for some $u = u_y \in P$, then q(y, z) $\preceq_E u$ whenever sequence $\{z_n\}$ in X converging to a point $z \in X$;
(QC_4) for all $c \in E$ with $\theta_E \ll c$, there exists $e \in E$ with $\theta_E \ll e$ such that $q(y_1, y_2) \ll e$ and $q(y_1, y_3) \ll e$ imply $d(y_2, y_3) \ll c$.

Remark 2 ([47]) In the Definition 4, consider (X, d) as a metric space with $E = \mathbb{R}_+$, $P = [0, +\infty)$ and replace the condition (QC_3) with the lower semi-continuity of $q(u, .)$ for all $u \in X$, then c-distance q becomes a w-distance on X. In fact, if $q(x, \cdot)$ is lower semi-continuous(lsc), then condition (QC_3) also holds. Thus, w-distance is a special case of c-distance. Hence, the definition of c-distance is more general.

Example 1 ([47]) Let (X, d) be a CMS with normal cone P. Put $q(y_1, y_2) = d(v, y_2)$ for all $y_1, y_2 \in X$, where $v \in X$ is a fixed point. Then, q is a c-distance.

Example 2 ([47]) Let (X, d) be a CMS with normal cone P. Put $q(y_1, y_2) = d(y_1, y_2)$ for all $y_1, y_2 \in X$. Then, q becomes a c-distance.

Remark 3 ([47]) For c-distance, we state two important remarks.
(R1) $q(y_1, y_2) = \theta_E$ is not necessarily equivalent to $y_1 = y_2$ for all $y_1, y_2 \in X$;
(R2) $q(y_1, y_2) = q(y_2, y_1)$ does not necessarily satisfy.

Next, we state a very important lemma.

Lemma 1 ([13]) *Let (X, d) be a CMS with q be a c-distance on X. Consider two sequences $\{u_n\}$ and $\{v_n\}$ in X with u, v, z $\in$ X. Let $\{\mu_n\}$ and $\{\sigma_n\}$ be two sequences in P converging to θ_E. Then, the following hold:*
(l_1) *If $q(u_n, v) \preceq_E \mu_n$ and $q(u_n, z) \preceq_E \sigma_n$ for $n \in \mathbb{N}$, then $v = z$. In fact, if $q(u, v) = \theta_E$ and $q(u, z) = \theta_E$, then $v = z$.*

(l_2) *A sequence* $\{v_n\}$ *converges to* $z(\in X)$ *if* $q(u_n, v_n) \preceq_E \mu_n$ *and* $q(u_n, z) \preceq_E \sigma_n$ *holds for* $n \in \mathbb{N}$.

(l_3) *A sequence* $\{u_n\}$ *is a Cauchy sequence in X if* $q(u_n, u_m) \preceq_E \mu_n$ *holds for* $m > n$.

(l_4) *A sequence* $\{u_n\}$ *is a Cauchy sequence in X if* $q(v, u_n) \preceq_E \mu_n$ *holds for* $n \in \mathbb{N}$.

From now, we write $(X, \lesssim)$ be a poset (means partially ordered set).

Definition 5 ([11]) Let $(X, \lesssim)$ be a poset. Suppose $g : X \mapsto X$ and $F : X^3 \mapsto X$ be two functions. The function F is said to satisfy g-MMP if for any $u, v, z \in X$

$$u_1, u_2 \in X, \ gu_1 \lesssim gu_2 \text{ implies } F(u_1, v, z) \lesssim F(u_2, v, z);$$

$$v_1, v_2 \in X, \ gv_1 \lesssim gv_2 \text{ implies } F(u, v_1, z) \gtrsim F(u, v_2, z);$$

$$z_1, z_2 \in X, \ gv_1 \lesssim gv_2 \text{ implies } F(u, v, z_1) \lesssim F(u, v, z_2).$$

If we take $g = i_X$ in the Definition 5, then F satisfies the MMP (see [9]).

Definition 6 ([27]) An element $(u, v, z) \in X^3$ is called a tripled coincidence point (in brief t.c.p) of the functions $g : X \mapsto X$ and $F : X^3 \mapsto X$ if $F(u, v, z) = gu$, $F(v, z, u) = gv$ and $F(z, u, v) = gz$.

If we consider $g = i_X$, then (u, v, z) is said to be a tripled fixed point (in brief t.f.p) of the function F (see [9]).

Definition 7 ([22]) Let $(X, \lesssim)$ be a poset. Suppose $g : X \mapsto X$ and $F : X^4 \mapsto X$ be two functions. The function F is said to satisfy g-MMP if for any $a, b, c, d \in X$

$$a_1, a_2 \in X, \ ga_1 \lesssim ga_2 \text{ implies } F(a_1, b, c, d) \lesssim F(a_2, b, c, d);$$

$$b_1, b_2 \in X, \ gb_1 \lesssim gb_2 \text{ implies } F(a, b_1, c, d) \gtrsim F(a, b_2, c, d);$$

$$c_1, c_2 \in X, \ gc_1 \lesssim gc_2 \text{ implies } F(a, b, c_1, d) \lesssim F(a, b, c_2, d);$$

$$d_1, d_2 \in X, \ gd_1 \lesssim gd_2 \text{ implies } F(a, b, c, d_1) \gtrsim F(a, b, c, d_2).$$

If we take $g = i_X$ in the Definition 7, then we say that F satisfies the MMP (see [23]).

Definition 8 ([27]) An element $(\xi_1, \xi_2, \xi_3, \xi_4) \in X^4$ is called a quadruple coincidence point (in brief q.c.p) of $g : X \mapsto X$ and $F : X^4 \mapsto X$ if

$$F(\xi_1, \xi_2, \xi_3, \xi_4) = g(\xi_1), \ F(\xi_2, \xi_3, \xi_4, \xi_1) = g(\xi_2), \ F(\xi_3, \xi_4, \xi_1, \xi_2) = g(\xi_3), \ F(\xi_4, \xi_1, \xi_2, \xi_3)$$

$= g(\xi_4)$ and $(g\xi_1, g\xi_2, g\xi_3, g\xi_4)(\in X^4)$ is called quadrupled point of coincidence (in brief q.p.c).

If we consider $g = i_X$, then $(\xi_1, \xi_2, \xi_3, \xi_4)(\in X^4)$ is said to be quadrupled fixed point of the function F (see [23]).

Definition 9 ([27]) An element $(\xi_1, \xi_2, \xi_3, \xi_4) \in X^4$ is called a common quadruple fixed point (in brief q.f.p) of $g : X \mapsto X$ and $F : X^4 \mapsto X$ if

$$F(\xi_1, \xi_2, \xi_3, \xi_4) = g(\xi_1) = \xi_1, \ F(\xi_2, \xi_3, \xi_4, \xi_1) = g(\xi_2) = \xi_2,$$

$$F(\xi_3, \xi_4, \xi_1, \xi_2) = g(\xi_3) = \xi_3, \ F(\xi_4, \xi_1, \xi_2, \xi_3) = g(\xi_4) = \xi_4.$$

Definition 10 ([22]) Let $g : X \mapsto X$ and $F : X^4 \mapsto X$ be two functions. Then two functions are said to be commutative if for all $\xi_1, \xi_2, \xi_3, \xi_4 \in X$

$$g(F(\xi_1, \xi_2, \xi_3, \xi_4)) = F(g\xi_1, g\xi_2, g\xi_3, g\xi_4).$$

Definition 11 ([51]) Let $g : X \mapsto X$ and $F : X^r \mapsto X$ $(r \geq 2)$ be two functions. Then two functions are called w-compatible if

$$F(g(\xi_1), g(\xi_2), ..., g(\xi_r)) = gF(\xi_1, \xi_2, ..., \xi_r),$$

whenever $(\xi_1, \xi_2, ..., \xi_r) \in C(F, g, r)$, where $C(F, g, r)$ is given by

$$g(\xi_1) = F(\xi_1, \xi_2, ..., \xi_{r-1}, \xi_r), \ g(\xi_2) = F(\xi_2, \xi_3, ..., \xi_r, \xi_1), \ \cdots, g(\xi_r) = F(\xi_r, \xi_1, ..., , \xi_{r-1}).$$

For a quadrupled case consider $r = 4$. The main aim of this paper is to extend the result of Berinde et al. [9] from t.f.p to q.f.p by applying c-distance. Motivated by the concepts of Sintunavarat et al. [47], we generalize the results. Finally, we provide an example and illustrate one application to justify our new findings.

3 Main Results

Theorem 1 *Let $(X, \lesssim)$ be a poset and suppose that (X, d) is a CMS with c-distance q on X. Let, further, $g : X \mapsto X$ and $F : X^4 \mapsto X$ be two functions satisfying $F(X^4) \subseteq g(X)$ and $(g(X), d)$ is a complete subspace of (X, d). Assume that F satisfy g-MMP and suppose that there exists four functions $\alpha, \beta, \gamma, \delta : X^4 \mapsto [0, 1)$ with satisfies*
(a_1) $\alpha(F(x_1^*, y_1^*, z_1^*, w_1^*), F(y_1^*, z_1^*, w_1^*, x_1^*), F(z_1^*, w_1^*, x_1^*, y_1^*), F(w_1^*, x_1^*, y_1^*, z_1^*)) \leq \alpha(gx_1^*, gy_1^*, gz_1^*, gw_1^*)$, this condition also holds for β, γ, δ;
(a_2) $\alpha(x_1^*, y_1^*, z_1^*, w_1^*) = \alpha(y_1^*, z_1^*, w_1^*, x_1^*) = \alpha(z_1^*, w_1^*, x_1^*, y_1^*) = \alpha(w_1^*, x_1^*, y_1^*, z_1^*)$, this condition also holds for β, γ, δ;
(a_3) $(\alpha + \beta + \gamma + \delta)(x_1^*, y_1^*, z_1^*, w_1^*) < 1$;
(a_4)

$$q(F(x_1^*, y_1^*, z_1^*, w_1^*), F(x_2^*, y_2^*, z_2^*, w_2^*)) \preceq_E \alpha(gx_1^*, gy_1^*, gz_1^*, gw_1^*)q(gx_1^*, gx_2^*)$$
$$+\beta(gx_1^*, gy_1^*, gz_1^*, gw_1^*)q(gy_1^*, gy_2^*)$$
$$+\gamma(gx_1^*, gy_1^*, gz_1^*, gw_1^*)q(gz_1^*, gz_2^*)$$
$$+\delta(gx_1^*, gy_1^*, gz_1^*, gw_1^*)q(gw_1^*, gw_2^*),$$

$$(1)$$

for all $x_i^*, y_i^*, z_i^*, w_i^* \in X$, $i \in \{1, 2\}$, and $(g(x_1^*) \lesssim g(x_2^*), g(y_2^*) \lesssim g(y_1^*), g(z_1^*) \lesssim g(z_2^*), g(w_2^*) \lesssim g(w_1^*))$ or $(g(x_2^*) \lesssim g(x_1^*), g(y_1^*) \lesssim g(y_2^*), g(z_2^*) \lesssim g(z_1^*), g(w_1^*) \lesssim g(w_2^*))$. If there exists $x_0, y_0, z_0, w_0 \in X$ such that the following hold

$$g(x_0) \lesssim F(x_0, y_0, z_0, w_0), \quad F(y_0, z_0, w_0, x_0) \lesssim g(y_0),$$

$$g(z_0) \lesssim F(z_0, w_0.x_0, y_0), \quad F(w_0, x_0, y_0, z_0) \lesssim g(w_0).$$

(a_5) Also, consider that X satisfies the following property

(i) if $\{a_n\}$ is a NIS with $\{a_n\} \to a$, then $a \lesssim a_n$, for all n,

(ii) if $\{b_n\}$ is a NDS with $\{b_n\} \to b$, then $b_n \lesssim b$, for all n.

Then, there exists $(h_1, h_2, h_3, h_4) \in X^4$ such that $h_1 = g\xi_1 = F(\xi_1, \xi_2, \xi_3, \xi_4)$, $h_2 = g\xi_2 = F(\xi_2, \xi_3, \xi_4, \xi_1)$, $h_3 = g\xi_3 = F(\xi_3, \xi_4, \xi_1, \xi_2)$, $h_4 = g\xi_4 = F(\xi_4, \xi_1, \xi_2, \xi_3)$ for some $\xi_1, \xi_2, \xi_3, \xi_4 \in X$, that is, $(h_1, h_2, h_3, h_4) \in X^4$ is a q.p.c of F and g with $q(h_i, h_i) = \theta_E$, for $i = 1, 2, 3, 4$.

Proof Define $gx_1 = F(x_0, y_0, z_0, w_0) \gtrsim gx_0$, $gy_1 = F(0, z_0, w_0, x_0) \lesssim gy_0$, $gz_1 = F(z_0, w_0, x_0, y_0) \gtrsim gz_0$, $gw_1 = F(w_0, x_0, y_0, z_0) \lesssim gw_0$. For $r \geq 1$ define $gx_n = F(x_{n-1}, y_{n-1}, z_{n-1}, w_{n-1})$, $gy_n = F(y_{n-1}, z_{n-1}, w_{n-1}, x_{n-1})$, $gz_n = F(z_{n-1}, w_{n-1}, x_{n-1}, y_{n-1})$, $gw_n = F(w_{n-1}, x_{n-1}, y_{n-1}, z_{n-1})$, we can construct such sequence, since $F(X^4) \subseteq g(X)$. As F satisfies g-MMP, we have $gx_1 = F(x_0, y_0, z_0, w_0) \lesssim F(x_1, y_1, z_1, w_1) = gx_2$, $gy_2 = F(y_1, z_1, w_1, x_1) \lesssim F(y_0, z_0, w_0, x_0) = gy_1$, $gz_1 = F(z_0, w_0, x_0, y_0) \lesssim F(z_1, w_1, x_1, y_1) = gz_2$, $gw_2 = F(w_1, x_1, y_1, z_1) \lesssim F(z_0, w_0, x_0, y_0) = gw_1$. Consequently, we get four sequences $\{gx_n\}$, $\{gy_n\}$, $\{gz_n\}$, $\{gw_n\}$ which satisfy the following relation

$$gx_n \lesssim gx_{n+1}, \quad gy_{n+1} \lesssim gy_n, \quad gz_n \lesssim gz_{n+1}, \quad gw_{n+1} \lesssim gw_n.$$

Now, our first target is to show that the sequences $\{gx_n\}$, $\{gy_n\}$, $\{gz_n\}$, $\{gw_n\}$ are Cauchy sequences in $(g(X), d)$. Thus, we have

$$q(gx_n, gx_{n+1}) = q(F(x_{n-1}, y_{n-1}, z_{n-1}, w_{n-1}), F(x_n, y_n, z_n, w_n))$$
$$\preceq_E \alpha(gx_{n-1}, gy_{n-1}, gz_{n-1}, gw_{n-1})q(gx_{n-1}, gx_n)$$
$$+\beta(gx_{n-1}, gy_{n-1}, gz_{n-1}, gw_{n-1})q(gy_{n-1}, gy_n)$$
$$+\gamma(gx_{n-1}, gy_{n-1}, gz_{n-1}, gw_{n-1})q(gz_{n-1}, gz_n)$$
$$+\delta(gx_{n-1}, gy_{n-1}, gz_{n-1}, gw_{n-1})q(gw_{n-1}, gw_n),$$

implies $q(gx_n, gx_{n+1}) \preceq_E \alpha(F(x_{n-2}, y_{n-2}, z_{n-2}, w_{n-2}), F(y_{n-2}, z_{n-2}, w_{n-2}, x_{n-2}),$
$\qquad F(z_{n-2}, w_{n-2}, x_{n-2}, y_{n-2}), F(w_{n-2}, x_{n-2}, y_{n-2}, z_{n-2}))q(gx_{n-1},$
$\qquad gx_n) + \beta(F(x_{n-2}, y_{n-2}, z_{n-2}, w_{n-2}), F(y_{n-2}, z_{n-2}, w_{n-2}, x_{n-2}),$
$\qquad F(z_{n-2}, w_{n-2}, x_{n-2}, y_{n-2}), F(w_{n-2}, x_{n-2}, y_{n-2}, z_{n-2}))q(gy_{n-1},$
$\qquad gy_n) + \gamma((F(x_{n-2}, y_{n-2}, z_{n-2}, w_{n-2}), F(y_{n-2}, z_{n-2}, w_{n-2}, x_{n-2}),$
$\qquad F(z_{n-2}, w_{n-2}, x_{n-2}, y_{n-2}), F(w_{n-2}, x_{n-2}, y_{n-2}, z_{n-2}))q(gz_{n-1},$
$\qquad gz_n) + \delta((F(x_{n-2}, y_{n-2}, z_{n-2}, w_{n-2}), F(y_{n-2}, z_{n-2}, w_{n-2}, x_{n-2}),$
$\qquad F(z_{n-2}, w_{n-2}, x_{n-2}, y_{n-2}), F(w_{n-2}, x_{n-2}, y_{n-2}, z_{n-2}))q(gw_{n-1}, gw_n),$

implies $q(gx_n, gx_{n+1}) \preceq_E \alpha(gx_{n-2}, gy_{n-2}, gz_{n-2}, gw_{n-2})q(gx_{n-1}, gx_n) + \beta(gx_{n-2}, gy_{n-2},$
$\qquad gz_{n-2}, gw_{n-2})q(gy_{n-1}, gy_n) + \gamma(gx_{n-2}, gy_{n-2}, gz_{n-2}, gw_{n-2})q(gz_{n-1},$
$\qquad gz_n) + \delta(gx_{n-2}, gy_{n-2}, gz_{n-2}, gw_{r-2})q(gw_{n-1}, gw_n),$

implies $q(gx_n, gx_{n+1}) \preceq_E \alpha(gx_{n-3}, gy_{n-3}, gz_{n-3}, gw_{n-3})q(gx_{n-1}, gx_n) + \beta(gx_{n-3}, gy_{n-3},$
$\qquad gz_{n-3}, gw_{n-3})q(gy_{n-1}, gy_n) + \gamma(gx_{n-3}, gy_{n-3}, gz_{n-3}, gw_{n-3})q(gz_{n-1},$
$\qquad gz_n) + \delta(gx_{n-3}, gy_{n-3}, gz_{n-3}, gw_{n-3})q(gw_{n-1}, gw_n).$

Proceeding in a similar way, we get,

$$q(gx_n, gx_{n+1}) \preceq_E \alpha(gx_0, gy_0, gz_0, gw_0)q(gx_{n-1}, gx_n) + \beta(gx_0, gy_0, gz_0, gw_0)q(gy_{n-1}, gy_n)$$
$$+\gamma(gx_0, gy_0, gz_0, gw_0)q(gz_{n-1}, gz_n) + \delta(gx_0, gy_0, gz_0, gw_0)q(gw_{n-1}, gw_n). \tag{2}$$

Similarly, by using (a_1), (a_2), we get

$$q(gy_n, gy_{n+1}) \preceq_E \alpha(gx_0, gy_0, gz_0, gw_0)q(gy_{n-1}, gy_n) + \beta(gx_0, gy_0, gz_0, gw_0)q(gz_{n-1}, gz_n)$$
$$+\gamma(gx_0, gy_0, gz_0, gw_0)q(gw_{n-1}, gw_n) + \delta(gx_0, gy_0, gz_0, gw_0)q(gx_{n-1}, gx_n), \tag{3}$$

$$q(gz_n, gz_{n+1}) \preceq_E \alpha(gx_0, gy_0, gz_0, gw_0)q(gz_{n-1}, gz_n) + \beta(gx_0, gy_0, gz_0, gw_0)q(gw_{n-1}, gw_n)$$
$$+\gamma(gx_0, gy_0, gz_0, gw_0)q(gx_{n-1}, gx_n) + \delta(gx_0, gy_0, gz_0, gw_0)q(gy_{n-1}, gy_n), \tag{4}$$

$$q(gw_n, gw_{n+1}) \preceq_E \alpha(gx_0, gy_0, gz_0, gw_0)q(gw_{n-1}, gw_n) + \beta(gx_0, gy_0, gz_0, gw_0)q(gx_{n-1}, gx_n)$$
$$+\gamma(gx_0, gy_0, gz_0, gw_0)q(gy_{n-1}, gy_n) + \delta(gx_0, gy_0, gz_0, gw_0)q(gz_{n-1}, gz_n). \tag{5}$$

Let us put $t = \alpha(gx_0, gy_0, gz_0, gw_0) + \beta(gx_0, gy_0, gz_0, gw_0) + \gamma(gx_0, gy_0, gz_0, gw_0) + \delta(gx_0, gy_0, gz_0, gw_0)$. Clearly, by (a_3), $t < 1$. Now, adding (2) to (5), we obtain

$$q(gx_n, gx_{n+1}) + q(gy_n, gy_{n+1}) + q(gz_n, gz_{n+1}) + q(gw_n, gw_{n+1})$$
$$\preceq_E t[q(gx_{n-1}, gx_n) + q(gy_{n-1}, gy_n) + q(gz_{n-1}, gz_n) + q(gw_{n-1}, gw_n)]. \tag{6}$$

Repeating (6), we obtain

$$q(gx_n, gx_{n+1}) + q(gy_n, gy_{n+1}) + q(gz_n, gz_{n+1}) + q(gw_n, gw_{n+1})$$
$$\preceq_E t^n[q(gx_0, gx_1) + q(gy_0, gy_1) + q(gz_0, gz_1) + q(gw_0, gw_1)]. \tag{7}$$

Put $L^* = [q(gx_0, gx_1) + q(gy_0, gy_1) + q(gz_0, gz_1) + q(gw_0, gw_1)]$.

Suppose $n < m$, then by (QC_2), we have

$$q(gx_n, gx_m) \preceq_E \frac{t^n}{1-t} L^*, \; q(gy_n, gy_m) \preceq_E \frac{t^n}{1-t} L^*,$$
$$q(gz_n, gz_m) \preceq_E \frac{t^n}{1-t} L^*, \; q(gw_n, gw_m) \preceq_E \frac{t^n}{1-t} L^*. \tag{8}$$

Thus, by applying Lemma 1 (l_3), we have that the sequences $\{gx_n\}$, $\{gy_n\}$, $\{gz_n\}$, $\{gw_n\}$ are Cauchy sequences in $(g(X), d)$. Since $(g(X), d)$ is complete, so there exists $\xi_1, \xi_2, \xi_3, \xi_4 \in X$ such that $gx_n \to g\xi_1$, $gy_n \to g\xi_2$, $gz_n \to g\xi_3$, $gw_n \to g\xi_4$. Now, applying (QC_3), (8) with the fact that $gx_n \to g\xi_1$, $gy_n \to g\xi_2$, $gz_n \to g\xi_3$, $gw_n \to g\xi_4$, we get,

$$q(gx_n, g\xi_1) \preceq_E \frac{t^n}{1-t} L^*, \; q(gy_n, g\xi_2) \preceq_E \frac{t^n}{1-t} L^*,$$
$$q(gz_n, g\xi_3) \preceq_E \frac{t^n}{1-t} L^*, \; q(gw_n, g\xi_4) \preceq_E \frac{t^n}{1-t} L^*. \tag{9}$$

Since $\{gx_n\}$, $\{gz_n\}$ are NDS and $\{gy_n\}$, $\{gw_n\}$ are NIS, thus applying (a_5), we obtain

$$q(gx_n, F(\xi_1, \xi_2, \xi_3, \xi_4) = q(F(x_{n-1}, y_{n-1}, z_{n-1}, w_{n-1}), F(\xi_1, \xi_2, \xi_3, \xi_4))$$
$$\preceq_E \alpha(gx_{n-1}, gy_{n-1}, gz_{n-1}, gw_{n-1})q(gx_{n-1}, g\xi_1)$$
$$+\beta(gx_{n-1}, gy_{n-1}, gz_{n-1}, gw_{n-1})q(gy_{n-1}, g\xi_2)$$
$$+\gamma(gx_{n-1}, gy_{n-1}, gz_{n-1}, gw_{n-1})q(gz_{n-1}, g\xi_3)$$
$$+\delta(gx_{n-1}, gy_{n-1}, gz_{n-1}, gw_{n-1})q(gw_{n-1}, g\xi_4),$$

implies $q(gx_n, F(\xi_1, \xi_2, \xi_3, \xi_4) \preceq_E \alpha(gx_{n-2}, gy_{n-2}, gz_{n-2}, gw_{n-2})q(gx_{n-1}, g\xi_1)$
$$+\beta(gx_{n-2}, gy_{n-2}, gz_{n-2}, gw_{n-2})q(gy_{n-1}, g\xi_2)$$
$$+\gamma(gx_{n-2}, gy_{n-2}, gz_{n-2}, gw_{n-2})q(gz_{n-1}, g\xi_3)$$
$$+\delta(gx_{n-2}, gy_{n-2}, gz_{n-2}, gw_{n-2})q(gw_{n-1}, g\xi_4).$$

Continuing in this manner, we obtain

$$q(gx_n, F(\xi_1, \xi_2, \xi_3, \xi_4) \preceq_E \alpha(gx_0, gy_0, gz_0, gw_0)q(gx_{n-1}, g\xi_1)$$
$$+\beta(gx_0, gy_0, gz_0, gw_0)q(gy_{n-2}, g\xi_2)$$
$$+\gamma(gx_0, gy_0, gz_0, gw_0)q(gz_{n-1}, g\xi_3)$$
$$+\delta(gx_0, gy_0, gz_0, gw_0)q(gw_{n-1}, g\xi_4).$$

Now, using (9), we have

$$q(gx_n, F(\xi_1, \xi_2, \xi_3, \xi_4)) \preceq_E \frac{t^{n+1}}{1-t} L^*. \tag{10}$$

Now, from (9–10) and Lemma 1 (l_1), we have $g\xi_1 = F(\xi_1, \xi_2, \xi_3, \xi_4)$. In a similar way we can show that $g\xi_2 = F(\xi_2, \xi_3, \xi_4, \xi_1)$, $g\xi_3 = F(\xi_3, \xi_4, \xi_1, \xi_2)$ and, $g\xi_4 = F(\xi_4, \xi_1, \xi_2, \xi_3)$ by the help of (a_1), (a_2). Put $h_1 = g\xi_1, h_2 = g\xi_2, h_3 = g\xi_3, h_4 = g\xi_4$.

Now,

$$q(h_1, h_1) = q(g\xi_1, g\xi_1) = q(F(\xi_1, \xi_2, \xi_3, \xi_4), F(\xi_1, \xi_2, \xi_3, \xi_4))$$
$$\preceq_E \alpha(g\xi_1, g\xi_2, g\xi_3, g\xi_4)q(g\xi_1, g\xi_1) + \beta(g\xi_1, g\xi_2, g\xi_3, g\xi_4)q(g\xi_2, g\xi_2) \tag{11}$$
$$+\gamma(g\xi_1, g\xi_2, g\xi_3, g\xi_4)q(g\xi_3, g\xi_3) + \delta(g\xi_1, g\xi_2, g\xi_3, g\xi_4)q(g\xi_4, g\xi_4).$$

Similarly, we calculate, $q(h_2, h_2)$, $q(h_3, h_3)$ and $q(h_4, h_4)$. Now, adding $q(h_1, h_1)$, $q(h_2, h_2)$, $q(h_3, h_3)$ and $q(h_4, h_4)$, we get,

$$\sum_{i=1}^{4} q(h_i, h_i) \preceq_E [\alpha + \beta + \gamma + \delta](g\xi_1, g\xi_2, g\xi_3, g\xi_4) \sum_{i=1}^{4} q(h_i, h_i).$$

By (a_3), $[\alpha + \beta + \gamma + \delta](g\xi_1, g\xi_2, g\xi_3, g\xi_4) < 1$ and applying Remark 1 (i), we get, $\sum_{i=1}^{4} q(h_i, h_i) = \theta_E$. But $\theta_E \preceq_E q(h_i, h_i)$ for $i = 1, 2, 3, 4$. Thus, we get $q(h_i, h_i) = \theta_E$ for $i = 1, 2, 3, 4$. Consequently, $(h_1, h_2, h_3, h_4) \in X^4$ is a q.p.c of F and g with $q(h_i, h_i) = \theta_E$ for $i = 1, 2, 3, 4$, and our claim is justified.

Corollary 1 *Let $(X, \precsim)$ be a poset and suppose that (X, d) is a complete CMS with c-distance q on X. Let $F : X^4 \mapsto X$ be a function satisfying $F(X^4) \subseteq X$. Assume that F satisfies MMP and there exist a scalar $\sigma \in [0, 1)$ with*
(a_6)

$$q(F(u_1^*, v_1^*, z_1^*, w_1^*), F(u_2^*, v_2^*, z_2^*, w_2^*)) \preceq_E \frac{\sigma}{4}(q(u_1^*, u_2^*) + q(v_1^*, v_2^*) + q(z_1^*, z_2^*) + q(w_1^*, w_2^*)), \tag{12}$$

for all $u_i^*, v_i^*, z_i^*, w_i^* \in X, i \in \{1, 2\}$ for which $(u_1^* \precsim u_2^*, v_2^* \precsim v_1^*, z_1^* \precsim z_2^*, w_2^* \precsim w_1^*)$ or $(u_2^* \precsim u_1^*, v_1^* \precsim v_2^*, z_2^* \precsim z_1^*, w_1^* \precsim w_2^*)$. If there exists $u_0, v_0, z_0, w_0 \in X$ such that the following hold

$$u_0 \precsim F(u_0, v_0, z_0, w_0), F(v_0, z_0, w_0, u_0) \precsim v_0, z_0 \precsim F(z_0, w_0.u_0, v_0), F(w_0, u_0, v_0, z_0) \precsim w_0.$$

(a_7) *Also, suppose that X satisfies the following property*
(i) if $\{a_n\}$ is a NIS with $\{a_n\} \to a$, then $a \precsim a_n$ for all n,
(ii) if $\{b_n\}$ is a NDS with $\{b_n\} \to b$, then $b_n \precsim b$ for all n.
Then there exists $(\xi_1, \xi_2, \xi_3, \xi_4) \in X^4$ such that $\xi_1 = F(\xi_1, \xi_2, \xi_3, \xi_4)$, $\xi_2 = F(\xi_2, \xi_3, \xi_4, \xi_1)$, $\xi_3 = F(\xi_3, \xi_4, \xi_1, \xi_2)$, $\xi_4 = F(\xi_4, \xi_1, \xi_2, \xi_3)$ for some $\xi_1, \xi_2, \xi_3, \xi_4 \in X$, that is, $(\xi_1, \xi_2, \xi_3, \xi_4) \in X^4$ is a q.f.p of F with $q(\xi_1, \xi_1) = q(\xi_2, \xi_2) = q(\xi_3, \xi_3) = q(\xi_4, \xi_4) = \theta_E$.

Proof It follows by taking $g = i_X$ (i_X =identity map), $\alpha(u_1^*, v_1^*, z_1^*, w_1^*) = \beta(u_1^*, v_1^*, z_1^*, w_1^*) = \gamma(u_1^*, v_1^*, z_1^*, w_1^*) = \delta(u_1^*, v_1^*, z_1^*, w_1^*) = \frac{\sigma}{4}$ for all $(u_1^*, v_1^*, z_1^*, w_1^*) \in X^4$ in Theorem 1.

Theorem 2 *Let $(X, \lesssim)$ be a poset and suppose that (X, d) is a complete CMS with c-distance q on X. Let $g : X \mapsto X$ and $F : X^4 \mapsto X$ be two functions such that $F(X^4) \subseteq g(X)$. Assume that F satisfies g-MMP and suppose there exists four functions $\alpha, \beta, \gamma, \delta : X^4 \mapsto [0, 1)$ with*

(a_8) $\alpha(F(u_1^*, v_1^*, z_1^*, w_1^*), F(v_1^*, z_1^*, w_1^*, u_1^*), F(z_1^*, w_1^*, u_1^*, v_1^*), F(w_1^*, u_1^*, v_1^*, z_1^*)) \leq \alpha(gu_1^*, gv_1^*, gz_1^*, gw_1^*)$, *this condition also holds for β, γ, δ;*

(a_9) $\alpha(u_1^*, v_1^*, z_1^*, w_1^*) = \alpha(v_1^*, z_1^*, w_1^*, u_1^*) = \alpha(z_1^*, w_1^*, u_1^*, v_1^*) = \alpha(w_1^*, u_1^*, v_1^*, z_1^*)$, *this condition also holds for β, γ, δ;*

(a_{10}) $(\alpha + \beta + \gamma + \delta)(u_1^*, v_1^*, z_1^*, w_1^*) < 1$;

(a_{11})

$$
\begin{aligned}
q(F(u_1^*, v_1^*, z_1^*, w_1^*), F(u_2^*, v_2^*, z_2^*, w_2^*)) \preceq_E\ & \alpha(gu_1^*, gv_1^*, gz_1^*, gw_1^*)q(gu_1^*, gu_2^*) \\
& + \beta(gu_1^*, gv_1^*, gz_1^*, gw_1^*)q(gv_1^*, gv_2^*) \\
& + \gamma(gu_1^*, gv_1^*, gz_1^*, gw_1^*)q(gz_1^*, gz_2^*) \\
& + \delta(gu_1^*, gv_1^*, gz_1^*, gw_1^*)q(gw_1^*, gw_2^*),
\end{aligned}
$$
(13)

for all $u_i^*, v_i^*, z_i^*, w_i^* \in X$, $i \in \{1, 2\}$ for which $(g(u_1^*) \lesssim g(u_2^*), g(v_2^*) \lesssim g(v_1^*), g(z_1^*) \lesssim g(z_2^*), g(w_2^*) \lesssim g(w_1^*))$ or $(g(u_2^*) \lesssim g(u_1^*), g(v_1^*) \lesssim g(v_2^*), g(z_2^*) \lesssim g(z_1^*), g(w_1^*) \lesssim g(w_2^*))$. If there exists $u_0, v_0, z_0, w_0 \in X$ such that the following hold

$$g(u_0) \lesssim F(u_0, v_0, z_0, w_0), F(v_0, z_0, w_0, u_0) \lesssim g(v_0),$$

$$g(z_0) \lesssim F(z_0, w_0.u_0, v_0), F(w_0, u_0, v_0, z_0) \lesssim g(w_0).$$

(a_{12}) *Suppose $g \in C(X)$ and $F \in C(X^4, X)$,*

(a_{13}) *g and F are commutative.*

Then, there exists $(h_1, h_2, h_3, h_4) \in X^4$ such that $h_1 = g\xi_1 = F(\xi_1, \xi_2, \xi_3, \xi_4)$, $h_2 = g\xi_2 = F(\xi_2, \xi_3, \xi_4, \xi_1)$, $h_3 = g\xi_3 = (\xi_3, \xi_4, \xi_1, \xi_2)$, $h_4 = g\xi_4 = F(\xi_4, \xi_1, \xi_2, \xi_3)$ for some $\xi_1, \xi_2, \xi_3, \xi_4 \in X$, that is, $(h_1, h_2, h_3, h_4) \in X^4$ is a q.p.c of F and g, with $q(h_i, h_i) = \theta_E$, for $i = 1, 2, 3, 4$.

Proof Let $u_0, v_0, z_0, w_0 \in X$ with $g(u_0) \lesssim F(u_0, v_0, z_0, w_0), F(v_0, z_0, w_0, u_0) \lesssim g(v_0), g(z_0) \lesssim F(z_0, w_0, u_0, v_0)$ and $F(w_0, u_0, v_0, z_0) \lesssim g(w_0)$. Since $F(X^4) \subseteq g(X)$ and F satisfy mixed g-monotone property, by using this we can construct four sequences $\{gu_n\}, \{gv_n\}, \{gz_n\}, \{gw_n\}$ with $gu_n \lesssim gu_{n+1}, gv_{n+1} \lesssim gv_n, gz_n \lesssim gz_{n+1}, gw_{n+1} \lesssim gw_n$. Again, applying the same technique of Theorem 1, we can show that $\{gu_n\}, \{gv_n\}, \{gz_n\}, \{gw_n\}$ are Cauchy sequences. Since X is complete, there exists $\xi_1, \xi_2, \xi_3, \xi_4 \in X$ such that $\lim_{n \to +\infty} gu_n = \xi_1, \lim_{n \to +\infty} gv_n =$

ξ_2, $\lim_{n \to +\infty} gz_n = \xi_3$, $\lim_{n \to +\infty} gw_n = \xi_4$. Again, g is continuous so $\lim_{n \to +\infty} ggu_n = g\xi_1$, $\lim_{n \to +\infty} ggv_n = g\xi_2$, $\lim_{n \to +\infty} ggz_n = g\xi_3$, $\lim_{n \to +\infty} ggw_n = g\xi_4$. As g and F commute, we have

$$gF(u_n, v_n, z_n, w_n) = F(gu_n, gv_n, gz_n, gw_n), \quad gF(v_n, z_n, w_n, u_n) = F(gv_n, gz_n, gw_n, gu_n)$$

$$gF(z_n, w_n, u_n, v_n) = F(gz_n, gw_n, gu_n, gv_n), \quad gF(w_n, u_n, v_n, z_n) = F(gw_n, gu_n, gv_n, gz_n).$$

Since $\lim_{n \to +\infty} gF(u_n, v_n, z_n, w_n) = \lim_{n \to +\infty} F(gu_n, gv_n, gz_n, gw_n)$, using continuity of F, we have $\lim_{n \to +\infty} ggu_{n+1} = F(\lim_{n \to +\infty} gu_n, \lim_{n \to +\infty} gv_n, \lim_{n \to +\infty} gz_n, \lim_{n \to +\infty} gw_n)$ implies $g\xi_1 = F(\xi_1, \xi_2, \xi_3, \xi_4)$. Similarly, one can show $g\xi_2 = F(\xi_2, \xi_3, \xi_4, \xi_1)$, $g\xi_3 = F(\xi_3, \xi_4, \xi_1, \xi_2)$, $g\xi_4 = F(\xi_4, \xi_1, \xi_2, \xi_3)$. Rest of the proof, that is, $q(g\xi_1, g\xi_1) = q(g\xi_2, g\xi_2) = q(g\xi_3, g\xi_3) = q(g\xi_4, g\xi_4) = \theta_E$ can be done from the calculation of Theorem 1.

Inspired by Nashine et al. [28], we introduce the concept of w^*-compatibility for the functions $F : X^4 \mapsto X$ and $g : X \mapsto X$.

Definition 12 Let $g : X \mapsto X$ and $F : X^4 \mapsto X$ be two functions. We say g and F are w^*-compatible if the following hold: $gF(a, a, a, a) = F(ga, ga, ga, ga)$, whenever $ga = F(a, a, a, a)$.

In Definition 11, authors [51] initiated the notion of w-compatibility for the functions $F : X^4 \mapsto X$, and $g : X \mapsto X$ (when $r = 4$). We claim every w-compatible functions are w^*-compatible but the converse is not true. To verify our claim, we consider the following example.

Example 3 Let $X = [0, +\infty)$. Suppose $g : X \mapsto X$ and $F : X^4 \mapsto X$ be two functions defined by,

$$F(x_1, x_2, x_3, x_4) = \begin{cases} 3 & \text{if } (x_1, x_2, x_3, x_4) = (1, 2, 3, 4), \\ 5 & \text{if } (x_1, x_2, x_3, x_4) = (2, 3, 4, 1), \\ 7 & \text{if } (x_1, x_2, x_3, x_4) = (3, 4, 1, 2), \\ 9 & \text{if } (x_1, x_2, x_3, x_4) = (4, 1, 2, 3), \\ 10 & \text{otherwise,} \end{cases}$$

$$g(x_1) = \begin{cases} 9 & \text{if } x_1 = 4, \\ 7 & \text{if } x_1 = 3, \\ 5 & \text{if } x_1 = 2, \\ 3 & \text{if } x_1 = 1, \\ 10 & \text{if } x_1 \in \{8, 10\}, \\ 12 & \text{otherwise.} \end{cases}$$

Then $g(1) = 3 = F(1, 2, 3, 4)$, $g(2) = 5 = F(2, 3, 4, 1)$, $g(3) = 7 = F(3, 4, 1, 2)$ and $g(4) = 9 = F(4, 1, 2, 3)$ but $gF(1, 2, 3, 4) = 7 \neq 10 = F(g(1), g(2), g(3), g(4))$, so g and F are not w-compatible. But $F(x_1, x_1, x_1, x_1) = gx_1$ holds only if $x_1 \in \{8, 10\}$ and in both the cases $gF(x_1, x_1, x_1, x_1) = 10 = F(gx_1, gx_1, gx_1, gx_1)$. So g and F are w^*-compatible, and our claim is justified.

Note that, if $(X, \lesssim)$ is a poset, then we give a partial order relation on the product space X^4 defined as follows: for (a, b, c, d), $(x, y, z, w) \in X^4$, $(a, b, c, d) \prec (x, y, z, w)$ if and only if $a \lesssim x$, $y \lesssim b$, $c \lesssim z$, $w \lesssim d$. We say that any two element of (a, b, c, d), $(x, y, z, w)(\in X^4)$ are comparable if $(a, b, c, d) \prec (x, y, z, w)$ or $(x, y, z, w) \prec (a, b, c, d)$ holds.

Theorem 3 *In addition to the hypotheses of Theorem 1, suppose that for every (a, b, c, d), $(a^*, b^*, c^*, d^*) \in X^4$ there exists $(k, l, r, s) \in X^4$ such that $(F(k, l, r, s), F(l, r, s, k), F(r, s, k, l), F(s, k, l, r))$ are comparable to both*

$$(F(a, b, c, d), F(b, c, d, a), F(c, d, a, b), F(d, a, b, c))$$

and

$$(F(a^*, b^*, c^*, d^*), F(b^*, c^*, d^*, a^*), F(c^*, d^*, a^*, b^*), F(d^*, a^*, b^*, c^*)).$$

Also, suppose F and g satisfie w^*-compatibility condition. Then, F and g have a unique common q.f.p and the structure of the q.f.p looks like (p, p, p, p), that is, the point $(p, p, p, p) \in X^4$ satisfies $p = gp = F(p, p, p, p)$.

Proof By Theorem 1, it is clear that the set of q.c.p is non-empty. Hence, our aim is to show that if there exists two q.c.p (a, b, c, d) and (a^*, b^*, c^*, d^*), that is, $ga = F(a, b, c, d)$, $gb = F(b, c, d, a)$, $gc = F(c, d, a, b)$, $gd = F(d, a, b, c)$, and $ga^* = F(a^*, b^*, c^*, d^*)$, $gb^* = F(b^*, c^*, d^*, a^*)$, $gc^* = F(c^*, d^*, a^*, b^*)$, $gd^* = F(d^*, a^*, b^*, c^*)$, then $(ga, gb, gc, gd) = (ga^*, gb^*, gc^*, gd^*)$ holds. By our assumption, there exists a quadrupled $(k, l, r, s) \in X^4$ such that $(F(k, l, r, s), F(l, r, s, k), F(r, s, k, l), F(s, k, l, r))$ is comparable to both $(F(a, b, c, d), F(b, c, d, a), F(c, d, a, b), F(d, a, b, c))$ and $(F(a^*, b^*, c^*, d^*), F(b^*, c^*, d^*, a^*), F(c^*, d^*, a^*, b^*), F(d^*, a^*, b^*, c^*))$. Put $k_0 = k$, $l_0 = l$, $r_0 = r$, $s_0 = s$ and choose $k_1, l_1, r_1, s_1 \in X$ such that $gk_1 = F(k_0, l_0, r_0, s_0)$, $gl_1 = F(l_0, r_0, s_0, k_0)$, $gr_1 = F(r_0, s_0, k_0, l_0)$, $gs_1 = F(s_0, k_0, l_0, r_0)$. Similarly, by using the same technique of Theorem 1, we can construct sequences $\{gk_n\}$, $\{gl_n\}$, $\{gr_n\}$, $\{gs_n\}$ with $gk_n = F(k_{n-1}, l_{n-1}, r_{n-1}, s_{n-1})$, $gl_n = F(l_{n-1}, r_{n-1}, s_{n-1}, k_{n-1})$, $gr_n = F(r_{n-1}, s_{n-1}, k_{n-1}, l_{n-1})$, $gs_n = F(s_{n-1}, k_{n-1}, l_{n-1}, r_{n-1})$, since $F(X^4) \subseteq g(X)$. In a similar way, we can construct sequences $\{ga_n\}$, $\{gb_n\}$, $\{gc_n\}$, $\{gd_n\}$ and sequences $\{ga_n^*\}$, $\{gb_n^*\}$, $\{gc_n^*\}$, $\{gd_n^*\}$, where $a_0 = a$, $b_0 = b$, $c_0 = c$, $d_0 = d$, and $a_0^* = a^*$, $b_0^* = b^*$, $c_0^* = c^*$, $d_0^* = d^*$. Since $(ga, gb, gc, gd) = (F(a, b, c, d), F(b, c, d, a), F(c, d, a, b), F(d, a, b, c)) = (ga_1, gb_1, gc_1, gd_1)$ and $(F(k, l, r, s), F(l, r, s, k), F(r, s, k, l), F(s, k, l, r)) = (gk_1, gl_1, gr_1, gs_1)$ are comparable, that is, $ga \lesssim gk_1$, $gl_1 \lesssim gb$,

$gc \precsim gr_1$ and $gs_1 \precsim gd$. By simple calculation, we can show that $ga \precsim gk_n, gl_n \precsim gb$, $gc \precsim gr_n$ and $gs_n \precsim gd$ for $n \geq 1$ (since g-MMP satisfies by F). Now, by using (a_1), (a_2) and (1), we get,

$$q(gk_{n+1}, ga) \preceq_E \alpha(gk_0, gl_0, gr_0, gs_0)q(gk_n, ga) + \beta(gk_0, gl_0, gr_0, gs_0)q(gl_n, gb)$$
$$+ \gamma(gk_0, gl_0, gr_0, gs_0)q(gr_n, gc) + (gk_0, gl_0, gr_0, gs_0)q(gs_r, gd).$$

In a similar way, one can calculate above type of inequality for $q(gl_{n+1}, gb)$, $q(gr_{n+1}, gc)$ and $q(gs_{n+1}, gd)$. Adding the terms $q(gk_{n+1}, ga)$, $q(gl_{n+1}, gb)$, $q(gr_{n+1}, gc)$, and $q(gs_{n+1}, gd)$, we obtain

$$q(gk_{n+1}, ga) + q(gl_{n+1}, gb) + q(gr_{n+1}, gc) + q(gs_{n+1}, gd)$$
$$\preceq_E \hat{\imath}^n[q(gk_0, ga) + q(gl_0, gb) + q(gr_0, gc) + q(gs_0, gd)],$$

where $\hat{\imath} = [\alpha + \beta + \gamma + \delta](gk_0, gl_0, gr_0, gs_0)$ with $\hat{\imath} < 1$, by (a_3).

Thus, we get,

$$q(gk_{n+1}, ga) \preceq_E \hat{\imath}^n \mu_1, \quad q(gl_{n+1}, gb) \preceq_E \hat{\imath}^n \mu_1, \tag{14}$$

$$q(gr_{n+1}, gc) \preceq_E \hat{\imath}^n \mu_1, \quad q(gs_{n+1}, gd) \preceq_E \hat{\imath}^n \mu_1. \tag{15}$$

In a similar way, we get,

$$q(gk_{n+1}, ga^*) \preceq_E \hat{\imath}^n \widetilde{L}, \quad q(gl_{n+1}, gb^*) \preceq_E \hat{\imath}^n \widetilde{L}, \tag{16}$$

$$q(gr_{n+1}, gc^*) \preceq_E \hat{\imath}^n \widetilde{L}, \quad q(gs_{n+1}, gd^*) \preceq_E \hat{\imath}^n \widetilde{L}, \tag{17}$$

where $\widetilde{L} = [q(gk_0, ga^*) + q(gl_0, gb^*) + q(gr_0, gc^*) + q(gs_0, gd^*)]$. Thus, by applying (14–17), and Lemma 1 (l_1) we get, $ga = ga^*$. Similarly, we can show that $gb = gb^*$, $gc = gc^*$, $gd = gd^*$. Consequently, (ga, gb, gc, gd) becomes the unique q.p.c of F and g. Note that, if we consider (ga, gb, gc, gd) as a q.p.c of F and g, then we can also view (gb, gc, gd, ga), (gc, gd, ga, gb) and (gd, ga, gb, gc) as a q.p.c of F and g. Since q.p.c of F and g is unique, thus we obtain $ga = gb = gc = gd$. Hence, (ga, ga, ga, ga) is the unique q.p.c of F and g. Next to show that F and g have unique q.f.p of the form (p, p, p, p).

To do this, put $p = ga$. Then we get $p = ga = F(a, a, a, a)$. Now, $gp = gga = gF(a, a, a, a) = F(ga, ga, ga, ga) = F(p, p, p, p)$, since F and g satisfies w^*-compatibility condition. Thus, $gp = F(p, p, p, p)$ is also a q.p.c of F and g. But by uniqueness, we obtain $gp = ga = p$. Thus, we get $p = gp = F(p, p, p, p)$, so (p, p, p, p) is a common q.f.p of F and g. Now we will check uniqueness of fixed point. Suppose (p^*, p^*, p^*, p^*) be another common q.f.p of F and g, then $p^* = gp^* = F(p^*, p^*, p^*, p^*)$. But by the uniqueness of q.p.c of F and g, we have $gp = gp^*$. Consequently, we obtain $p = p^*$ and our claim is justified.

Corollary 2 *Let $(X, \lesssim)$ be a poset and suppose that (X, d) be a complete CMS with c-distance q on X. Let $F : X^4 \mapsto X$ be a function satisfying $F(X^4) \subseteq X$. Assume that F satisfies the MMP and there exist a scalars $\sigma_1, \sigma_2, \sigma_3, \sigma_4 \in [0, 1)$ such that $\sigma_1 + \sigma_2 + \sigma_3 + \sigma_4 < 1$ with*
(a_{14})

$$q(F(u_1^*, v_1^*, z_1^*, w_1^*), F(u_2^*, v_2^*, z_2^*, w_2^*)) \preceq_E \sigma_1 q(u_1^*, u_2^*) + \sigma_2 q(v_1^*, v_2^*) + \sigma_3 q(z_1^*, z_2^*) + \sigma_4 q(w_1^*, w_2^*), \tag{18}$$

for all $u_i^*, v_i^*, z_i^*, w_i^* \in X, i \in \{1, 2\}$ for which $(u_1^* \lesssim u_2^*, v_2^* \lesssim v_1^*, z_1^* \lesssim z_2^*, w_2^* \lesssim w_1^*)$ or $(u_2^* \lesssim u_1^*, v_1^* \lesssim v_2^*, z_2^* \lesssim z_1^*, w_1^* \lesssim w_2^*)$. If there exists $u_0, v_0, z_0, w_0 \in X$ such that the following hold

$$u_0 \lesssim F(u_0, v_0, z_0, w_0), \ F(v_0, z_0, w_0, u_0) \lesssim v_0,$$

$$z_0 \lesssim F(z_0, w_0.u_0, v_0), \ F(w_0, u_0, v_0, z_0) \lesssim w_0.$$

(a_{15}) *Also, suppose that X has the following property*
   (i) *if $\{a_n\}$ is a NIS with $\{a_n\} \to a$, then $a \lesssim a_n$, for all n,*
   (ii) *if $\{b_n\}$ is a NDS with $\{b_n\} \to b$, then $b_n \lesssim b$, for all n.*

(a_{16}) *Further, assume that for every $(u_1^*, v_1^*, z_1^*, w_1^*)$, $(u_2^*, v_2^*, z_2^*, w_2^*) \in X^4$ there exists $(k, l, r, s) \in X^4$ such that $(F(k, l, r, s), F(l, r, s, k), F(r, s, k, l), F(s, k, l, r))$ is comparable to both*

$$(F(u_1^*, v_1^*, z_1^*, w_1^*), F(v_1^*, z_1^*, w_1^*, u_1^*), F(z_1^*, w_1^*, u_1^*, v_1^*), F(w_1^*, u_1^*, v_1^*, z_1^*))$$

and

$$(F(u_2^*, v_2^*, z_2^*, w_2^*), F(v_2^*, z_2^*, w_2^*, u_2^*), F(z_2^*, w_2^*, u_2^*, v_2^*), F(w_2^*, u_2^*, v_2^*, z_2^*)).$$

Then, F has a unique q.f.p $(\tilde{a}, \tilde{b}, \tilde{c}, \tilde{d})$ with $\tilde{a} = \tilde{b} = \tilde{c} = \tilde{d}$.

Proof We can prove it by considering $g = i_X$, $i_X =$ identity map. Consider $\alpha(x, y, z, w) = \sigma_1$, $\beta(x, y, z, w) = \sigma_2$, $\gamma(x, y, z, w) = \sigma_3$, and $\delta(x, y, z, w) = \sigma_4$ in Theorem 3.

Theorem 4 *Let $(X, \lesssim)$ be a poset. Suppose that (X, d) be a CMS with a c-distance q in X. Also suppose that E is a Banach space and $P \subseteq E$ be a given normal cone, with normal constant M. Let $g : X \mapsto X$ and $F : X^4 \mapsto X$ be two functions satisfying $F(X^4) \subseteq g(X)$ with $(g(X), d)$ is a complete subspace of (X, d). Assume that F satisfy g-MMP and there exists four functions $\alpha, \beta, \gamma, \delta : X^4 \mapsto [0, 1)$ with*
(a_{17}) $\alpha(F(u, v, z, w), F(v, z, w, u), F(z, w, u, v), F(w, u, v, z)) \leq \alpha(gu, gv, gz, gw)$, *this condition also holds for β, γ, δ*
(a_{18}) $\alpha(u, v, z, w) = \alpha(v, z, w, u) = \alpha(z, w, u, v) = \alpha(w, u, v, z)$, *this condition also holds for β, γ, δ*
(a_{19}) $(\alpha + \beta + \gamma + \delta)(u, v, z, w) < 1$

(a_{20})

$$
\begin{aligned}
q(F(u, v, z, w), F(a, b, c, d)) \preceq_E\ & \alpha(gu, gv, gz, gw)q(gu, ga) \\
& +\beta(gu, gv, gz, gw)q(gv, gb) \\
& +\gamma(gu, gv, gz, gw)q(gz, gc) \\
& +\delta(gu, gv, gz, gw)q(gw, gd),
\end{aligned}
\tag{19}
$$

(a_{21})

$$
\begin{aligned}
\inf\{\| & q(F(u, v, z, w), ga) \| + \| q(F(v, z, w, u), gb) \| + \| q(F(z, w, u, v), gc) \| \\
& + \| q(F(w, u, v, z), gd) \| + \| q(gu, ga) \| + \| q(gv, gb) \| + \| q(gz, gc) \| \\
& + \| q(gw, gd) \| + \| q(gu, F(u, v, z, w)) \| + \| q(gv, F(v, z, w, u)) \| \\
& + \| q(gz, F(z, w, u, v)) \| + \| q(gw, F(w, u, v, z)) \| \} > 0,\ \textit{for all } ga \neq F(a, b, c, d) \\
& \textit{or } gb \neq F(b, c, d, a)\, \textit{or } gc \neq F(c, d, a, b)\, \textit{or } gd \neq F(d, a, b, c),
\end{aligned}
$$

for all $u, v, z, w, a, b, c, d \in X$, *for which* $(g(u) \lesssim g(a),\ g(b) \lesssim g(v),\ g(z) \lesssim g(c),$ $g(d) \lesssim g(w))$ *or* $(g(a) \lesssim g(u),\ g(v) \lesssim g(b),\ g(c) \lesssim g(z),\ g(w) \lesssim g(d))$. *If there exists* $u_0, v_0, z_0, w_0 \in X$ *such that the following hold*

$$
g(u_0) \lesssim F(u_0, v_0, z_0, w_0),\ F(v_0, z_0, w_0, u_0) \lesssim g(v_0),
$$

$$
g(z_0) \lesssim F(z_0, w_0.u_0, v_0),\ F(w_0, u_0, v_0, z_0) \lesssim g(w_0).
$$

Then there exists $(h_1, h_2, h_3, h_4) \in X^4$ *such that* $h_1 = ga = F(a, b, c, d)$, $h_2 = gb = F(b, c, d, a)$, $h_3 = gc = F(c, d, a, b)$, $h_4 = gd = F(d, a, b, c)$ *for some* $a,$ $b, c, d \in X$, *that is,* $(h_1, h_2, h_3, h_4) \in X^4$ *is a q.p.c of* F *and* g *with* $q(ga, ga) = q(gb, gb) = q(gc, gc) = q(gd, gd) = \theta_E$.

Proof From the proof of Theorem 1, we can easily construct sequences $\{gu_n\}, \{gv_n\},$ $\{gz_n\}$, and $\{gw_n\}$ by induction in such a way $gu_n = F(u_{n-1}, v_{n-1}, z_{n-1}, w_{n-1})$, $gv_n = F(v_{n-1}, z_{n-1}, w_{n-1}, u_{n-1})$, $gz_n = F(z_{n-1}, w_{n-1}, u_{n-1}, v_{n-1})$, $gw_n = F(w_{n-1}, u_{n-1}, v_{n-1}, z_{n-1})$. Similarly, we can show $\{gu_n\}, \{gv_n\}, \{gz_n\}$, and $\{gw_n\}$ are Cauchy sequences in $(g(X), d)$. Since $(g(X), d)$ is a complete CMS, so there exists $a, b, c, d \in X$ such that $gu_n \to ga,\ gv_n \to gb,\ gz_n \to gc$ and $gw_n \to gd$. Now we can obtain the following equations by doing same calculation that have already done to obtain inequalities (8) to (11),

$$
q(gu_n, gu_m) \preceq_E \frac{\tilde{t}^n}{1 - \tilde{t}} N^*,\ q(gv_n, gv_m) \preceq_E \frac{\tilde{t}^n}{1 - \tilde{t}} N^*,
\tag{20}
$$

$$
q(gz_n, gz_m) \preceq_E \frac{\tilde{t}^n}{1 - \tilde{t}} N^*,\ q(gw_n, gw_m) \preceq_E \frac{\tilde{t}^n}{1 - \tilde{t}} N^*,
\tag{21}
$$

where $N^* = q(gu_0, gu_1) + q(gv_0, gv_1) + q(gz_0, gz_1) + q(gw_0, gz_1)$ and $\tilde{t} = [\alpha + \beta + \gamma + \delta](gu_0, gv_0, gz_0, gw_0),\ \tilde{t} < 1$. Since $gu_n \to ga,\ gv_n \to gb,\ gz_n \to gc$ and

$gw_n \to gd$, by using $(QC3)$, inequalities (20) and (21), we obtain

$$q(gu_n, ga) \preceq_E \frac{\tilde{t}^n}{1-\tilde{t}} N^*, \quad q(gv_n, gb) \preceq_E \frac{\tilde{t}^n}{1-\tilde{t}} N^*, \tag{22}$$

$$q(gz_n, gc) \preceq_E \frac{\tilde{t}^n}{1-\tilde{t}} N^*, \quad q(gw_n, gd) \preceq_E \frac{\tilde{t}^n}{1-\tilde{t}} N^*. \tag{23}$$

As P is normal cone with normal constant M, thus we obtain

$$\| q(gu_n, ga) \| \le M \frac{\tilde{t}^n}{1-\tilde{t}} \| N^* \|, \quad \| q(gv_n, gb) \| \le M \frac{\tilde{t}^n}{1-\tilde{t}} \| N^* \|, \tag{24}$$

$$\| q(gz_n, gc) \| \le M \frac{\tilde{t}^n}{1-\tilde{t}} \| N^* \|, \quad \| q(gw_n, gd) \| \le \frac{\tilde{t}^n}{1-\tilde{t}} \| N^* \|. \tag{25}$$

Now put $m = n + 1$ in (20), (21), and again using the fact that P is a normal cone with normal constant M, we obtain

$$\| q(gu_n, gu_{n+1}) \| \le M \frac{\tilde{t}^n}{1-\tilde{t}} \| N^* \|, \quad \| q(gv_n, gv_{n+1}) \| \le M \frac{\tilde{t}^n}{1-\tilde{t}} \| N^* \|, \tag{26}$$

$$\| q(gz_n, gz_{n+1}) \| \le M \frac{\tilde{t}^n}{1-\tilde{t}} \| N^* \|, \quad \| q(gw_n, gw_{n+1}) \| \le \frac{\tilde{t}^n}{1-\tilde{t}} \| N^* \|. \tag{27}$$

Now assume that $ga \ne F(a, b, c, d)$, $gb \ne F(b, c, d, a)$, $gc \ne F(c, d, a, b)$, and $gd \ne F(d, a, b, c)$. Then, we have

$$
\begin{aligned}
0 < &\inf\{\| q(F(u, v, z, w), ga) \| + \| q(F(v, z, w, u), gb) \| + \| q(F(z, w, u, v), gc) \| \\
&+ \| q(F(w, u, v, z), gd) \| + \| q(gu, ga) \| + \| q(gv, gb) \| + \| q(gz, gc) \| + \| q(gw, gd) \| \\
&+ \| q(gu, F(u, v, z, w)) \| + \| q(gv, F(v, z, w, u)) \| + \| q(gz, F(z, w, u, v)) \| \\
&+ \| q(gw, F(w, u, v, z)) \| \} \text{ for all } u, v, z, w \in X \\
\le &\inf\{\| q(F(u_n, v_n, z_n, w_n), ga) \| + \| q(F(v_n, z_n, w_n, u_n), gb) \| + \| q(F(z_n, w_n, \\
&u_n, v_n), gc) \| + \| q(F(w_n, u_n, v_n, z_n), gd) \| + \| q(gu_n, ga) \| + \| q(gv_n, gb) \| \\
&+ \| q(gz_n, gc) \| + \| q(gw_n, gd) \| + \| q(gu_n, F(u_n, v_n, z_n, w_n)) \| + \| q(gv_n, \\
&F(v_n, z_n, w_n, u_n)) \| + \| q(gz_n, F(z_n, w_n, u_n, v_n)) \| + \| q(gw_n, F(w_n, u_n, v_n, z_n)) \| \} \\
= &\inf\{\| q(u_{n+1}, ga) \| + \| q(v_{n+1}, gb) \| + \| q(z_{n+1}, gc) \| + \| q(w_{n+1}, gd) \| \\
&+ \| q(gu_n, ga) \| + \| q(gv_n, gb) \| + \| q(gz_n, gc) \| + \| q(gw_n, gd) \| \\
&+ \| q(gu_n, gu_{n+1}) \| + \| q(gv_n, gv_{n+1}) \| + \| q(gz_n, gz_{n+1}) \| \\
&+ \| q(gw_n, gw_{n+1}) \| \} \\
\le &\inf\{12M \frac{\tilde{t}^n}{1-\tilde{t}} \| N^* \| \} = 0 \text{ as } n \to +\infty.
\end{aligned}
$$

Rest of the proof that is to show $q(ga, ga) = q(gb, gb) = q(gc, gc) = q(gd, gd) = \theta_E$ can be done by using the same technique that have already done in Theorem 1.

4   Some Comparison Between Berinde-Borcut's Result with Our Result

- Berinde and Borcut's results were established in the context of ordered metric space whereas our results are in the context of ordered c-distance which is more general.
- In the Berinde and Borcut's result all the $\alpha, \beta, \gamma, \delta$ are constant, whereas in our case these are functions.
- Berinde and Borcut's results were established for tripled fixed point, whereas our result establish for quadruple fixed point.
- As a special case Berinde and Borcut's results were obtained from our results.

Example 4 Let $X = [0, 1)$, suppose $d : X^2 \to E$ and $q : X^2 \to E$ be two functions given by $d(x, y) = | x - y |$, $q(x, y) = d(x, y)$, where $E = \mathbb{R}$ and $P = \{x \in E : x \geq 0\}$. Then (X, d) is a partially ordered CMS with usual ordering and q is a c-distance defined on X. Let $F : X^4 \to X$ and $g : X \to X$ be two functions given by $F(x, y, z, w) = \frac{x^2}{1944}$ for all $x, y, z, w \in X$ and $gx = \frac{x}{9}$ for all $x \in X$. Let us consider four functions $\alpha, \beta, \gamma, \delta$ from X^4 to $[0, 1)$ as

$$\alpha(x, y, z, w) = \beta(x, y, z, w) = \gamma(x, y, z, w) = \delta(x, y, z, w) = \frac{1 + x + y + z + w}{24} \text{ for all } x, y, z, w \in X.$$

Now notice the followings:

- $F(X^4) \subseteq g(X)$;
- F satisfies g-MMP;
- $[\alpha + \beta + \gamma + \delta](x) < 1$, for all $x \in X^4$;
- $\alpha(x, y, z, w) = \alpha(y, z, w, x) = \alpha(z, w, x, y) = \alpha(w, x, y, z)$, this also holds for β, γ, and δ;
-

$$\alpha(F(x, y, z, w), F(y, z, w, x), F(z, w, x, y), F(w, x, y, z))$$

$$= \alpha\left(\frac{x^2}{1944}, \frac{y^2}{1944}, \frac{z^2}{1944}, \frac{w^2}{1944}\right)$$

$$= \frac{1 + \frac{x^2}{1944} + \frac{y^2}{1944} + \frac{z^2}{1944} + \frac{w^2}{1944}}{24}$$

$$\leq \frac{1 + \frac{x^2}{9} + \frac{y^2}{9} + \frac{z^2}{9} + \frac{w^2}{9}}{24}$$

$$\leq \alpha(gx, gy, gz, gw).$$

This also holds for β, γ, and δ also. Next, we calculate

$$
\begin{aligned}
q(F(x, y, z, w), F(a, b, c, d)) &= d(F(x, y, z, w), F(a, b, c, d)) \\
&= d(\frac{x^2}{1944}, \frac{a^2}{1944}) \\
&= \frac{|x^2 - a^2|}{1944} \\
&= \frac{(x + a)|x - a|}{1944} \\
&\leq \frac{(1 + x)|x - a|}{1944} \\
&= \frac{81(\frac{1}{9} + \frac{x}{9})|\frac{x}{9} - \frac{a}{9}|}{1944} \\
&\leq \frac{(1 + \frac{x^2}{9} + \frac{y^2}{9} + \frac{z^2}{9} + \frac{w^2}{9})|\frac{x}{9} - \frac{a}{9}|}{24} \\
&\leq \alpha(gx, gy, gz, gw)q(gx, ga).
\end{aligned}
$$

Thus, we have

$$
\begin{aligned}
q(F(x, y, z, w), F(a, b, c, d)) &\leq \alpha(gx, gy, gz, gw)q(gx, ga) + \beta(gx, gy, gz, gw)q(gy, gb) \\
&+ \gamma(gx, gy, gz, gw)q(gz, gc) + \delta(gx, gy, gz, gw)q(gw, gd),
\end{aligned}
$$

for all $x, y, z, w, a, b, c, d \in X$. Hence, all the conditions of Theorem 1 are satisfied and $(0, 0, 0, 0)$ is a q.p.c of F and g with $q(0, 0) = 0$.

5 Application

Now, we show one application to investigate under which condition the following system of non-linear integral equations of fredholm type give unique solution.

$$F(u_1, u_2, u_3, u_4)(t) = \int_0^\Omega R(t, \tau) f(\tau, u_1(\tau), u_2(\tau), u_3(\tau), u_4(\tau)) d\tau + h(t),$$

$$F(u_2, u_3, u_4, u_1)(t) = \int_0^\Omega R(t, \tau) f(\tau, u_2(\tau), u_3(\tau), u_4(\tau), u_1(\tau)) d\tau + h(t),$$

$$F(u_3, u_4, u_1, u_2)(t) = \int_0^\Omega R(t, \tau) f(\tau, u_3(\tau), u_4(\tau), u_1(\tau), u_2(\tau)) d\tau + h(t),$$

$$F(u_4, u_1, u_2, u_3)(t) = \int_0^\Omega R(t, \tau) f(\tau, u_4(\tau), u_1(\tau), u_2(\tau), u_3(\tau)) d\tau + h(t),$$

$$(28)$$

where $\Omega \in \mathbb{R}$ with $\Omega > 0$, $t \in [0, \Omega)$, and $u, v, z, w \in C[0, \Omega)$. Now we define a function d on X by,

$$d(x, y) = \sup_{t \in [0, \Omega)} |x(t) - y(t)|,$$

for all $x, y \in X$, where $X = C([0, \Omega], \mathbb{R})$ (the space of all real valued continuous functions on $[0, \Omega)$). Then, (X, d) is a complete metric space. Also, suppose the space X has the following partial relation,

$$x \lesssim y \text{ if and only if } x(t) \leq y(t), \text{ for all } t \in [0, \Omega),$$

where $x, y \in X$. Note that, if $(X, \lesssim)$ is a partially ordered set, then we can endow the product space $X \times X \times X \times X$ by the ordering $(u_1, v_1, z_1, w_1) \prec (u_2, v_2, z_2, w_2)$ if and only if $u_1 \lesssim u_2, v_2 \lesssim v_1, z_1 \lesssim z_2, w_2 \lesssim w_1$ where $(u_i, v_i, z_i, w_i) \in X^4$ for $i = \{1, 2\}$. Now suppose that the following conditions are satisfied:

(b_1) $R \in C([0, \Omega] \times [0, \Omega], [0, +\infty))$ such that $\sup_{0 \leq s, \tau \leq \Omega} R(s, \tau) = H < \frac{1}{\Omega}$;

(b_2) $h \in C([0, \Omega], \mathbb{R})$;

(b_3) $f \in C([0, \Omega] \times \mathbb{R} \times \mathbb{R} \times \mathbb{R} \times \mathbb{R}, \mathbb{R})$;

(b_4) for all $u_i, v_i \in X$, with $v_1 \lesssim u_1, u_2 \lesssim v_2, v_3 \lesssim u_3, u_4 \lesssim v_4$ where $i = 1, 2, 3, 4$ we have

$$0 \leq [f(\tau, u_1(\tau), u_2(\tau), u_3(\tau), u_4(\tau)) - f(\tau, v_1(\tau), v_2(\tau), v_3(\tau), v_4(\tau))]$$
$$\leq \rho_1 |u_1(\tau) - v_1(\tau)| + \rho_2 |u_2(\tau) - v_2(\tau)| + \rho_3 |u_3(\tau) - v_3(\tau)| + \rho_4 |u_4(\tau) - v_4(\tau)|, \text{ for all } \tau \in [0, \Omega)$$

where $\rho_i \geq 0$, $i = 1, 2, 3, 4$ and $\sum_{i=1}^4 \rho_i < 1$;

(b_5) suppose that f is monotone increasing in it's second and fourth components and monotone decreasing in it's third and fifth components;

(b_6) suppose that there exists functions $u_0, v_0, z_0, w_0 \in C([0, \Omega), \mathbb{R})$ with

$$x_0(t) \le \int_0^\Omega R(t, \tau) f(\tau, x_0(\tau), y_0(\tau), z_0(\tau), w_0(\tau)) d\tau + h(t),$$

$$y_0(t) \ge \int_0^\Omega R(t, \tau) f(\tau, y_0(\tau), z_0(\tau), w_0(\tau), x_0(\tau)) d\tau + h(t),$$

$$\tag{29}$$

$$z_0(t) \le \int_0^\Omega R(t, \tau) f(\tau, z_0(\tau), w_0(\tau), x_0(\tau), y_0(\tau)) d\tau + h(t),$$

$$w_0(t) \ge \int_0^\Omega R(t, \tau) f(\tau, w_0(\tau), x_0(\tau), x_0(\tau), z_0(\tau)) d\tau + h(t).$$

Theorem 5 *Under the assumptions* (b_1)–(b_6), *the system of Eq. 28 has a unique solution in* $C([0, \Omega), \mathbb{R})$.

Proof Let us consider an operator $F : X^4 \to X$ given by:

$$F(u_1, u_2, u_3, u_4)(t) = \int_0^\Omega R(t, \tau) f(\tau, u_1(\tau), u_2(\tau), u_3(\tau), u_4(\tau)) d\tau + h(t), \ t \in [0, \Omega).$$

It is easy to see that (u_1, u_2, u_3, u_4) is a solution of the system of non-linear integral Eq. 28 if and only if (u_1, u_2, u_3, u_4) is the common q.f.p of the operator F. Now we show that F satisfies mixed monotone property. To check this let us consider $u_1, u_2 \in X$ with $u_1 \lesssim u_2$. Then, by (b_5) f is monotone increasing in it's second component, we have $f(\tau, u_1(\tau), v(\tau), y(\tau), z(\tau)) \le f(\tau, u_2(\tau), v(\tau), y(\tau), z(\tau))$ for all $\tau \in [0, \Omega)$ implies $F(u_1, v, y, z)(t) \le F(u_2, v, y, z)(t)$ implies $F(u_1, v, y, z) \lesssim F(u_2, v, y, z)$, where $v, y, z \in X$. Consequently, the first component of F is monotone increasing. Similarly, by (b_5) we can show that the third component of F is increasing and second, fourth component of F are decreasing. Hence, F satisfies MMP. Next, we calculate $d(F(u_1, u_2, u_3, u_4), F((v_1, v_2, v_3, v_4))$ for $(v_1, v_2, v_3, v_4) \prec (u_1, u_2, u_3, u_4)$. Since $v_1 \lesssim u_1$, $u_2 \lesssim v_2, v_3 \lesssim u_3, u_4 \lesssim v_4$, thus we have

$$| F(u_1, u_2, u_3, u_4)(t) - F(v_1, v_2, v_3, v_4)(t) |$$

$$= \int_0^\Omega R(t, \tau)[f(\tau, u_1(\tau), u_2(\tau), u_3(\tau), u_4(\tau)) - f(\tau, v_1(\tau), v_2(\tau), v_3(\tau), v_4(\tau))] d\tau$$

$$\le \int_0^\Omega R(t, \tau)[\rho_1 | u_1(\tau) - v_1(\tau) | + \rho_2 | u_2(\tau) - v_2(\tau) | + \rho_3 | u_3(\tau) - v_3(\tau) |$$

$$+ \rho_4 | u_4(\tau) - v_4(\tau) |] d\tau$$

$$\le \left(\int_0^\Omega R(t, \tau) d\tau \right) [\rho_1 d(u_1, v_1) + \rho_2 d(u_2, v_2) + \rho_3 d(u_3, v_3) + \rho_4 d(u_4, v_4)].$$

Hence, we have

$$d(F(u_1, u_2, u_3, u_4), F(v_1, v_2, v_3, v_4)) \le H\Omega[\rho_1 d(u_1, v_1) + \rho_2 d(u_2, v_2) + \rho_3 d(u_3, v_3) + \rho_4 d(u_4, v_4)].$$

Clearly, $H\Omega \sum_{i=1}^4 \rho_i < 1$. Again, from inequality (29), we have $x_0 \lesssim F(x_0, y_0, z_0, w_0)$, $F(y_0, z_0, w_0, x_0) \lesssim y_0, z_0 \lesssim F(z_0, w_0.x_0, y_0), F(w_0, x_0, y_0, z_0) \lesssim w_0$. Now for any $u, v \in X$, $\max\{u(\tau), v(\tau)\}$, $\min\{u(\tau), v(\tau)\} \in X$, for all $\tau \in [0, \Omega]$. Using this concept for any $(u_1, u_2, u_3, u_4), (v_1, v_2, v_3, v_4) \in X^4$, we have

$(\max\{u_1(\tau), v_1(\tau)\}, \min\{u_2(\tau), v_2(\tau)\}, \max\{u_3(\tau), v_3(\tau)\}, \min\{u_4(\tau), v_4(\tau)\})$ is comparable to (u_1, u_2, u_3, u_4) and (v_1, v_2, v_3, v_4). Thus, by applying Corollary 2 we can conclude that the system of non-linear integral Eq. 28 has a unique solution $(u_1, u_2, u_3, u_4) \in X^4$ with $u_1 = u_2 = u_3 = u_4$.

6 Conclusion

In this paper, we initiate the notion of w^*-compatibility for the functions F and g. We prove quadruple fixed point results in the context of c-distance, which generalize many well-known fixed point results that are available in literature. In our new findings, we use a much improved technique to obtain the fixed point results. As a by product, we apply our results to study the existence and uniqueness solution of a system of non-linear integral equations. More well-known metric fixed point results will be studied in the context of c-distance, which orient the future study of the authors.

Acknowledgements The authors are very grateful to the editor and the anonymous reviewers for the valuable comments and several useful suggestions which improved the presentation of the paper. The first author (SKG) would like to thank University Grants Commission, Govt. of India (Sr. No. 2121540966, Ref. No: 20/12/2015(ii)EU-V), for financial support.

References

1. Agarwal, R.P., Karapinar, E., O'Regan, D., Rolda'n-Lo'pez-de-Hierro, A.F.: Fixed Point Theory in Metric Type Spaces. Springer, Switzerland (2015)
2. Abbas, M., Jungck, G.: Common fixed point results for noncommuting mappings without continuity in cone metric spaces. J. Math. Anal. Appl. **341**(1), 416–420 (2008)
3. Abbas, M., Khan, M.A., Radenovic, S.: Common coupled fixed point theorems in cone metric spaces for w-compatible mappings. Appl. Math. Comput. **217**(1), 195–202 (2010)
4. Abbas, M., Rhoades, B.E., Nazir, T.: Common fixed points for four maps in cone metric spaces. Appl. Math. Comput. **216**(1), 80–86 (2010)
5. Aleksić, S., Kadelburg, Z., Mitrović, Z.D., Radenović, S.: A new survey: cone metric spaces. J. Int. Math. Virtual Inst. **9**, 93–121 (2019)
6. Alegre, C., Marin, J., Romaguera, S.: A fixed point theorem for generalized contractions involving w-distances on complete quasi-metric spaces. Fixed Point Theory Appl. **2014** (4) (2014)
7. Aydi, H., Karapinar, E., Mustafa, Z.: Coupled coincidence point results on generalized distance in ordered cone metric spaces. Positivity **17**(4), 979–993 (2013)
8. Banach, S.: Sur les opérations dans les ensembles abstraits et leur application aux équations intégrales. Fund. Math. **3**(1), 133–181 (1922)
9. Berinde, V., Borcut, M.: Tripled fixed point theorems for contractive type mappings in partially ordered metric spaces. Nonlinear Anal. **74**(15), 4889–4897 (2011)
10. Bhaskar, T.G., Lakshmikantham, V.: Fixed point theorems in partially ordered metric spaces and applications. Nonlinear Anal. **65**(7), 1379–1393 (2006)
11. Borcut, M.: Tripled coincidence theorems for contractive type mappings in partially ordered metric spaces. Appl. Math. Comput. **218**(14), 7339–7346 (2012)

12. Cho, Y.J., Saadati, R., Wang, S.: Common fixed point theorems on generalized distance in ordered cone metric spaces. Comput. Math. Appl. **61**(4), 1254–1260 (2011)
13. Dordevic, M., Doric, D., Kadelburg, Z., Radenovic, S., Spasic, D.: Fixed point results under c-distance in tvs-cone metric spaces. Fixed Point Theory Appl., 29 (2011)
14. Fréchet, M.: La notion décart et le calcul fonctionnel. CR Acad. Sci. Paris **140**, 772–774 (1905)
15. Harjani, J., Sadarangani, K.: Generalized contractions in partially ordered metric spaces and applications to ordinary differential equations. Nonlinear Anal. **72**(3–4), 1188–1197 (2010)
16. Huang, L.G., Zhang, X.: Cone metric spaces and fixed point theorems of contractive mappings. J. Math. Anal. Appl. **332**(2), 1468–1476 (2007)
17. Ilić, D., Rakočević, V.: Quasi-contraction on a cone metric space. Appl. Math. Lett. **22**(5), 728–731 (2009)
18. Ilic, D., Rakocevic, V.: Common fixed points for maps on metric space with $w-$distance. Appl. Math. Comput. **199**(2), 599–610 (2008)
19. Kada, O., Suzuki, T., Takahashi, W.: Nonconvex minimization theorems and fixed point theorems in complete metric spaces. Sci. Math. Jpn. **44**(2), 381–391 (1996)
20. Kadelburg, Z., Pavlovic, M., Radenovic, S.: Common fixed point theorems for ordered contractions and quasicontractions in ordered cone metric spaces. Comput. Math. Appl. **59**(9), 3148–3159 (2010)
21. Kadelburg, Z., Radenovic, S., Rakocevic, V.: Remarks on Quasi-contraction on a cone metric space. Appl. Math. Lett. **22**(11), 1674–1679 (2009)
22. Karapinar, E., Berinde, V.: Quadruple fixed point theorems for nonlinear contractions in partially ordered metric spaces. Banach J. Math. Anal. **6**(1), 74–89 (2012)
23. Karapinar, E., Van Luong, N.: Quadruple fixed point theorems for nonlinear contractions. Comput. Math. Appl. **64**(6), 1839–1848 (2012)
24. Lakshmikantham, V., Ciric, L.: Coupled fixed point theorems for nonlinear contractions in partially ordered metric spaces. Nonlinear Anal. **70**(12), 4341–4349 (2009)
25. Lakzian, H., Gopal, D., Sintunavarat, W.: New fixed point results for mappings of contractive type with an application to nonlinear fractional differential equations. J. Fixed Point Theory Appl. **18**(2) (2016)
26. Luong Van , N., Thuan, N.X.: Coupled fixed point theorems for mixed monotone mappings and an application to integral equations. Comput. Math. Appl. **62**(11), 4238–4248 (2011)
27. Mustafa, Z., Aydi, H., Karapinar, E.: Mixed g-monotone property and quadruple fixed point theorems in partially ordered metric spaces. Fixed Point Theory Appl., 71 (2012)
28. Nashine, H.K., Kadelburg, Z., Radenovic, S.: Coupled common fixed point theorems for w^{*}-compatible mappings in ordered cone metric spaces. Appl. Math. Comput. **218**(9), 5422–5432 (2012)
29. Nieto, J.J., Rodríguez-López, R.: Existence and uniqueness of fixed point in partially ordered sets and applications to ordinary differential equations. Acta Math. Sin. (Engl. Ser.) **23**(12), 2205–2212 (2007)
30. Paunović, L.R.: Teorija Apstraktnih Metrickih Prostora-Neki Novi Rezultati. University of Pristina, Leposavic, Serbia (2017)
31. Radenovic, S., Rhoades, B.E.: Fixed point theorem for two non-self mappings in cone metric spaces. Comput. Math. Appl. **57**(10), 1701–1707 (2009)
32. Radenović, S., Vetro, P., Nastasi, A., Quan, L.T.: Coupled fixed point theorems in C*-algebra-valued b-metric spaces. Scientific publications of the state University of Novi Pazar, Ser. A: Appl. Math. Inform. Mech. **9**(1), 81–90 (2017)
33. Rahimi, H., Radenović, S., Rad, G.S., Kumam, P.: Quadrupled fixed point results in abstract metric spaces. Comput. Appl. Math. **33**(3), 671–685 (2014)
34. Ran, A.C., Reurings, M.C.: A fixed point theorem in partially ordered sets and some applications to matrix equations. Proc. Amer. Math. Soc., 1435–1443 (2004)
35. Rao, N.S., Kalyani, K.: Generalized contractions to coupled fixed point theorems in partially ordered metric spaces. J. Sib. Fed. Univ. Math. Phys. **13**(4), 492–502 (2020)
36. Rao, N.S., Kalyani, K.: Coupled fixed point theorems with rational expressions in partially ordered metric spaces. J. Anal. **28**(4), 1085–1095 (2020)

37. Rao, N.S., Kalyani, K.: Unique fixed point theorems in partially ordered metric spaces. Heliyon **6**(11), e05563 (2020)
38. Rao, N.S., Kalyani, K., Khatri, K.: Contractive mapping theorems in Partially ordered metric spaces. Cubo (Temuco) **22**(2), 203–214 (2020)
39. Rao, N.S., Kalyani, K., Mitiku, B.: Fixed point theorems for nonlinear contractive mappings in ordered b-metric space with auxiliary function. BMC Res. Notes **13**(1), 1–8 (2020)
40. Razani, A., Nezhad, Z., Boujary, M.: A fixed point theorem for $w-$distance. Appl. Sci. **11**, 114–117 (2009)
41. Rezapour, S., Hamlbarani, R.: Some notes on the paper Cone metric spaces and fixed point theorems of contractive mappings. J. Math. Anal. Appl. **345**(2), 719–724 (2008)
42. Sabetghadam, F., Masiha, H.P., Sanatpour, A.H.: Some coupled fixed point theorems in cone metric spaces. Fixed Point Theory Appl., Article ID 125426, 8 p. (2009)
43. Sang, Y.: Existence and uniqueness of fixed points for mixed monotone operators with perturbations. Electron. J. Differ. Equ. **233**, 1–16 (2013)
44. Sang, Y., Meng, Q.: Fixed point theorems with generalized altering distance functions in partially ordered metric spaces via w-distances and applications. Fixed Point Theory Appl. **1**, 1–25 (2015)
45. Shatanawi, W.: Partially ordered cone metric spaces and coupled fixed point results. Comput. Math. Appl. **60**(8), 2508–2515 (2010)
46. Shatanawi, W., Karapinar, E., Aydi, H.: Coupled coincidence points in partially ordered cone metric spaces with a c-distance. J. Appl. Math., Article ID 312078, 15 p. (2012)
47. Sintunavarat, W., Cho, Y.J., Kumam, P.: Common fixed point theorems for c-distance in ordered cone metric spaces. Comput. Math. Appl. **62**(4), 1969–1978 (2011)
48. Vetro, P.: Common fixed points in cone metric spaces. Rend. Circ. Mat. Palermo (2) **56** (3), 464–468 (2007)
49. Wang, S., Guo, B.: Distance in cone metric spaces and common fixed point theorems. Appl. Math. Lett. **24**(10), 1735–1739 (2011)
50. Zabrejko, P.P.: K-metric and K-normed linear spaces: survey. Collect. Math. **48**(4), 825–859 (1997)
51. Zhu, L., Zhu, C.X., Chen, C.F., Stojanovic, Ž.: Multidimensional fixed points for generalized ψ-quasi-contractions in quasi-metric-like spaces. J. Inequal. Appl., 27 (2014)

Dice Similarity Measure for Fuzzy Numbers and its Applications in Multi-critera Decision Making and Pattern Recognition

Palash Dutta[ID] **and Bornali Saikia**[ID]

Abstract Owing to the uncertainty classical decision-making process becomes more multifarious and reprehensible, and, consequently, the fuzzy decision-making process was developed. In the fuzzy decision-making process, measuring similarity plays a crucial role in discriminating between two uncertain parameters or sets of objects. Fortunately, different similarity measures have been developed in the literature to manage uncertain factors in our real-life problems; however, some do not yield desirable outcomes. In this regard, an effort has been made to construct a new dice similarity measure of fuzzy numbers based on the exponential area of expected intervals of membership functions. Subsequently, some discussion on essential properties of the proposed similarity measure is presented. To demonstrate the proposed approach's validity and novelty, a comparison between the proposed method and existing methods is performed by adopting some pairs of fuzzy numbers from previous works. At last, multi-criteria decision-making problems and pattern recognition through the concept of proposed dice similarity measure of fuzzy numbers have been demonstrated to show its applicability in our real-life problems.

Keywords Fuzzy set · Trapezoidal fuzzy number · Triangular fuzzy number · Dice similarity measure · Multi-criteria decision making · Pattern recognition

1 Introduction

The similarity measure is an effective tool to measure the closeness between two objects. The main goal of constructing a similarity measure is to capture the commonality between two objects. Many similarity measures have been developed and

P. Dutta (✉) · B. Saikia
Dibrugarh University, Dibrugarh 786004, India
e-mail: palash.dtt@gmail.com; palashdutta@dibru.ac.in

© The Author(s), under exclusive license to Springer Nature Singapore Pte Ltd. 2023   63
P. Gyei-Kark et al. (eds.), *Engineering Mathematics and Computing*,
Studies in Computational Intelligence 1042,
https://doi.org/10.1007/978-981-19-2300-5_5

getting more and more attention among a chunk of researchers due to their wide applications in real-life problem. Historically, Jaccard [1], Dice [2] and Cosine similarity measures [3] are frequently applied to determine the similarity between two objects. However, measuring similarity is entirely complicated when available data or information is vague or imprecise in nature. In 1965, Zadeh [4], the father of Fuzzy Logic, introduced fuzzy set theory (FST) as an extension of conventional set theory to represent uncertain human behavior in mathematical form. Mathematically, it is a set of elements having a degree of membership in the unit closed interval. Afterward, FST was broadly applied in various areas such as decision making, pattern recognition, cluster analysis, medical diagnosis, taxonomy, engineering and technology, and many other areas.

In Multicriteria decision making (MCDM) problems and pattern recognition, the study of similarity measure of fuzzy numbers (FNs) play a significant role. MCDM is an act of choosing the best option from the available options in the presence of multiple, conflicting, and incommensurable criteria. In daily life, MCDM problems are familiar to everyone, such as selecting a candidate in an interview, choosing the best book from the available books, buying a house or car, or many other things. There are two MCDM problems in literature: multiple attribute decision making (MADM) and multiple objective decision making (MODM). MADM is a process of choosing the best alternative from a finite set of alternatives based on multiple attributes. On the other hand, MODM is a process of selecting the best alternative from a finite set of alternatives based on several conflicting criteria. Pattern recognition is an essential part of science for observing the patterns collected from the surroundings and classify them into known patterns. In other words, pattern recognition is a process of matching the received data/information with the data/information already stored in the knowledge base. The concept of similarity measure becomes a more crucial issue in pattern recognition while matching a new sample pattern with a stored pattern to which it closely resembles.

Many similarity measures have been established in the literature. Chen et al. [5, 6], Chen [9, 10], Lee [11], Hsieh and Chen [12], Yong et al. [13] proposed some similarity measures for trapezoidal fuzzy numbers (TrFNs). Further, Wen et al. [14] introduced a new similarity measure with the help of center of gravity, the area, the perimeter, the height of generalized trapezoidal fuzzy number (GTrFNs). In this work, some pairs of generalized fuzzy numbers (GFNs) are used to compare this method with the other existing similarity measures. This method can overcome the drawbacks of the existing similarity measures and applied in pattern recognition and fuzzy risk analysis problems. Similarly, Patra and Modal [16] proposed a similarity measure by using the concepts of center of gravity, the area, the perimeter, the height of GTrFNs. Rezvani [17] proposed a similarity measure between GTrFNs based on left and right apex angles and after that Rezvani and Mousavi [18] continued this work and proposed new similarity measure by combining the concept of the center of gravity, the area, the perimeter, the height of GTrFNs. Also some examples are depicted to compare with the other existing similarity measures. Zhou [19] proposed a similarity measure to calculate the degree of similarity between GTrFNs. In this work, some pairs of different GTrFNs have been illustrated to compare the proposed similarity measure with the other existing similarity measures and finally applied in risk analysis. Khorshidi and Nikfalazar [21] developed a similarity measure to evaluate the degree of similarity between GTrFNs. This similarity measure is proposed by using the concepts of geometric distance, center of gravity (COG), area, perimeter, and height. In this work, this method is compared with the other existing methods by using twenty different sets of GTrFNs and applied in risk analysis. Dhivya and Sridevi [22] introduced a similarity measure for FNs by using the concept of center of gravity (COG) points and intuitionistic fuzzy difference of distance of points of FNs and successfully applied in fingerprint matching. These existing similarity measures sometimes provide identical results for two non-identical FNs. Therefore, Chen and Wen [23] suggested a new similarity measure based on quadratic mean operator and Bevel-projection Difference Averaging operator (BPDA) to solve the difficulties in the existing similarity measures of FNs. A comparison between the new one and the other existing approaches has been done by adopting some pairs of FNs. This approach overcomes all the difficulties in the existing approaches. Recently, Sen [24] proposed a similarity measure for GTrFNs and applied in the field of risk analysis. Dutta [25] proposed a DSM for GFNs with different heights and applied in MCDM problems. Ulucay [31] proposed similarity function of trapezoidal fuzzy multi-numbers and applied in multi-criteria decision making.

Although many SMs such as Chen [6], Yong [13], Farhadinia [15], Patra and Mondal [16] Dhivya and Sridevi [22], Khorshidi and Nikfalazar [21], Rezvani and Mousavi [18], Chen [6], Farhadinia [15], Patra and Mondal [16], Yong [13], Wen et al. [14], Dhivya and Sridevi [22], Khorshidi and Nikfalazar [21], Rezvani and Mousavi [18] are available in the literature, despite having advantages, some disadvantages and limitations are encountered in the approaches. The drawbacks and limitations of the existing approaches motivate us in devising a new similarity measure. The contribution of this work is that the devised SM is an advanced one that can overcome the drawbacks of the earlier works. Furthermore, the present SM can solve not only difficult MCDM and pattern recognition problems but also other real-world problems under uncertainty.

The structure of this work is organized as follows. Section 2 introduces some definitions and basic concepts about Fuzzy sets [4], DSM [2], expected interval (EI) [27]. Section 3 presents the new DSM. Some properties of the proposed DSM have been discussed in this section and also 10 pairs of FNs are adopted from [23] to compare the proposed similarity measure with the other existing similarity measures. In Sects. 4 and 5, we discuss a MCDM problem and pattern recognition through the proposed DSM. Finally, Sect. 6 draws some final results and remarks of the new introduced DSM measure.

2   Preliminaries

In this section discuss some basic definitions related to fuzzy set and similarity measure.

Definition 1 Fuzzy set [4].
   Let X be the non-void set of real numbers. A fuzzy set P on X is defined as

$$P = \{(x, \mu_P(x)) : x \in X\}$$

where $\mu_P(x) : X \rightarrow [0, 1]$ is the degree of membership function(MF) of P.

Definition 2 Intuitionistic Fuzzy set (IFS) [29].
   Let X be the non-void set of real numbers. An IFS P on X is defined as

$$P = \{(x, \mu_P(x), \nu_P(x)) : x \in X\}$$

where $\mu_P(x)$ and $\nu_P(x)$ are the degree of membership function (MF) and non-membership function (NMF) of the element x of P and the following conditions are satisfied

$$0 \leq \mu_P(x) + \nu_P(x) \leq 1$$

and $\pi = 1 - (\leq \mu_P(x) + \nu_P(x))$ is called degree of hesitancy.

Definition 3 Picture Fuzzy set (PFS) [28].
   Let X be the non-void set of real numbers. A PFS P on X is defined as

$$P = \{(x, \mu_P(x), \eta_P(x), \nu_P(x)) : x \in X\}$$

where $\mu_P(x)$, $\eta_P(x)$ and $\nu_P(x)$ are the degree of MF, neutral MF and NMF of the element x of P and the following conditions are satisfied

$$0 \leq \mu_P(x) + \eta_P(x) + \nu_P(x) \leq 1$$

and $\pi = 1 - (\leq \mu_P(x) + \eta_P(x) + \nu_P(x))$ is called degree of refusal membership.

Definition 4 α-cut [25].
   An α-cut set of P is a crisp subset X, which is defined as follows

$$^\alpha P = \{x \mid \mu_P(x) \geq \alpha\}$$

α-cut is a closed interval denoted by $^\alpha P = [L_P(\alpha), R_P(\alpha)]$.

Definition 5 GTrFN [18].

Let a GTrFN $P = <p_1, q_1, r_1, s_1; w_1>$ whose MF $\mu_P(x)$ is defined by

$$\mu_P(x) = \begin{cases} w_1 \frac{x-p_1}{q_1-p_1}, & x \in [p_1, q_1] \\ w_1, & x \in [q_1, r_1] \\ w_1 \frac{s_1-x}{s_1-r_1}, & x \in [r_1, s_1] \\ 0, & otherwise \end{cases}$$

For normal FNs $w_1 = 1$

Definition 6 GTFN [18].

Let a GTFN $P =< p_1, q_1, s_1; w_1 >$ whose MF $\mu_P(x)$ is defined by

$$\mu_P(x) = \begin{cases} w_1 \frac{x-p_1}{q_1-p_1}, & x \in [p_1, q_1] \\ w_1, & x \in [q_1, q_1] \\ w_1 \frac{s_1-x}{s_1-q_1}, & x \in [q_1, s_1] \\ 0, & otherwise \end{cases}$$

For normal fuzzy numbers $w_1 = 1$

Definition 7 Trapezoidal fuzzy positive ideal solution (TrFPIS) [26].

TrFPIS is defined by $I^+ =< 1, 1, 1, 1; 1 >$

Definition 8 Expected interval (EI) [27].

EI of fuzzy set P on X of real numbers R is given by

$$EI(P) = [E_l(P), E_u(P)] = [\int_0^{w_1} P_l(\alpha)d\alpha, \int_0^{w_1} P_u(\alpha)d\alpha]$$

where $P_l(\alpha) = inf\{x \in X | \mu_P(x) \geq \alpha\}$ and $P_u(\alpha) = sup\{x \in X | \mu_P(x) \geq \alpha\}$ are α−cuts of a fuzzy number P.

Definition 9 Dice similarity [2].

Let two vectors $U = (u_1, u_2, ..., u_n)$ and $V = (v_1, v_2, ..., v_n)$ of lenght n where all coordinates are positive real numbers. Then Dice similarity measure between two vectors U and V is defined by

$$D(U, V) = \frac{2U.V}{||U||_2^2 + ||V||_2^2} = \frac{2\sum_{i=1}^n u_i v_i}{\sum_{i=1}^n u_i^2 + \sum_{i=1}^n v_i^2} \tag{1}$$

where $U.V$ is the inner product of the vectors U and V and $||U||_2$ and $||V||_2$ are known as Euclidean or L_2 norms of U and V.

It is also defined that if $u_i = v_i = 0$, for $i = 1, 2, ...n$ then measure value be zero.

The DSM for two vectors is also satisfied the following properties

*P*1. $0 \le D(U, V) \le 1$;

*P*2. $D(U, V) = D(V, U)$;

*P*3. $D(U, V) = 1$ if and only if $U = V$ i.e. $u_i = v_i = 0$, for $i = 1, 2, \ldots n$.

3 New DSM for FNs

Historically Jaccard, Dice and Cosine similarity measures are well known similarity measures and widely applied in the field of decision science. However, it is not possible to define Cosine similarity measure if we consider one vector as zero. For this purpose, Dice proposed DSM to overcome the difficulties in Cosine similarity measure. Therefore, the idea of DSM is used in this work.

In this section, we propose a new DSM. In this approach first we find out the $\alpha-$cuts of FNs and then we take exponential of $\alpha-$cuts. After that exponential area of FNs has been evaluated. Finally the proposed similarity measure has been formulated by using the exponential area of EIs and the existing DSM (1). In addition, we also generalized the proposed dice similarity measure for IFNs and PFNs.

Suppose A and B be two fuzzy set whose $\alpha-$cuts are

$$^{\alpha}A = [L_\alpha(A), \ R_\alpha(A)], \quad ^{\alpha}B = [L_\alpha(B), \ R_\alpha(B)]$$

Taking exponential of $\alpha-$cuts as given below

$$e^{\alpha A} = [e^{L_\alpha(A)}, \ e^{R_\alpha(A)}], \quad e^{\alpha B} = [e^{L_\alpha(B)}, \ e^{R_\alpha(B)}]$$

EIs $EI(A)$ and $EI(B)$ of A and B are evaluated as given below

$$EI(A) = \left[ln\left(\int_0^{w_1} e^{L_\alpha(A)} d\alpha \right), \ ln\left(\int_0^{w_1} e^{R_\alpha(A)} d\alpha \right) \right] = \left[\Delta_A(x_1), \Delta_A(x_2) \right]$$

$$EI(B) = \left[ln\left(\int_0^{w_2} e^{L_\alpha(B)} d\alpha \right), \ ln\left(\int_0^{w_2} e^{R_\alpha(B)} d\alpha \right) \right] = \left[\Delta_B(x_1), \Delta_B(x_2) \right]$$

Dice similarity measure $S_D F(A, B)$ of A and B is defined by using above EIs and Eq. (1)

$$S_{DF}(A, B) = \sum_{i=1}^{2} \frac{2\{\Delta_A(x_i)\Delta_B(x_i)\}}{\{\Delta_A(x_i)\}^2 + \{\Delta_B(x_i)\}^2} \tag{2}$$

Similarly, we can generalize this proposed DSM for IFNs $S_{DIF}(A, B)$ and PFNs $S_{DPF}(A, B)$ by evaluating exponential area of NMFs and NFs.

$$S_{DIF}(A, B) = \sum_{i=1}^{2} \frac{2\{\Delta_A(x_i)\Delta_B(x_i) + \Delta'_A(x_i)\Delta'_B(x_i)\}}{\{\Delta_A(x_i)\}^2 + \{\Delta_B(x_i)\}^2 + \{\Delta'_A(x_i)\}^2 + \{\Delta'_B(x_i)\}^2}$$

$$S_{DPF}(A, B) = \sum_{i=1}^{2} \frac{2\{\Delta_A(x_i)\Delta_B(x_i) + \Delta'_A(x_i)\Delta'_B(x_i) + \Delta''_A(x_i)\Delta''_B(x_i)\}}{\{\Delta_A(x_i)\}^2 + \{\Delta_B(x_i)\}^2 + \{\Delta'_A(x_i)\}^2 + \{\Delta'_B(x_i)\}^2}$$

where, $\Delta'_A(x_i)$, $\Delta'_B(x_i)$ are exponential area of EIs of NMFs and $\Delta''_A(x_i)$, $\Delta''_B(x_i)$ are exponential area of EIs of NFs A and B.

In this paper we will work on DSM of FNs, its properties and applications only.

Remark

(1) Suppose that $P =< p_1, p_2, p_3, p_4; w_1 >$ and $Q =< q_1, q_2, q_3, q_4; w_2 >$ be two GTrFNs. Then the DSM of P and Q is defined by

$$S_{DF}(P, Q) = \frac{2[ln\{w_1(\frac{e^{p_2}-e^{p_1}}{p_2-p_1})\}ln\{w_2(\frac{e^{q_2}-e^{q_1}}{q_2-q_1})\} + ln\{w_1(\frac{e^{p_4}-e^{p_3}}{p_4-p_3})\}ln\{w_2(\frac{e^{q_4}-e^{q_3}}{q_4-q_3})\}]}{\{ln\{w_1(\frac{e^{p_2}-e^{p_1}}{p_2-p_1})\}\}^2 + \{ln\{w_2(\frac{e^{q_2}-e^{q_1}}{q_2-q_1})\}\}^2 + \{ln\{w_1(\frac{e^{p_4}-e^{p_3}}{p_4-p_3})\}\}^2 + \{ln\{w_2(\frac{e^{q_4}-e^{q_3}}{q_4-q_3})\}\}^2} \quad (3)$$

For $p_3 = p_2$ and $q_3 = q_2$, the the above GTrFNs will transform into GTFNs is as defined below

$$S_{DF}(P, Q) = \frac{2[ln\{w_1(\frac{e^{p_2}-e^{p_1}}{p_2-p_1})\}ln\{w_2(\frac{e^{q_2}-e^{q_1}}{q_2-q_1})\} + ln\{w_1(\frac{e^{p_4}-e^{p_2}}{p_4-p_2})\}ln\{w_2(\frac{e^{q_4}-e^{q_2}}{q_4-q_2})\}]}{\{ln\{w_1(\frac{e^{p_2}-e^{p_1}}{p_2-p_1})\}\}^2 + \{ln\{w_2(\frac{e^{q_2}-e^{q_1}}{q_2-q_1})\}\}^2 + \{ln\{w_1(\frac{e^{p_4}-e^{p_2}}{p_4-p_2})\}\}^2 + \{ln\{w_2(\frac{e^{q_4}-e^{q_2}}{q_4-q_2})\}\}^2} \quad (4)$$

(2) If $S_{DF}(A, B) = 0$ then there is no similarity between A and B and $S_{DF}(A, B) = 1$ means A and B are completely similar. Suppose, $S_{DF}(A, B)$ is the similarity between A and B and $S_{DF}(P, Q)$ is the similarity between P and Q and if the value of $S_{DF}(A, B)$ is more than $S_{DF}(P, Q)$ then the similarity between A and B will be more than P and Q.

(3) For $P =< p_1, p_1, p_1, p_1; w_1 >$, logarithm of exponential area

$$ln\left(\int_0^{w_1} e^{L_\alpha(A)}d\alpha\right) = p_1, \quad ln\left(\int_0^{w_1} e^{R_\alpha(A)}d\alpha\right) = p_1.$$

3.1 Properties

The proposed DSM given in (2) satisfies the following properties
(p1) $0 \le S_{DF}(A, B) \le 1$
(p2) $S_{DF}(A, B) = 1 \implies A = B$
(p3) $S_{DF}(A, B) = S_D(B, A)$

Proof (p1) From the definition, it is obvious that $S_{DF}(A, B) \ge 0$.

$$2[ln(\int_0^{w_1} e^{L_\alpha(A)} d\alpha) \, ln(\int_0^{w_2} e^{L_\alpha(B)} d\alpha) + ln(\int_0^{w_1} e^{R_\alpha(A)} d\alpha) \, ln(\int_0^{w_2} e^{R_\alpha(B)} d\alpha)]$$

$$\leq \{ln(\int_0^{w_1} e^{L_\alpha(A)} d\alpha)\}^2 + \{ln(\int_0^{w_1} e^{R_\alpha(A)} d\alpha)\}^2 + \{ln(\int_0^{w_2} e^{L_\alpha(B)} d\alpha)\}^2 + \{ln(\int_0^{w_2} e^{R_\alpha(B)} d\alpha)\}^2$$

Consequently, $0 \leq S_{DF}(A, B) \leq 1$

(p2) From (2) we have

$$S_{DF}(A, B) = \frac{2[ln(\int_0^{w_1} e^{L_\alpha(A)} d\alpha) \, ln(\int_0^{w_2} e^{L_\alpha(B)} d\alpha) + ln(\int_0^{w_1} e^{R_\alpha(A)} d\alpha) \, ln(\int_0^{w_2} e^{R_\alpha(B)} d\alpha)]}{\{ln(\int_0^{w_1} e^{L_\alpha(A)} d\alpha)\}^2 + \{ln(\int_0^{w_1} e^{R_\alpha(A)} d\alpha)\}^2 + \{ln(\int_0^{w_2} e^{L_\alpha(B)} d\alpha)\}^2 + \{ln(\int_0^{w_2} e^{R_\alpha(B)} d\alpha)\}^2}$$

Suppose that $A = B$ then rewrite the above equation,

$$S_{DF}(A, A) = \frac{2[\{ln(\int_0^{w_1} e^{L_\alpha(A)} d\alpha)\}^2 + \{ln(\int_0^{w_1} e^{R_\alpha(A)} d\alpha)\}^2]}{2[\{ln(\int_0^{w_1} e^{L_\alpha(A)} d\alpha)\}^2 + \{ln(\int_0^{w_1} e^{R_\alpha(A)} d\alpha)\}^2]}$$

$$= 1$$

Conversely, suppose that $S_{DF}(A, A) = 1$. Then

$$\implies \frac{2[ln(\int_0^{w_1} e^{L_\alpha(A)} d\alpha) \, ln(\int_0^{w_2} e^{L_\alpha(B)} d\alpha) + ln(\int_0^{w_1} e^{R_\alpha(A)} d\alpha) \, ln(\int_0^{w_2} e^{R_\alpha(B)} d\alpha)]}{\{ln(\int_0^{w_1} e^{L_\alpha(A)} d\alpha)\}^2 + \{ln(\int_0^{w_1} e^{R_\alpha(A)} d\alpha)\}^2 + \{ln(\int_0^{w_2} e^{L_\alpha(B)} d\alpha)\}^2 + \{ln(\int_0^{w_2} e^{R_\alpha(B)} d\alpha)\}^2} = 1$$

$$\implies \{ln(\int_0^{w_1} e^{L_\alpha(A)} d\alpha) - ln(\int_0^{w_2} e^{L_\alpha(B)} d\alpha)\}^2 + \{ln(\int_0^{w_1} e^{R_\alpha(A)} d\alpha) - ln(\int_0^{w_2} e^{R_\alpha(B)} d\alpha)\}^2 = 0$$

$$\implies \{ln(\int_0^{w_1} e^{L_\alpha(A)} d\alpha - \int_0^{w_2} e^{L_\alpha(B)} d\alpha)\}^2 = 0, \quad \{ln(\int_0^{w_1} e^{R_\alpha(A)} d\alpha - \int_0^{w_2} e^{R_\alpha(B)} d\alpha)\}^2 = 0$$

Here, $ln(\int_0^{w_1} e^{L_\alpha(A)} d\alpha) - ln(\int_0^{w_2} e^{L_\alpha(B)} d\alpha) = 0$ and $ln(\int_0^{w_1} e^{R_\alpha(A)} d\alpha) - ln(\int_0^{w_2} e^{R_\alpha(B)} d\alpha) = 0$ if $e^{L_\alpha(A)} = e^{L_\alpha(B)}$, $e^{R_\alpha(A)} = e^{R_\alpha(B)}$ and $w_1 = w_2$, otherwise, $S_D(A, B) \neq 1$ which is a contradiction.

Therefore, $L_\alpha(A) = L_\alpha(B)$ and $R_\alpha(A) = R_\alpha(B)$

$$\implies A = B$$

(P3) $S_{DF}(A, B) = S_{DF}(B, A)$

Since

$$S_{DF}(A, B) = \frac{2[ln(\int_0^{w_1} e^{L_\alpha(A)} d\alpha) \, ln(\int_0^{w_2} e^{L_\alpha(B)} d\alpha) + ln(\int_0^{w_1} e^{R_\alpha(A)} d\alpha) \, ln(\int_0^{w_2} e^{R_\alpha(B)} d\alpha)]}{\{ln(\int_0^{w_1} e^{L_\alpha(A)} d\alpha)\}^2 + \{ln(\int_0^{w_1} e^{R_\alpha(A)} d\alpha)\}^2 + \{ln(\int_0^{w_2} e^{L_\alpha(B)} d\alpha)\}^2 + \{ln(\int_0^{w_2} e^{R_\alpha(B)} d\alpha)\}^2}$$

$$= \frac{2[ln(\int_0^{w_2} e^{L_\alpha(B)} d\alpha) \, ln(\int_0^{w_1} e^{L_\alpha(A)} d\alpha) + ln(\int_0^{w_2} e^{R_\alpha(B)} d\alpha) \, ln(\int_0^{w_1} e^{R_\alpha(A)} d\alpha)]}{\{ln(\int_0^{w_2} e^{L_\alpha(B)} d\alpha)\}^2 + \{ln(\int_0^{w_2} e^{R_\alpha(B)} d\alpha)\}^2 + \{ln(\int_0^{w_1} e^{L_\alpha(A)} d\alpha)\}^2 + \{ln(\int_0^{w_1} e^{R_\alpha(A)} d\alpha)\}^2}$$

$$= S_{DF}(B, A)$$

Hence the proof.

3.2   Comparative Study

In order to show the superiority of the proposed DSM, comparison between proposed DSM and other established similarity measures has been performed by adopting some pairs of FNs from [23] as shown in the following figures. Here we compare the proposed DSM with sixteen existing similarity measures presented by Chen [6], Wen and Chen [9], Lee [11], Hsieh and Chen [12], Yong [13], Wen et al. [14], Frahadinia [15], Patra and Mondal [16], Rezvani and Mousavi [18], Zhou [19], Chen and Huang [20], Khorshidi and Nikfalazar [21], Dhivya and Sridevi [22], Chen and Wen [23], Sen et al. [24], Dutta [25]with the help of ten pairs of FNs.

In the above Fig. 1, set 01 and set 02 are two different sets of FNs. Here, we observe that FNs in set 02 are more similar than those are in set 1. But according to the Table 1, Chen [6], Yong [13], Farhadinia [15], Patra and Mondal [16] methods provide same degree of similarity. Also, Dhivya and Sridevi [22], Khorshidi and Nikfalazar [21], Rezvani and Mousavi [18] give us incorrect result as shown in the Table 1.

From above figure, set 03 and set 04 are clearly two different sets of FNs. In the Fig. 1, it is observed that the similarity of FNs in set 04 are more than those FNs in set 3. However, the Table 1 shows that Chen [6], Farhadinia [15], Patra and Mondal [16] methods provide same degree of similarity. The methods developed by Yong [13], Wen et al. [14], Dhivya and Sridevi [22], Khorshidi and Nikfalazar [21], Rezvani and Mousavi [18] provided incorrect result as shown in the Table 1.

Set 05 and set 06 are two different sets of FNs. From the Fig. 1, it is clear that FNs in set 5 is more similar as compared to those FNs in set 06. But Table 1 shows that Chen [6], Farhadinia [15], Patra and Mondal [16] methods provide same degree of similarity. Table 1 also shows that Chen and Wen [9], Hsieh and Chen [12], Lee [11], Cen and Huang [20], Zhou [19], Dhivya and Sridevi [22], Chen and Wen [23], Dutta [25] methods can not provide satisfactory outcomes.

Set 07 and set 08 are two different pairs of fuzzy numbers. It is observed in that the degree of similarity of fuzzy numbers in set 07 is more than those fuzzy numbers in set 08. However, Table 1 shows that Chen [6], Lee [11], Hsieh and Chen [12] methods provide identical results and Yong [13], Dhivya and Sridevi [22] methods can not provide proper result.

Set 09 and set 10 are two different sets of FNs. Figure 1 shows that similarity measure of FNs in set 10 is more than those FNs in set 09. But Wen et al. [14], Chen and Wen [9], Rezvan and Mousavi [18], Patra and Mondal [16], Dhivya and Sridevi [22], Farhadinia [15] methods yield incorrect result that the similarity between FNs in set 09 is more than the similarity between FNs in set 10.

Set 08 and set 05 are different pairs of FNs. From human intuition point of view it is clear that similarity of FNs in set 05 is more than those FNs in set 08. But Sen et al. [24] provides incorrect result that the similarity between FNs in set 08 is more than the similarity between FNs in set 05.

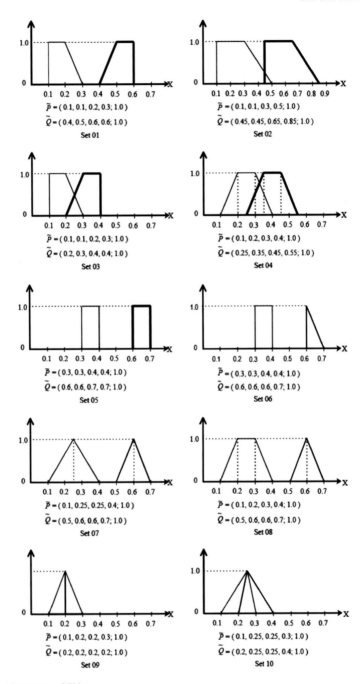

Fig. 1 10 patterns of FNs

Table 1 Comparative table of the proposed DSM with the other existing methods

Existing methods	Set 01	Set 02	Set 03	Set 04	Set 05	Set 06	Set 07	Set 08	Set 09	Set 10
S_C [6]	0.6500	0.6500	0.8500	0.8500	0.9500	0.9500	0.6500	0.6500	0.8000	0.8500
S_{WC} [9]	0.5818	0.6500	0.7577	0.8500	0.8364	0.9500	0.6073	0.5753	0.8000	0.4447
S_L [11]	0.7317	0.7407	0.8571	0.8696	0.9375	0.9524	0.7407	0.7407	0.8333	0.8571
S_{HC} [12]	0.3000	0.5333	0.5000	0.6667	0.5000	0.6667	0.4167	0.4167	0.6000	0.6500
S_Y [13]	0.5394	0.5394	0.7898	0.7875	0.9344	0.9364	0.5382	0.4397	0.7214	0.4933
S_W [14]	0.6583	0.6500	0.8583	0.8500	0.9667	0.9500	0.6073	0.5753	0.8000	0.7744
S_F [15]	0.6500	0.6500	0.8500	0.8500	0.9500	0.9500	0.5871	0.5419	0.8000	0.7632
S_{PM} [16]	0.6500	0.6500	0.8500	0.8500	0.9500	0.9500	0.6338	0.6175	0.8000	0.6503
S_{RM} [18]	0.6583	0.6500	0.8583	0.8500	0.7000	0.6715	0.6073	0.5753	0.8658	0.8333
S_Z [19]	0.7364	0.7533	0.8699	0.8859	0.9454	0.9538	0.7609	0.7591	0.8530	0.8601
S_{CH} [20]	0.6209	0.6500	0.8109	0.8500	0.9060	0.9500	0.5782	0.5488	0.8000	0.3868
S_{KN} [21]	0.8930	0.8904	0.9810	0.9799	0.9195	0.8906	0.8612	0.8348	0.9280	0.9353
S_{DS} [22]	0.6580	0.6500	0.8580	0.8580	0.0915	0.6600	0.1451	0.5500	0.6600	0.9330
S_{CW} [23]	0.5588	0.5909	0.7298	0.8500	0.7000	0.9048	0.5493	0.5488	0.8000	0.3674
S_{SKS} [24]	0.7046	0.1738	0.2231	0.3679	0.025	0.0002	0.0848	0.0969	0.0899	0.6025
S_D [25]	0.6094	0.7435	0.8333	0.9013	0.8364	0.8502	0.7179	0.7159	0.9412	0.9636
Proposed method	0.4869	0.7390	0.8989	0.9074	0.8364	0.4279	0.7156	0.7131	0.1526	0.4815

After analysis the above Table 1 and Fig. 1, it is clear that the proposed DSM provides proper results for all 10 patterns of FNs. So this new method will be more helpful as compared to the other existing methods.

4 MCDM via Proposed DSM

In this section, we will apply the proposed DSM of FNs to solve a MCDM problems where all the available information are expressed in terms of TrFNs. The methodologies and a numerical examples are discussed in the following text.

4.1 Methodology

Step 1: First consider a set of k alternatives $P = (P_1, P_2, ..., P_k)$ and a set of l criteria $C = (C_1, C_2, ..., C_l)$. The preference value of each alternative $P_i(i = 1, 2, ..., k)$ on given criterion $C_j(j = 1, 2, ..., l)$ are in the form of TrFNs $p_{ij} =$

$(p_{ij1}, p_{ij2}, p_{ij3}, p_{ij4})$ $(i = 1, 2, ..., k; j = 1, 2, ..., l)$ and the decision matrix $P = (p_{ij})_{k \times l}$ is constructed as given below,

$$
P = \begin{array}{c} \\ P_1 \\ P_2 \\ \vdots \\ P_k \end{array}
\begin{array}{cccc}
C_1 & C_2 & \cdots & C_l \\
\left[\begin{array}{cccc}
(p_{111}, p_{112}, p_{113}, p_{114}) & (p_{121}, p_{122}, p_{123}, p_{124}) & \cdots & (p_{1l1}, p_{1l2}, p_{1l3}, p_{1l4}) \\
(p_{211}, p_{212}, p_{213}, p_{214}) & (p_{221}, p_{222}, p_{223}, p_{224}) & \cdots & (p_{2l1}, p_{2l2}, p_{2l3}, p_{2l4}) \\
\cdots & \cdots & \cdots & \cdots \\
(p_{k11}, p_{k12}, p_{k13}, p_{k14}) & (p_{k21}, p_{k22}, p_{k23}, p_{k24}) & \cdots & (p_{kl1}, p_{kl2}, p_{kl3}, p_{kl4})
\end{array}\right]
\end{array}
$$

Step 2: The criteria are generally not proportionate and to transform them into comparable values, the decision matrix needs to be normalized by using the following equation

$$
n_{ij} = \frac{p_{ij}}{\max\limits_{i=1,...,k} (p_{ij})}
\tag{5}
$$

$$
R = \begin{array}{c} \\ P_1 \\ P_2 \\ \vdots \\ P_k \end{array}
\begin{array}{cccc}
C_1 & C_2 & \cdots & C_l \\
\left[\begin{array}{cccc}
(n_{111}, n_{112}, n_{113}, n_{114}) & (n_{121}, n_{122}, n_{123}, n_{124}) & \cdots & (n_{1l1}, n_{1l2}, n_{1l3}, n_{1l4}) \\
(n_{211}, n_{212}, n_{213}, n_{214}) & (n_{221}, n_{222}, n_{223}, n_{224}) & \cdots & (n_{2l1}, n_{2l2}, n_{2l3}, n_{2l4}) \\
\cdots & \cdots & \cdots & \cdots \\
(n_{k11}, n_{k12}, n_{k13}, n_{k14}) & (n_{k21}, n_{k22}, n_{k23}, n_{k24}) & \cdots & (n_{kl1}, n_{kl2}, n_{kl3}, n_{kl4})
\end{array}\right]
\end{array}
$$

Step 3: Finally, the similarity between the ideal alternative I^+ and given alternatives P_i is defined by using normalized matrix and proposed DSM (3) as shown below

$$
S_{DF}(P_i, I^+) = \sum_{j=1}^{l} \frac{2\left\{ln\left(\dfrac{e^{n_{ij2}} - e^{n_{ij1}}}{n_{ij2} - n_{ij1}}\right) + ln\left(\dfrac{e^{n_{ij4}} - e^{n_{ij3}}}{n_{ij4} - n_{ij3}}\right)\right\}}{\left\{ln\left(\dfrac{e^{n_{ij2}} - e^{n_{ij1}}}{n_{ij2} - n_{ij1}}\right)\right\}^2 + \left\{ln\left(\dfrac{e^{n_{ij4}} - e^{n_{ij3}}}{n_{ij4} - n_{ij3}}\right)\right\}^2 + 2}
\tag{6}
$$

where $i = 1, 2, ..., k$

by using the above Eq. (6), rank of alternatives can be determined. The bigger value of $S_{DF}(P_i, I^+)$, the better the rank of P_i.

4.1.1 Numerical Example

Suppose a person wants to buy medical insurance from the available insurance companies P_1, P_2, P_3, P_4 and P_5 on the basis of four criteria say C_1, C_2, C_3 and C_4. The ratings of criteria are in term of TrFNs as in the Table 2.

Table 2 Ratings of criteria

	C_1	C_2	C_3	Cr_4
P_1	(0.40, 0.45, 0.55, 0.60)	(0.35, 0.55, 0.65, 0.80)	(0.32, 0.42, 0.57, 0.69)	(0.03, 0.08, 0.13, 0.18)
P_2	(0.90, 0.95, 1.05, 1.10)	(0.40, 0.60, 0.70, 0.85)	(0.40, 0.50, 0.65, 0.70)	(0.05, 0.10, 0.15, 0.20)
P_3	(0.40, 0.45, 0.55, 0.60)	(0.20, 0.50, 0.60, 0.75)	(0.15, 0.30, 0.40, 0.55)	(0.01, 0.06, 0.11, 0.16)
P_4	(0.85, 0.90, 1.00, 1.05)	(0.35, 0.55, 0.65, 0.80)	(0.40, 0.50, 0.65, 0.70)	(0.04, 0.08, 0.13, 0.18)
P_5	(0.55, 0.60, 0.70, 0.75)	(0.25, 0.45, 0.55, 0.70)	(0.19, 0.26, 0.40, 0.50)	(0.02, 0.07, 0.12, 0.17)

Let $P = (p_{ij})_{k \times l}$ be the decision matrix formed by using Table 2

$$P = \begin{bmatrix} (0.40, 0.45, 0.55, 0.60) & (0.35, 0.55, 0.65, 0.80) & (0.32, 0.42, 0.57, 0.69) & (0.03, 0.08, 0.13, 0.18) \\ (0.90, 0.95, 1.05, 1.10) & (0.40, 0.60, 0.70, 0.85) & (0.40, 0.50, 0.65, 0.70) & (0.05, 0.10, 0.15, 0.20) \\ (0.40, 0.45, 0.55, 0.60) & (0.20, 0.50, 0.60, 0.75) & (0.15, 0.30, 0.40, 0.55) & (0.01, 0.06, 0.11, 0.16) \\ (0.85, 0.90, 1.00, 1.05) & (0.35, 0.55, 0.65, 0.80) & (0.40, 0.50, 0.65, 0.70) & (0.04, 0.08, 0.13, 0.18) \\ (0.55, 0.60, 0.70, 0.75) & (0.25, 0.45, 0.55, 0.70) & (0.19, 0.26, 0.40, 0.50) & (0.02, 0.07, 0.12, 0.17) \end{bmatrix}$$

After constructing the decision matrix, we check whether the decision matrix is in the normalized form or not. If it is not in the normalized form, we will normalized the whole matrix by using Eq. (5).

$$R = \begin{bmatrix} (0.44, 0.47, 0.52, 0.59) & (0.88, 0.92, 0.93, 0.94) & (0.80, 0.84, 0.88, 0.98) & (0.60, 0.80, 0.86, 0.90) \\ (1.00, 1.00, 1.00, 1.00) & (1.00, 1.00, 1.00, 1.00) & (1.00, 1.00, 1.00, 1.00) & (1.00, 1.00, 1.00, 1.00) \\ (0.44, 0.47, 0.52, 0.59) & (0.50, 0.83, 0.86, 0.88) & (0.38, 0.60, 0.62, 0.79) & (0.20, 0.60, 0.73, 0.80) \\ (0.94, 0.95, 0.95, 0.95) & (0.88, 0.92, 0.93, 0.94) & (1.00, 1.00, 1.00, 1.00) & (0.80, 0.80, 0.87, 0.9) \\ (0.61, 0.63, 0.66, 0.68) & (0.62, 0.75, 0.79, 0.82) & (0.48, 0.52, 0.62, 0.71) & (0.40, 0.70, 0.80, 0.85) \end{bmatrix}$$

From Table 3, it is observed that P_2 is the best insurance company among the five insurance companies and it is also observed that the ranking results of proposed method is similar with Patra [16], Dutta [25].

4.2 Methodology

Step 1: A company considers a set of k investment alternatives $Q = (Q_1, Q_2, ..., Q_k)$ and each alternative is evaluated from a set of l criteria $C = (C_1, C_2, ..., C_l)$. Due to the presence of uncertainty in MCDM problems, preference value of each alternative $Q_i (i = 1, 2, ..., k)$ on given criterion $C_j (j = 1, 2, ..., l)$ are represented in the

Table 3 Comapring results with Patra [16], Dutta [25]

Alternatives	Proposed method	[16]	[25]
P_1	3.7986	2.7917	3.7523
P_2	4.0000	4.0000	4.0000
P_3	3.4895	2.0508	3.4575
P_4	3.9968	3.6827	3.9795
P_5	3.7912	2.5414	3.6509
Ranking order	$P_2 > P_4 > P_1 > P_5 > P_3$	$P_2 > P_4 > P_1 > P_5 > P_3$	$P_2 > P_4 > P_1 > P_5 > P_3$

form of TFNs, $q_{ij} = (q_{ij1}, q_{ij2}, q_{ij3})$ $(i = 1, 2, ..., k; j = 1, 2, ..., l)$ and the judgment matrix $J = (q_{ij})_{k \times l}$ is constructed based on alternatives and criteria as given below,

$$
J = \begin{array}{c} \\ \varrho_1 \\ \varrho_2 \\ \vdots \\ \varrho_k \end{array}
\begin{array}{cccc}
C_1 & C_2 & .... & C_l \\
\left[\begin{array}{cccc}
(q_{111}, q_{112}, q_{113}) & (q_{121}, q_{122}, q_{123}) & .... & (q_{1l1}, q_{1l2}, q_{1l3}) \\
(q_{211}, q_{212}, q_{213}) & (q_{221}, q_{222}, q_{223}) & .... & (q_{2l1}, q_{2l2}, q_{2l3}) \\
.... & & .... & .... \\
(q_{k11}, q_{k12}, q_{k13}) & (q_{k21}, q_{k22}, q_{k23}) & .... & (q_{kl1}, q_{kl2}, p_{kl3})
\end{array} \right]
\end{array}
$$

Step 2: Generally, there are two categories of criteria containing benefit criterion Δ^{BC} and cost criterion Δ^{CC} in MCDM problem. If the given criteria are not proportionate and to convert them into comparable values, the judgment matrix needs to be normalized by using [30] as given below

For benefit criteria $(C_l \in \Delta^{BC})$

$$
\begin{cases}
n_{ij1} = q_{ij1}/max(q_{ij3}) \\
n_{ij2} = q_{ij2}/max(q_{ij2}) \\
n_{ij3} = q_{ij3}/max(q_{ij1}) \wedge 1 \\
w'_{ij} = w_{ij}/max(w_{ij})
\end{cases}
\tag{7}
$$

For cost criteria $(C_l \in \Delta^{CC})$

$$
\begin{cases}
n_{ij1} = max(q_{ij3})/q_{ij1} \\
n_{ij2} = max(q_{ij2})/q_{ij2} \\
n_{ij3} = max(q_{ij1})/q_{ij2} \wedge 1 \\
w'_{ij} = max(w_{ij})/w_{ij}
\end{cases}
\tag{8}
$$

Here, we consider $w_{ij} = 1$ for $i = 1, 2, ..., k$ and $j = 1, 2, ..., l$

Then the normalized judgment matrix will be of the form

$$J' = \begin{array}{c} \\ Q_1 \\ Q_2 \\ \vdots \\ Q_k \end{array} \begin{bmatrix} \overset{C_1}{(n_{111}, n_{112}, n_{113})} & \overset{C_2}{(n_{121}, n_{122}, n_{123})} & \overset{\cdots}{\cdots\cdots} & \overset{C_l}{(n_{1l1}, n_{1l2}, n_{1l3})} \\ (n_{211}, n_{212}, n_{213}) & (n_{221}, n_{222}, n_{223}) & \cdots\cdots & (n_{2l1}, n_{2l2}, n_{2l3}) \\ & \cdots\cdots & \cdots\cdots & \cdots\cdots & \cdots\cdots \\ (n_{k11}, n_{k12}, n_{k13}) & (n_{k21}, n_{k22}, n_{k23}) & \cdots\cdots & (n_{kl1}, n_{kl2}, n_{kl3}) \end{bmatrix}$$

Step 3: Finally, we evaluate the similarity between the ideal alternative I^+ and given alternatives Q_i by using normalized matrix and proposed DSM of TFN as shown below

$$S_{DF}(Q_i, I^+) = \sum_{j=1}^{l} \frac{2\left\{ ln\left(\frac{e^{n_{ij2}} - e^{n_{ij1}}}{n_{ij2} - n_{ij1}}\right) + ln\left(\frac{e^{n_{ij3}} - e^{n_{ij2}}}{n_{ij3} - n_{ij2}}\right) \right\}}{\left\{ ln\left(\frac{e^{n_{ij2}} - e^{n_{ij1}}}{n_{ij2} - n_{ij1}}\right) \right\}^2 + \left\{ ln\left(\frac{e^{n_{ij3}} - e^{n_{i23}}}{n_{ij3} - n_{ij2}}\right) \right\}^2 + 2} \tag{9}$$

where $i = 1, 2, ..., k$

With the help of above Eq. (9), ordering of alternatives can be determined. The bigger value of $S_{DF}(Q_i, I^+)$, the better the rank of Q_i.

4.2.1 Numerical Example

Suppose a company 'Y' considers four available investment alternatives Q_1, Q_2, Q_3 and Q_4 and each alternative is chosen from four criteria C_1, C_2, C_3 and C_4. Here C_2 & C_3 profit criteria and C_1 & C_4 are cost criteria. Let $J = (q_{ij})_{k \times l}$ be the judgment matrix is constructed based on alternatives and criteria as shown below

$$J = \begin{bmatrix} (5, 6, 7) & (3, 4, 5) & (4, 5, 6) & (0.4, 0.5, 0.6) \\ (9, 10, 11) & (5, 5.5, 6) & (5, 5.5, 6) & (1.4, 1.7, 2) \\ (5, 5.5, 6) & (4, 4.5, 5) & (3, 3.5, 4) & (0.8, 0.9, 1) \\ (8, 9, 10) & (3, 3.5, 4) & (3, 3.5, 4) & (0.5, 0.6, 0.7) \end{bmatrix}$$

After constructing the judgment matrix, we check whether the matrix is in the normalized form or not. If it is not in the normalized form, then we will make it normalized by using Eqs. (7) and (8) as shown below.

$$J' = \begin{bmatrix} (0.71, 0.92, 1.00) & (0.50, 0.73, 1.00) & (0.67, 0.91, 1.00) & (0.67, 1.00, 1.00) \\ (0.45, 0.55, 0.67) & (0.83, 1.00, 1.00) & (0.83, 1.00, 1.00) & (0.20, 0.29, 0.43) \\ (0.83, 1.00, 1.00) & (0.67, 0.82, 1.00) & (0.50, 0.64, 0.80) & (0.40, 0.56, 0.75) \\ (0.50, 0.61, 0.75) & (0.50, 0.64, 0.80) & (0.50, 0.64, 0.80) & (0.57, 0.83, 1.00) \end{bmatrix}$$

Now evaluate $S_{DF}(Q_i, I^+)$ with the help of normalized judgment matrix J' and the Eq. (9), we have

$S_{DF}(Q_1, I^+) = 3.8499$, $S_{DF}(Q_2, I^+) = 3.3968$, $S_{DF}(Q_3, I^+) = 3.7393$ and $S_{DF}(Q_4, I^+) = 3.6792$

Form the above it is observed that the degree of similarity between Q_1 and the ideal alternative I^+ is maximum, so Q_1 is the best alternative among the available alternatives and also the required ordering of alternatives is $Q_1 > Q_3 > Q_4 > Q_2$.

5 Pattern Recognition via Proposed DSM

Again this section describes about the application of proposed DSM in the field of pattern recognition. In this section we adopt some data from Yong et al. [13] to show the significance of our proposed DSM.

Suppose that, there exist l patterns which are in the form of generalized or normal TrFNs P_i, $1 \leq i \leq l$. Again, suppose that there be a sample pattern to recognized which is in the form of generalized or normal TrFN P.

$$S_{DF}(P_j, P) = \max_{1 \leq i \leq l} \{S_D(P_i, P)\} \tag{10}$$

According to the above formula (5), the pattern sample P belongs to that pattern P_j whose have maximum $S_D(P_j, P)$.

5.1 Numerical Example

Consider, five patterns P_1, P_2, P_3, P_4 and P_5 which are of the form of TrFNs as shown in the following Table 4. Suppose, a new sample pattern P = (0.1614, 0.2683, 0.7052, 1.095; 1.0) which will be recognized.

From the above Table 5, it is observed that pattern P belongs to the pattern P_4 as $S_D(P_4, P)$ has maximum value than the others. This numerical example shows the applicability of proposed DSM and provides similar result as [6–8, 13].

Table 4 Patteren description in terms of FNs

S. no.	Patterns	TrFNs
1	P_1	(0.0, 0.0, 0.02, 0.07; 1.0)
2	P_2	(0.04, 0.1, 0.18, 0.23; 1.0)
3	P_3	(0.17, 0.22, 0.36, 0.42; 1.0)
4	P_4	(0.32, 0.41, 0.58, 0.65; 1.0)
5	P_5	(1.0, 1.0, 1.0, 1.0, 0.8)

Table 5 Comparision the outputs of the proposed measure with the other existing measures

S. no.	Patterns	[6]	[7]	[8]	[13]	Proposed measure
1	$S_D(P_1, P)$	0.4650	0.5995	0.1962	0.1854	0.0765
2	$S_D(P_2, P)$	0.5800	0.7332	0.3226	0.3079	0.4399
3	$S_D(P_3, P)$	0.7307	0.8557	0.5092	0.4947	0.6442
4	$S_D(P_4, P)$	0.7824	0.9353	0.7056	0.6876	0.7444
5	$S_D(P_5, P)$	0.5100	0.6664	0.1958	0.2046	0.3828

6 Conclusion

In this work, a new DSM based on the exponential area of EIs is proposed for TrFNs and TFNs. The proposed DSM is an extended and straightforward form of DSM of vectors. In addition, we generalized this new DSM through the concept of IFNs and PFNs. Some essential properties of the proposed DSM are discussed in this work. Despite many similarity measures, no one can discriminate against FNs adequately in all situations. Here, we adopted some numerical examples from [23] for comparative study to illustrate the advantages of the proposed DSM. Moreover, to study the application of the proposed DSM, an MCDM problem and pattern recognition has been performed using numerical examples where all available information are expressed in the form of TrFNs. Although the present SM is quite well and has tremendous advantages, it has a limitation that it could not evaluate similarity value for bell-shaped fuzzy numbers. In the future, we will work on MCDM problems through the concept of intuitionistic and picture fuzzy DSMs.

References

1. Jaccard, P.: Distribution de la flore alpine dans le Bassin des Drouces et dans quelques regions voisines. Bulletin de la Societe Vaudoise des Sciences Naturelles **37**(140), 241–272 (1901)
2. Dice, L.R.: Measures the amount of ecologic association between species. Ecology **26**, 297–302 (1945)
3. Salton, G., Mcgill, M.J.: Introduction to Modern Information Retrieval. Auckland, McGrae-hill (1983)
4. Zadeh L. A.: Fuzzy sets. Inform. Control **8**, 338–356 (1965). https://doi.org/10.1016/S0019-9958(65)90241-X
5. Chen, S.M., Yeh, M.S., Hsiao, P.Y.: A comparison of similarity measures of fuzzy values. Fuzzy Sets Syst. **72**(1), 79–89 (1995)
6. Chen, S.M.: New method for subjective mental workload assessment and fuzzy risk analysis. Cybern. Syst. **27**(5), 449–472 (1996)
7. Lee, H.S.: Optimal consensus of fuzzy opinions under group decision making environment. Fuzzy Sets Syst. **132**, 303–315 (2002)
8. Chen, S.J., Chen, S.M.: Fuzzy risk analysis based on similarity measures of generalized fuzzy numbers. IEEE Trans. Fuzzy Syst. **11**, 45–56 (2003)

9. Chen, S.J.: A new similarity measures of generalized fuzzy numbers based on geometric-mean averaging operator. Fuzzy Systems, IEEE International Conference on 2006 (2006)
10. Chen, S.J.: A new similarity measures between fuzzy numbers using quadratic-mean operator in intelligent information technology application,. IITA'08. Second International Symposium on 2008, IEEE (2008)
11. Lee, H.S.: An optimal aggregation method for fuzzy opinions of group decision. In: Systems, Man, and Cybernetics, 1999. IEEE SMC'99 Conference Proceedings. 1999 IEEE International Conference on, Vol. 3, pp. 314–319. IEEE(1999)
12. Hsieh, C.H., Chen, S.H.: Similarity of generalized fuzzy numbers with graded mean integration representation. In: Proceeding of 8th International Fuzzy Systems Association World Congress, Taipei, Taiwan, Republic of Chaina, Vol. 2, pp. 551–55
13. Yong, D., Wenkang, S., Feng, D., Qi, L.: A new similarity measure of generalized fuzzy numbers and its application to pattern recognition. Pattern Recogn. Lett. 25(8), 875–883 (2004)
14. Wen, J., Fana, X., Duanmua, D., Yong, D.: A modified similarity measures of generalized fuzzy numbers. Procedia Eng. 15, 2773–2777 (2011)
15. Farhadinia, B.: On the Similarity Measure of Generalized Fuzzy Numbers Based on the Geometric Distance and the Perimeter Concepts (2012)
16. Patra, K., Mondal, S.K.: Fuzzy risk analysis using area and height based similarity measure on generalized trapezoidal fuzzy numbers and its application. Appl. Soft Comput. 28, 276–284 (2015)
17. Rezvani, S.: A new similarity measure of generalized fuzzy numbers based on left and right apex angles (I). Palestine J. Math. 4(1), 117U126 (2015)
18. Rezvani, S., Mousavi, M.: A new similarity measure of generalized fuzzy numbers based on left and right apex angles (II). Palestine J. Math. 4(2) (2015)
19. Zhou, Y.: A novel similarity measure for generalized trapezoidal fuzzy numbers and its application to decision making. Int. J. u-and e-Serv. Sci. Technol. 9(3), 131–148 (2016)
20. Chen, S.J., Huang, S.J.: Similarity measure between generalized fuzzy numbers based on quadratic mean. Eng. Technol. Comput. Basic Appl. Sci. (ECBA), vol. 407, Hong Kong (2017)
21. Khorshidi, H., Nikfalazar, S.: An improved similarity measure for generalized fuzzy numbers and its application in fuzzy risk analysis. Appl. Soft Comput. 52, 478–486 (2017)
22. Dhivya, J., Sridevi, B.: Intuitionistic fuzzy similarity measure for generalized fuzzy numbers and its application in fingerprint matching. IETE J. Res. (2018). https://doi.org/10.1080/03772063.2018.1433081
23. Chen, S.J., Wen, C.C.: New method for measuring the similarity of generalized fuzzy numbers. Int. J. Adv. Electron. Comput. Sci. (2018) ISSN: 2393-2835
24. Sen, S., Patra, K., Mondal, S.K.: A new approach to similarity measure for generalized trapezoidal fuzzy numbers and its application to fuzzy risk analysis. Granul. Comput. (2020). https://doi.org/10.1007/s41066-020-00227-1
25. Dutta, P.: An advanced dice similarity measure of generalized fuzzy numbers and its application in multicriteria decision making. Arab J. Basic Appl. Sci. 27(1), 75–92 (2020)
26. Saini, N., Bajaj, R.K., Gandotra, N., Dwivedi, R.P.: Multi-criteria decision making with triangular intuitionistic fuzzy number based on distance measure & parametric entropy approach. Procedia Comput. Sci. 125, 34–41 (2018)
27. Ye, J.: Multicriteria decision-making method using the Dice similarity measure between expected intervals of trapezoidal fuzzy numbers. J. Decis. Syst. 21(4), 307–317 (2012)
28. Joshi, D., Kumar, S.: An approach to multi-criteria decision making problems using dice similarity measure for picture fuzzy sets. In: International Conference on Mathematics and Computing, pp. 135–140. Springer, Singapore (2018)
29. Singh, A., Kumar, S.: A novel dice similarity measure for IFSs and its applications in pattern and face recognition. Expert Syst. Appl. 149, 113245 (2020)
30. Xu, Z., Shang, S., Qian, W., Shu, W.: A method for fuzzy risk analysis based on the new similarity of trapezoidal fuzzy numbers. Expert Syst. Appl. 37(3), 1920–1927 (2010)
31. Ulucay, V.: A new similarity function of trapezoidal fuzzy multi-numbers based on multi-criteria decision making. J. Inst. Sci. Technol. 10(2), 1233–1246 (2020)

Identifying Cyberspace Users' Tendency in Blog Writing Using Machine Learning Algorithms

Samah W. G. AbuSalim, Salama A. Mostafa, Aida Mustapha, Rosziati Ibrahim, and Mohd Helmy Abd Wahab

Abstract A blog is a form of direct interactive communication technology, which allows users to interact and communicate with each other through posting comments and sharing links as well. A blog is a platform where a writer or group of writers gives their opinion on a specific topic. Many issues and topics that are in a certain country being censored and controlled by the government from being presented through the mass media. Nevertheless, blogs have the space to provide a wide platform for exchanging ideas and opinions on various issues. There is a specific proportion between blog features and bloggers' tendency to social, political, and cultural patterns of different countries and nations that create trends among the bloggers in these countries. In this paper, we use an existing data set from previous research, which has 100 records of data, and manipulate the data by applying three machine learning algorithms for implementing classification and regression tasks. The algorithms are Decision Tree (c4.5), Linear Regression (LR), and Decision Forest (DF) with a 10-fold cross-validation method for training and testing. The results showed that C4.5

S. W. G. AbuSalim
Department of Computer Information Sciences, Universiti Teknologi PETRONAS (UTP), 32610, Seri Iskandar, Perak, Malaysia
e-mail: samah_21000332@utp.edu.my

S. A. Mostafa (✉) · R. Ibrahim
Faculty of Computer Science and Information Technology, Universiti Tun Hussin Onn Malaysia, 86400 Batu Pahat, Johor, Malaysia
e-mail: salama@uthm.edu.my

R. Ibrahim
e-mail: rosziati@uthm.edu.my

A. Mustapha
Faculty of Computer Applied Science, Universiti Tun Hussein Onn Malaysia, 84500 Panchor, Johor, Malaysia
e-mail: aidam@uthm.edu.my

M. H. A. Wahab
Faculty of Electrical and Electronic Engineering, Universiti Tun Hussein Onn Malaysia, 86400 Batu Pahat, Johor, Malaysia
e-mail: helmy@uthm.edu.my

© The Author(s), under exclusive license to Springer Nature Singapore Pte Ltd. 2023
P. Gyei-Kark et al. (eds.), *Engineering Mathematics and Computing*,
Studies in Computational Intelligence 1042,
https://doi.org/10.1007/978-981-19-2300-5_6

achieves the best overall results of 81% accuracy, 83% precision, and 91% recall, compared with the other two algorithms.

1 Introduction

Nowadays, the internet has become a huge part of human life. During free time or working, most of the people will connect their devices to the internet. Internet becomes a huge platform for most of the people to search for or share knowledge because the internet is limitless. Social media network such as Instagram, Facebook, Snapchat, and YouTube are examples of the current platform for sharing ideas, thought, and spreading the news and this includes blogging. This social media has become one of the things that cannot be ignored in our daily life [1]. Different from the other social media listed before, blogging has more deep content, and the number of people who care about blogging increases by reading and writing blogs on a daily basis [2]. Blogs are one of the most common web resources to provide virtual space contexts [3].

Bloggers usually address specific issues that they want to share and focus on, and the blog site includes topics that provide useful information to the readers. Style and pattern in blogging can be different for each person and each region or country. The environment, that the writers live in and the events surrounding them mostly affect their blogs content. Moreover, many factors affect the tendency toward pattern for blogging. Some patterns are highlighted based on the tendency with the social, political and cultural patterns of different nations and countries [4]. Identifying a pattern of blogging in a country or region is important. The pattern will show the current issues that evolve in that country or region. The discussion on the blogging site shows the thought of the citizen in that area, toward the issue. Observing the pattern of writing and discussion can be a useful data for some organization and parties to read and recognize the interests, strengths, and weaknesses of the region. The trend in blogging can also help organizations within the region to plan a strategic program and help to determine policies and planning for social, political, economic, and cultural pathologies and lead to related strategies.

This project aims to identify the users' tendency in blogs. It proposes different kinds of algorithms and data mining tasks to do blog tendency prediction. The algorithms and data mining tasks were run using Microsoft Azure Studio [5]. Microsoft Azure Studio provides a cloud computing service that contains a different type of data mining algorithm and task that helps in manipulating the data. The blog's user in Kohkiloye and Boyer Ahmad was chosen as the sample for the data set based on their regional context of the province, the population, and social and political structure. The data was collected and analyzed from the information database, social networks, blogs, websites, and virtual communities which are used by them.

For this project, the data mining tasks used is regression. The algorithm for regression are Decision Tree (C4.5), Linear Regression (LR), Decision Forest (DF) [6, 7]. The algorithms are chosen based on the ability to predict this type of work. The results from the three algorithms were compared with each other and compared with the results of the related work. The dataset used contain attribute such as degree, caprice, topic, local media turnover, local, political, and social space, and pro blogger. The dataset will be explained in Sect. 3.

The remaining of this paper is organized as follows. Section 2 reviews work related to the usage of the machine learning algorithms of other similar research papers as a benchmark. Also to study the methodology used in these research papers. Section 3 and subsection for it presents the methodology used which is how to use the algorithm to perform the regression. It also presents the dataset and the evaluation metrics. Section 4 presents the results obtained from applying the algorithms and discussing of their results. Finally, in Sect. 5, the findings of the project are concluded then mentioning future research works.

2 Related Work

The paper with the related work that has been chosen uses the recent method or technology to approve the study of this project that uses the latest method.

Alghobiri [8] compared three diverse classification algorithms, namely: Naïve Bayes, C4.5, and Support Vector Machine (SVM) algorithms. They choose the data based on the size or number of data and the nature of the attributes. The comparison results showed that the SVM algorithm is better than the other classification algorithms based on accuracy, precision, and F-measure.

Asim et al. [9] evaluated various machine learning techniques including Lazy learning algorithms, Eager learning algorithms, and Ensemble techniques for data classification. They identified the best approaches by which to improve the accuracy of data classification among the techniques used. The results show that the Random Forest and Nearest Neighbor Classifier (IB1) can classify high-precision nominal data with 85% accuracy and 84.8% precision for classification.

Masetic et al. [10] evaluate detecting malicious Web sites by using C4.5 decision tree classifier. The classifier is assessed by a variety of performance assessment parameters, including precision, sensitivity, specificity, and region under the ROC curve. The results show that the C4.5 decision tree classifier with 96.5% accuracy has achieved considerable success in the identification of malicious websites.

Samsudin et al. [11] suggested using the Random Forest algorithm to implement a blog classification using the same set of data. The key concern is to determine the accuracy in classifying a blogger who chooses to write a blog as a professional rather than seasonal. The results show that the ROC Area of the Random Forest algorithm is 11% better than C4.5 and 6% better than K-NN algorithms in blog classification. Also, the Random Forest algorithm has 97% recall better than C4.5 and K-NN algorithms.

Dias and Dias [12] look at the possibility of finding an appropriate machine learning technique by comparing a linear regression, neural network regression, and decision forest regression approaches to forecast the monthly ad revenue that a blog can generate with greater accuracy, using statistics from Google Analytics and Google AdSense. In conclusion, the Decision Forest Regression model came out as the best fit with an accuracy of over 70%.

Chen et al. [13] improved the recurrent neural network model which is used to classify Chinese micro-blog text. A feature fusion method based on shallow and deep learning is used to extract the micro-blog text features. Moreover, considering the internal correlation of the text sequence, a recurrent neural network model based on LSTM is used. The final classification precision rate is 85.04%, which is 3.17% higher than the traditional SVM method based on shallow learning features. The results confirm the effectiveness of the proposed method.

Simaki et al. [14] examined the suitability of many regression algorithms for the age estimation function. A variety of machine learning algorithms have been tested to analyze and verify their suitability for performing this function. Forty-two text features have been used in experimenting. The results showed that the Bagging algorithm with the RepTreebase learner provided the best value for estimating the age of the site users with a Mean Absolute Error (MAE) of 5.44, while the Root-Mean-Square Error (RMSE) was approximately 7.14.

Yang et al. [15] suggest a new model called DUAPM for discovering, modeling, and predicting microblogging user behavior and activities in the Cyber-Physical-Social System (CPSS) applications in order to detect spam and fake accounts. Three important characteristics have been relied upon in implementing their approach: personal information, social relationship, and user interaction. As part of the assessment of a sample data set containing 3,621 users of Sina Weibo over 20 weeks, the results show that the accuracy of the DUAPM model is higher than other models used for predicting user behavior in various social media which used both logical regression and random forest algorithms.

Mostafa et al. [16] try to diagnose Parkinson's disease by implementing three classification methods independently which are Decision Tree, Naïve Bayes, and Neural Network. Then identify the best method among the three. They intend to solve the problem by measuring the efficiency of the three methods. The results show that the Decision Tree with 91.63% accuracy and Neural Network, with 91.01% accuracy is better than Naïve Bayes with 89.46% accuracy rate and they recommend using both Decision Tree and the Neural Network for datasets with similar properties.

Woo et al. [17] identify Korean keywords for the identification of outbreaks of influenza from social media info. The following steps were carried out: initial keyword selection; Keyword time series generation using a pre-processing approach; optimal selection of characteristic; model creation and validation using the least absolute shrinkage and selection operator; Support Vector Machine (SVM), and Random Forest Regression (RFR). The results show that a total of 15 keywords optimally detected epidemiological influenza, uniformly distributed across Twitter and blog sources. Their model can be used in other countries, languages, infectious diseases, and social media sources. Table 1 summarizes the literature review related to different classification algorithms.

Table 1 The comparison of classifications algorithms

Authors and year	Objectives	Algorithms	Result
Alghobiri [8]	Comparative analysis between three classification algorithms that are used to classify various data sets	– NAÏVE Bayes algorithm, C4.5 Algorithm and Support vector machine algorithm (SVM)	SVM outperforms the other classification algorithms
Asim et al. [9]	Investigates the factors that influence professionalism in blogging	Decision tree algorithms, Lazy learning algorithms and Ensembling methods	Nearest-neighbor classifier (IB1) and Random Forest have results with 85% accuracy and 84.8% precision
Masetic et al. [10]	Assess the ability of the C4.5 decision tree classifier to identify malicious websites	C4.5 decision tree classifier algorithm	C4.5 decision tree algorithm achieved significant success in detecting malicious Web sites
Samsudin et al. [11]	Suggested an alternative algorithm to present a case study of professional and seasonal bloggers from Kohkiloye and Boyer Ahmad Province	Random Forest algorithm, C4.5 algorithm and K-Nearest Neighbor algorithm	Random Forest performed better compared to single classifier C4.5 and K-Nearest Neighbor (ROC Area 92%, Recall 97%, Precision 88%)
Dias and Dias [12]	Forecast the monthly ad revenue that a blog can generate to greater accuracy, using statistics from Google Analytics and Google AdSense	Linear regression, neural network regression and decision forest regression	Decision Forest Regression model came out as the best fit with an accuracy of over 70%
Simaki et al. [14]	Evaluation of regression algorithms for estimating the age of bloggers	Regression algorithms	The RepTree algorithm proved to outperform all the evaluated regression algorithms and provided accurate age estimations

3 Methodology

This study will use classification and regression as data mining tasks. Classification in machine learning represents the decision of differentiating between entities, things, objects, or events and identifying their corresponding groups. Regression is basically a statistical approach to find the relationship between variables. In machine learning, this is used to predict the outcome of an event based on the relationship between variables obtained from the data set. This technique is used for forecasting, time series modeling, climate prediction, and finding the causal effect relationship between the variables [18]. One of the key benefits that regression analysis offers is that it indicates the strength of the impact of multiple independent variables on a dependent variable [19]. The simplest regression technique is linear regression and the advanced regression technique is multiple regression [20]. For the classification and regression task, the algorithms that are used are Decision Tree (c4.5), Linear Regression (LR), and Decision Forest (DF). The Azure Machine Learning tool was used to conduct our experiments with a 10-fold cross-validation method for training and testing.

3.1 Data set

A data set is defined as a set of data, which is often incomplete and inconsistent, obtained from a specific source, such as the Internet [21]. This research used data sets from the UCI Machine Learning Repository [22]. This data set contains 100 instances and 6 attributes. The data set was compiled from Gharehchopogh et al. [23].

3.2 Algorithms

Classification and regression are two techniques used in machine learning for decision-making [24]. This section provides a summary of three algorithms used as classification and regression techniques which are Decision Tree (c4.5), Linear Regression (LR), and Decision Forest (DF) as follows.

3.2.1 Decision Tree (C4.5)

It is known as the decision tree algorithm and it is considered one of the most powerful and famous classification methods. This algorithm is used in Data Mining as a Decision Tree Classifier that can be used to produce a decision based on a given data sample (univariate or multivariate predictors). The C4.5 tree operates by recursively partitioning the training data set to check the capacity of the function values when separating the classes [25]. Some advantages of the C4.5 Classification algorithm are (a) Ease of understanding the results of the tree diagram analysis. (b) Ease of creating less experimental data and recovering it from other classification algorithms. (c) Calculation time is comparatively quicker than other classification techniques [26].

3.2.2 Linear Regression (LR)

Linear regression is a statistical model that analyzes the linear relationship between a dependent variable and a specific set of independent variables. It is considered one of the simplest machine learning algorithms within the category of supervised learning, which model the concept of regression. There are two types of linear regression: The first is the Simple Linear Regression and it contains one independent variable, so there will be a linear relationship between the Independent Variable and Dependent Variables. The second type is Multiple Linear Regression and contains several independent variables. The advantages of this algorithm are: (a) It is easier to implement in various programming languages and simpler to understand. (b) Specifically used for predicting numeric values.

3.2.3 Decision Forest (DF)

A decision forest is an ensemble method in which a classifier is constructed by combining several different Independent base classifiers. Decision forest aims to improve the predictive performance of a single decision tree by training multiple trees and combining their predictions [27]. A decision forest can be created by two main means: (1) using a general ensemble method (such as AdaBoost) that can virtually be used with any base learning method, including decision trees and (2) ensemble methods that were designed specifically for creating a decision forest (such as Random Forest) [27]. Random Decision Forest has been proven fast and efficient for handling different data analysis problems such as classification, regression, clustering, and dimensionality reduction [28]. It can be applied to graphics processing units (GPUs) and provides real-time efficiency. Random Decision Forest ensembles enhance the capabilities of Decision Trees providing better generalization [29].

3.3 Evaluation Metrics

The evaluation metrics for classification tasks used in the experiments are accuracy, precision, recall, and F-Measure. For regression tasks, the evaluation metrics used are mean absolute error, root mean squared error, relative absolute error, relative squared error, and coefficient of determination. The following measures are used to find the results of the given classifiers:

Accuracy is the ratio of the number of accurate observations of the total number of samples entered Eq. 1 shows the accuracy formula.

$$\text{Accuracy} = \frac{\text{True Positives } + \text{False Negatives}}{\text{Total Number of Samples}} \tag{1}$$

Precision is defined as the number of true positives divided by the summation of the number of true positives and the number of false positives. The formula for calculating precision is shown in Eq. 2.

$$\text{Precision} = \frac{\text{truePositives}}{\text{truePositives} + \text{falsePositives}} \tag{2}$$

The recall is defined as the number of true positives divided by the summation of the number of true positives and the number of false negatives. The formula for calculating recall is shown in Eq. 3.

$$\text{recall} = \frac{\text{truePositives}}{\text{truePositives} + \text{fasleNegatives}} \tag{3}$$

The weighted harmonic mean of precision and recall score. (F1) is another evaluation criterion. The formula for calculating the F1 score is shown in Eq. 4.

$$F1 = 2 * \frac{\text{precision} * \text{recall}}{\text{precision} + \text{recall}} \tag{4}$$

Mean Absolute Error (MAE) is a linear score which means that all the individual differences are weighted equally in the average. The formula for calculating mean absolute error is shown in Eq. 5.

$$\text{MAE} = \frac{1}{N} \sum_{i=1}^{N} |y_i - \hat{y}_i| \tag{5}$$

Root Mean Squared Error (RMSE) is the square root of the mean square error to make the scale of the errors to be the same as the scale of targets. The formula for calculating the root mean squared error is shown in Eq. 6.

$$\text{RMSE} = \sqrt{\frac{1}{N} \sum_{i=1}^{N} (y_i - \hat{y}_i)^2} = \sqrt{\text{MSE}} \tag{6}$$

Relative Absolute Error (RAE) can be compared between models whose errors are measured in different units. The formula for calculating relative absolute error is shown in Eq. 7.

$$\text{RAE} = \frac{\sum_{i=1}^{n} |p_i - a_i|}{\sum_{i=1}^{n} |\bar{a} - a_i|} \tag{7}$$

Relative Squared Error (RSE) can be compared between models whose errors are measured in different units. The formula for calculating relative squared error is shown in Eq. 8.

$$RSE = \frac{\sum_{i=1}^{n} (p_i - a_i)^2}{\sum_{i=1}^{n} (\overline{a} - a_i)^2} \qquad (8)$$

Coefficient of determination (R^2) summaries the explanatory of the regression model and is computed from the sums-of-square terms. The formula for calculating the coefficient of determination is shown in Eq. 9.

$$R^2 = \frac{\sum_{i=1}^{n} (\hat{y} - \overline{Y})^2}{\sum_{i=1}^{n} (Y - \overline{Y})^2} \qquad (9)$$

4 Results and Discussion

Our experiments aim to compare the Decision Tree (c4.5), Linear Regression (LR), and Decision Forest (DF) in terms of performance using different evaluation metrics. It can provide correct anticipation of bloggers' behaviors and see the role of each factor and their importance in a professional approach. For c4.5, LR, and DF, the data split using a tenfold cross-validation method for training and testing. The result is shown in Table 2. The results in Table 2 showed that LR has the lowest accuracy which is 69%. The lowest MAE of 0.1834 is achieved by the c4.5 in which the MAE measure shows how close the prediction is to the actual outcomes on blogger prediction as the lower score represents the best result. Besides, the R^2 of c4.5 near to 1 which is better than the DF and LR.

The results above also showed that the F-Score of the c4.5 is 0.873 which is the closest to the ideal F-Score value of 1. The results showed that C4.5 achieves the best overall classification results of 81% accuracy, 83% precision, and 91% recall. On the other hand, the results showed that C4.5 achieves the best overall regression

Table 2 Comparison between algorithms

Algorithm metric	c4.5	LR	DF
Accuracy	0.8113	0.6997	0.7334
Precision	0.8389	0.6627	0.6962
Recall	0.9125	0.8855	0.9413
F1-Score	0.8731	0.7445	0.8001
MAE	0.1834	0.2386	0.2381
RMSE	0.1387	0.1451	0.1442
RAE	0.1419	0.1787	0.1980
RSE	0.0052	0.0544	0.0104
R^2	0.9998	0.9459	0.9895

results of MAE, RMSE, RAE, RSE, and R^2. Ultimately, the decision tree algorithm is better and more reliable in predicting the bloggers's tendency in writing blogs.

5 Conclusions and Future Work

In conclusion, many algorithms can be used to present the case study of bloggers' behavior and tendency in blog writing using the same data set. From the result of the classification and regression tasks which are performed by the Decision Tree (c4.5), Linear Regression (LR) and Decision Forest (DF), the LR algorithm has the poorest performance compared to the c4.5 and DF algorithms toward the data set. The F1-Score of c4.5 has the highest value which is 0.873. Moreover, the results showed that the C4.5 algorithm achieves the best overall classification results of 81% accuracy, 83% precision, and 91% recall, compared with the other two algorithms. On the other hand, the results showed that the C4.5 algorithm achieves the best overall regression results of MAE, RMSE, RAE, RSE, and R^2. The algorithm might perform differently depending on the sample size of the data set. Some algorithms can perform well using small data rather than a bigger data set and vice versa. Based on this project, the most suitable algorithm for the data set that contains 100 samples of bloggers' behavior and tendency is the c4.5 algorithm. In future research, a bigger sample size will be used to get more robust results.

Acknowledgements This research is supported by Universiti Tun Hussein Onn Malaysia.

References

1. Bandorski, D., Kurniawan, N., Baltes, P., Hoeltgen, R., Hecker, M., Stunder, D., Keuchel, M.: Contraindications for video capsule endoscopy World. J. Gastroenterol. **22**, 9898–9908 (2016)
2. Alsamadani, H.A.: The effectiveness of using online blogging for students' individual and group writing. Int. Educ. Stud. **11**(1), 44 (2017). https://doi.org/10.5539/ies.v11n1p44
3. Gharehchopogh, F.S., Khaze, S.R., Maleki, I.: A new approach in bloggers classification with hybrid of K-nearest neighbor and artificial neural network algorithms. Indian J. Sci. Technol. **8**(3), 237 (2015). https://doi.org/10.17485/ijst/2015/v8i3/59570
4. Hand, D.J.: Principles of data mining. Drug-Safety **30**, 621–622 (2007). https://doi.org/10.2165/00002018-200730070-00010
5. Cloud Computing Services: Microsoft Azure. (n.d.). https://azure.microsoft.com/en-in/. Accessed April 21, 2020
6. Dalatu, P.I., Fitrianto, A., Mustapha, A.: A comparative study of linear and nonlinear regression models for outlier detection. In: Recent Advances on Soft Computing and Data Mining, pp 316–326 (2016).https://doi.org/10.1007/978-3-319-51281-5_32
7. Geetha, M.C.S., Shanthi, I.E., Raman, S.S.: A survey and analysis on regression data mining techniques in agriculture. Int. J. Pure Appl. Math. **118**(8), 341–347 (2018). ISSN: 1311-8080 (printed version); ISSN: 1314-3395 (on-line version)
8. Alghobiri, M.: A comparative analysis of classification algorithms on diverse datasets. Eng. Technol. Appl. Sci. Res. **8**(2), 2790–2795 (2018)

9. Asim, Y., Shahid, A.R., Malik, A.K., Raza, B.: Significance of machine learning algorithms in professional blogger's classification. Comput. Electr. Eng. **65**, 461–473 (2018). https://doi.org/10.1016/j.compeleceng.2017.08.001

10. Masetic, Z., Subasi, A., Azemovic, J.: Malicious web sites detection using C4.5 decision tree. Southeast Eur. J. Soft Comput. **5**(1) (March 2016). ISSN 2233–1859

11. Samsudin, N.A., Mustapha, A., Wahab, M.H.A.: Ensemble classification of cyber space users tendency in blog writing using random forest. In: 2016 12th International Conference on Innovations in Information Technology (IIT) (2016). https://doi.org/10.1109/innovations.2016.788 0046

12. Diasa, D.S., Diasb, N.G.J.: Forecasting monthly ad revenue from blogs using machine learning. In: The 3rd International Conference on Advances in Computing and Technology, ICACT 2018 (2018)

13. Chen, Q., Guo, Z., Sun, C., Li, W.: Research on chinese micro-blog sentiment classification based on recurrent neural network. In: 2017 2nd International Conference on Computer Science and Technology (CST 2017) (2017) ISBN: 978-1-60595-461-5

14. Simaki, V., Aravantinou, C., Mporas, I., Megalooikonomou, V.: Automatic estimation of web bloggers' age using regression models. In: Ronzhin, A., Potapova, R., Fakotakis, N. (eds.) Speech and Computer: 17th International Conference, SPECOM 2015, Athens, Greece, September 20–24, 2015, Proceedings. Lecture Notes in Computer Science, vol. 9319, pp. 113–120. Springer (2015). https://doi.org/10.1007/978-3-319-23132-7_14

15. Yang, P., Yang, G., Liu, J., Qi, J., Yang, Y., Wang, X., Wang, T.: DUAPM: an effective dynamic micro-blogging user activity prediction model towards cyber-physical-social systems. IEEE Trans. Industr. Inf., 1–1 (2019).https://doi.org/10.1109/tii.2019.2959791

16. Mostafa, S.A., Mustapha, A., Khaleefah, S.H., Ahmad, M.S., Mohammed, M.A.: Evaluating the performance of three classification methods in diagnosis of Parkinson's disease. In: International Conference on Soft Computing and Data Mining, pp. 43–52. Springer, Cham (February 2018)

17. Woo, H., Sung Cho, H., Shim, E., Lee, J.K., Lee, K., Song, G., Cho, Y.: Identification of keywords from twitter and web blog posts to detect influenza epidemics in Korea. Disaster Med. Public Health Prep. **12**(03), 352–359 (2017). https://doi.org/10.1017/dmp.2017.84

18. Geetha, M.C.S., Shanthi, I., Raman, S.: A survey and analysis on regression data mining techniques in agriculture. Int. J. Pure Appl. Math. **118**, 341–346 (2018)

19. Rui, L.T., Afif, Z.A., Saedudin, R.D.R., Mustapha, A., Razali, N.: A regression approach for prediction of Youtube views. Bull. Electr. Eng. Inform. **8**(4), 1502–1506 (December 2019). ISSN: 2302-9285. https://doi.org/10.11591/eei.v8i4.1630

20. Bini, B.S., Mathew, T.: Clustering and regression techniques for stock prediction. Procedia Technol. **24**, 1248–1255 (2016). https://doi.org/10.1016/j.protcy.2016.05.104

21. Dali, A.D., Omar, N.A., Mustapha, A.: Data mining approach to herbs classification (2018)

22. Dua, D., Graff, C.: UCI machine learning repository http://archive.ics.uci.edu/ml. University of California, School of Information and Computer Science, Irvine, CA (2019)

23. Gharehchopogh, F.S., Khaze, S.R.: Data mining application for cyber space users tendency in blog writing: a case study. Int. J. Comput. Appl. (0975–888) **47**(18) (June 2012)

24. Nafi, S.N.M.M., Mustapha, A., Mostafa, S.A., Khaleefah, S.H., Razali, M.N.: Experimenting two machine learning methods in classifying river water quality. In: Communications in Computer and Information Science, pp. 213–222. Springer, Cham (September 2019)

25. Rahim, R., Zufria, I., Kurniasih, N., Simargolang, M.Y., Hasibuan, A., Sutiksno, D.U., et al.: C4.5 classification data mining for inventory control. Int. J. Eng. Technol. 7(2.3), 68 (2018). https://doi.org/10.14419/ijet.v7i2.3.12618

26. Novaković, J., Strbac, P., Bulatović, D.: Toward optimal feature selection using ranking methods and classification algorithms. Yugoslav J. Op. Res. **21**(1), 119–135 (2011). https://doi.org/10.2298/YJOR1101119N

27. Rokach, L.: Decision forest: twenty years of research. Inform. Fusion **27**, 111–125 (2016). https://doi.org/10.1016/j.inffus.2015.06.005

28. Abobakr, A., Hossny, M., Nahavandi, S.: A skeleton-free fall detection system from depth images using random decision forest. IEEE Syst. J., 1–12 (2018). https://doi.org/10.1109/jsyst.2017.2780260
29. Akram, B.A., Akbar, A.H., Shafiq, O.: HybLoc: hybrid indoor Wi-Fi Localization using soft clustering based random decision forest ensembles. IEEE Access, 1–1 (2018). https://doi.org/10.1109/access.2018.2852658

An Intelligent Intrusion Detection System Using a Novel Combination of PCA and MLP

Ratul Chowdhury, Arindam Roy, Banani Saha, and Samir Kumar Bandyopadhyay

Abstract Cyber threats are diversified in both volume and variety as most of the organizations develop and accept emerging technologies related to big data, cloud computing, and Internet of Things. In the area of cyber security, intrusion detection system (IDS) plays a significant role to identify the existing attack from network traffic. Over the past few decades, a lot of research works have been performed in this area. Researchers have used DARPA, KDD98, KDD99, and NSL-KDD datasets as a benchmark for their experiments. However, in the current network scenario, these datasets do not intuitively reflect proper network traffic and modern low footprint attacks. In this regard, this paper proposes a novel intrusion detection technique, where principal component analysis (PCA) has been used for dimensionality reduction, and multilayer perceptron (MLP) has been applied for classification. All the experiments have been conducted over the current UNSW-NB15 dataset consisting of a total of 2540,044 records with 9 different types of modern low footprint attacks. The experimental results demonstrate that the proposed misuse-based technique achieved a higher detection rate and low false alarm rate in comparison with other existing methods in the literature.

Keywords Intrusion detection system · UNSW-NB15 · Principal component analysis · Multilayer perceptron · Weka

R. Chowdhury (✉)
Future Institute of Engineering and Management, Kolkata, India
e-mail: ratul.cse87@gmail.com

A. Roy · B. Saha · S. K. Bandyopadhyay
University of Calcutta, Kolkata, India

1 Introduction

Internet is one of the greatest revolutions of the twentieth century. It has played a significant role in various sectors like communication, research, education, financial transaction, and real-time update. This rapid use of internet has created some serious issues in front of us:

1. Privacy and Information Security: The increasing rate of sensitive data and the present big data concept enhances the privacy and security issues in a higher semantic level.
2. Various Third-Party Risk: Many financial organizations collaborate with the third party to reduce the cost made by outsourcing services. Internal structure and data sharing with third parties further increase the risk factor.
3. Emerging and Advanced Threats: An advanced persistent threat (APT) is a network attack, in which, an unauthorized person gains access to a network and stays there undetected for a long period of time.
4. Exploration of New Signature-Based Threat: Day by day, the nature of the attack is updated by the changing trends of their signature.

In this aspect, cyber security plays an important role to protect our computer, network, program, and data from unauthorized access. This paper mainly focuses on network security through IDS. It is an important area of research for the past few decades. The crucial aspect of IDS is to impartially identify different attacks from network packets. Researchers have adopted various machine learning approaches for the construction of intelligent IDS. The learning methods of IDS are broadly classified into 3 types [1]: supervised learning, unsupervised learning, and semi supervised learning. In supervised learning, all the instances in the dataset have a specific level and in unsupervised learning, there is no level. Semi supervised learning is a mixture of both.

Based on the nature of attack, again IDSs are of three types: [4]

 i. Misuse-based,
 ii. Anomal-based, and
 iii. Hybrid.

Misuse-based or signature-based IDS is used to identify known attacks only. So, zero day attack is undetected by these kind of IDS. Anomaly-based technique, on the other hand, creates a model according to the system behavior and identifies anomalies as deviations from the normal behavior. Increasing false alarm rate is the main drawback of anomaly-based IDS. Hybrid IDS is a combination of both misuse-based and anomaly-based IDS. Snort is a well-known IDS that is freely available and mostly used in UNIX or Linux operating system. In the current scenario, the nature of the attack has been completely changed. So, in this paper, we have used the current UNSW-NB15 dataset [13] for experimental purpose. It contains 9 different types of more realistic attacks with normal behavior. The extraction of appropriate features from the preprocessed dataset plays an important role to identify attacks. Various wrapper-based, filter-based, and embedded methods are available for feature reduction.

In this paper, we have implemented PCA as a feature of the reduction technique. PCA [16] achieves dimension reduction by creating new and artificial variables called principal components. Each principal component is a linear combination of the observed variables. After feature reduction and preprocessing steps, MLP has been used for classification purpose. So the paper is structured as follows. Section 2 reflects the related work, Sect. 3 describes the detailed PCA-MLP model, Sect. 4 highlights experimental set up, Sect. 5 analyzes the results, and conclusion is presented in Sect. 6.

2 Related Work

Many authors have adopted different machine learning approaches like Artificial neural network (ANN), Genetic algorithm (GA), Support vector machine (SVM), Multilayer perceptron (MLP), and deep neural network approaches such as Convolution neural network (CNN) and Recurrent neural network (RNN) in their work. Canady [5] first used ANN as a multi-category classifier to identify the attack. He has used the data generated by a RealSecure^{TM} network monitor. 10% of the data was selected randomly for testing, but here, ANN is used as a complete black box. Lippmann et al. [11] proposed a keyword-based technique to identify unknown attacks. The popular DARPA 1998 dataset was used for experimental purposes. The experimental result shows that the proposed method was good for user-to-root attack only. GA is another approach which has been used in different ways for the construction of IDS. In [8], the author has used GA in his work. He has mainly focused on representing the datasets through GA but doesn't use any standard dataset for implementation purpose. In their work, Shaveta et al. [15] identified that the effectiveness of the GA generally depends on three factors, namely: selection of fitness function, representation of individuals, and values of the GA parameters. They have proposed a good fitness function and applied it to KDD99 dataset. For binary classification, the result was good, but the analysis for individual attack was not mentioned. SVM is another classification technique, used for higher dimension and it produces higher detection rate if the data set is not linearly separable. Jha et al. [7] used SVM for intrusion detection. Since the number of dimensions affects the performance of SVM, therefore, after applying information gain, they have used SVM. The accuracy was quite impressive. In another work, Enache et al. [6] proposed a method where swarm intelligence algorithm had been used for feature selection and the resultant feature set is fed into a SVM classifier. They used the NSL-KDD dataset, but randomly chose only 9,566 records for training and testing purposes which was too small compared to the actual data size. The authors in [9] proposed a completely new intrusion detection technique by gradually removing the features. With the combination of clustering methods such as ant colony optimization and SVM, they have developed a classifier which worked efficiently with a minimum number of feature sets. MLP is another type of classifier based on feed forward artificial neural network. It contains more than one hidden layer that are fully connected, and backpropagation algorithm is used in between the hidden and output layer for weight update. Amato et al. [2] applied MLP on KDD dataset. They have used the greedy stepwise approach for dimension reduction and reduced the number of features from 41 to 11. The overall accuracy was up to the mark, but the accuracy for user-to-root attack and root-to-local attack was not satisfactory. After Professor Hinton introduced the concept of deep learning, a new platform was opened for the researchers. Li et al. [10] have used convolution neural network for the construction of IDS. Considering the significant advantages of CNN in image classification, a range-specific one-hot encoding is applied to convert the NSL-KDD dataset into image format. Though the idea was very impressive, but the accuracy level was not up to the mark. Wang et al. [18] introduced a new malware traffic classification technique, where automatic feature extraction had been performed through CNN. According to the authors, the main limitation of the work was that they used only spatial feature of traffic and they didn't perform any CNN parameter tuning. Yin et al. [19] defined RNN for intrusion detection. KDD dataset was used for this purpose and achieved the highest accuracy among different classifiers. But, as compared to bidirectional RNN and LSTM-RNN, the accuracy was not up to the mark. Wang et al. [17] had also proposed a combined concept of CNN and RNN, the IDS extracted the low level spatial features by using CNN and high level temporal features using LSTM networks. The dataset used was DARPA 1998 and ISCX 2012. The accuracy level of their model outperforms other published approaches. Belouch et al. [3] used a rep tree algorithm for intrusion detection. They converted the dataset into protocol-based, and finally compared the result with NSL-KDD dataset. Their experimental results reflect that for UDP protocol, the accuracy level is not satisfactory.

After analyzing the above works, it is concluded that in spite of some research attempts, still there is a lot of room for improvement.

Most of the research attempts have been done on various old datasets, but in current scenario, the nature of the attack is changing day by day, therefore, the construction of intelligent IDS is a big challenge for us. UNSW-NB15 is the latest dataset which contains 175,341 records for training and 82,332 number of records for testing. The whole dataset contains 9 modern types of low footprint attacks. In [14], the complete feature set of UNSW-NB15 dataset was described. It contains 6 different types of features: flow feature, basic feature, content feature, time feature, additional generated feature, and label feature. They have also presented a complete statistical analysis of UNSW-NB15 dataset and a critical comparison has been done with KDD 99 dataset. The experimental result shows that UNSW-NB15 is more complex than KDD99 and is considered to be a new benchmark data set for evaluating NIDSs.

3   Proposed Methodologies

The proposed model as illustrated in Fig. 2 comprises of 3 primary sections.

  i. Preprocessing,
 ii. Dimension Reduction, and
iii. Train-Test Split.

   In preprocessing phase, the normalization and numerization have been carried out, whereas in dimension reduction phase, the 47 dimensional feature vectors have been reduced to 22 dimensions. Finally, the model has been constructed based on the trained data and the accuracy has been achieved through testing. The model construction has been performed through MLP. MLP is a simple feed forward network with one or more hidden layer. Here, the data flows from the input to output layer and it follows backpropagation algorithm to train the network. MLP model is basically used to solve those problems that are linearly inseparable. Figure 1 shows the basic MLP model which contains 3 layers, namely: input, output, and a single hidden layer.
   The MLP algorithm works in two passes. In the forward pass, the predicted output is evaluated based on the input and its corresponding weight. Similarlry, in the backward pass, the back propagation algorithm is used for weight updation.

3.1   Dataset Description

The UNSW-NB15 [3] dataset was created by the IXIA PerfectStorm tool in the Cyber Range Lab of the Australian Centre for Cyber Security (ACCS). It consists of

Fig. 1 Basic MLP model with one hidden layer

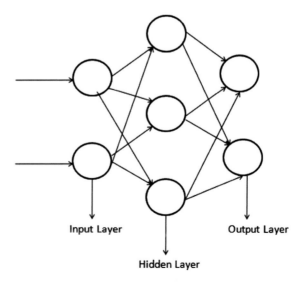

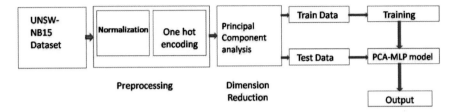

Fig. 2 Basic PCA-MLP model

2,540,044 records which are stored in four CSV files. Apart from this, a part from this dataset is separated for training and testing purposes, which consists of 175,341 and 82,332 number of records, respectively. The training and testing set have 9 different types of low footprint attack like Fuzzer, Analysis, Backdoor, Dos, Exploit, Generic, Reconnaissance, Shellcode, and Worm. The UNSW-NB15 dataset has a total of 49 features with the class level and it has been categorized into 6 groups, namely: flow features, basic features, content features, time features, additional generated features, and labeled features.

3.2 Data Preprocessing and Normalization

There are a total of 46 numeric and 3 non-numeric features available in UNSW-NB15 dataset. As the PCA-MLP model takes only numeric value, therefore, the non-numeric features need to be converted into numeric form. In this dataset, the non-numeric features are: 'protocol', 'service', and 'state'. The state field contains 4 different types of states: INT, FIN, CON, and REQ, which are converted into binary vectors $(1, 0, 0, 0)$, $(0, 1, 0, 0)$, $(0, 0, 1, 0)$, and $(0, 0, 0, 1)$ by using one hot encoding method. Similarly, the features protocol and service are encoded into binary form. The second part is normalization. Since the difference between the maximum and minimum values of some attributes has a very large scope, therefore, we have normalized it between the range [0–1] by using Eq. 1, where MAX and MIN are the maximum and minimum values of each feature, respectively.

$$X_i = \frac{X_i - MIN}{MAX - MIN} \qquad (1)$$

3.3 Dimension Reduction

Dimension reduction plays an important role in intrusion detection. The principal component analysis is an unsupervised method used to reduce the dimensionality of the feature space. PCA is a well established mathematical technique which creates new artificial variables called principal components. Each principal component is a linear combination of different feature set. In this work, we have used PCA on UNSW-NB15 dataset, which reduces 49 features to 22 features. These 22 features are specially used for further classification. The basic steps of PCA for two-dimensional data x_1 and x_2 is described as follows:

Step 1: Computation of covariance matrix. The covariance matrix is defined as follows:

$$\begin{pmatrix} varience(x_1) & covarience(x_1, x_2) \\ covarence(x_1, x_2) & varience(x_2) \end{pmatrix} \tag{2}$$

The covariance is computed by the following formula:

$$\sum (x_1 - \overline{x_1}) * (x_2 - \overline{x_2}) \tag{3}$$

Step 2: Obtain the Eigenvalues by solving

$$(A - \lambda I) = 0 \tag{4}$$

Step 3: Obtain the Eigenvector.
Step 4: Obtain Coordinates of data points in direction of Eigenvector. The estimated points are the first principal components.

3.4 Methodology

The resultant output of PCA is fed into the MLP model. In the MLP structure, an error has been calculated in the output layer and back propagated into the hidden layer for weight optimization. The forward propagation and weight update algorithm for PCA-MLP model is defined as follows:

i. x_i (i = 1, 2, …,n), defines n number of input samples.
ii. h_i (i = 1, 2, …,m), defines m number of hidden states.
iii. y_i (i = 1, 2, …,n) defines n number of output samples.
iv. o_i (i = 1, 2, …,l) defines l number of output states., where l is defined as the number of classes.
v. w_k defines the weight from input to hidden layer.
vi. w_o defines the weight from hidden to output layer.

Algorithm 1 Forward propagation from input to hidden layer

1: **for** (i in 1 to m) **do**
2:    $U_{hi} \leftarrow \sum_{j=1}^{n} w_{kj}$
3:    $V_{hi} \leftarrow U_{hi} + b_i$
4:    $h_i \leftarrow sigmoidal(V_{hi})$
5: **end for**

Algorithm 2 Forward propagation from hidden to O/P layer

1: **for** (i in 1 to l) **do**
2:    $U_{oi} \leftarrow \sum_{j=1}^{m} h_i w_{oj}$
3:    $V_{oi} \leftarrow U_{oi} + b_i$
4:    $o_i \leftarrow sigmoidal(V_{oi})$
5: **end for**

Algorithm 3 Weight update between hidden to output layer

1: **for** (i in 1 to m) **do**
2:    **for** (j in 1 to l) **do**
3:       $W_{oij} \leftarrow W_{oij} - \alpha * \Delta_{ol}$ ▷ Where $\Delta_{ol} = \delta E/\delta W_{oij}$, E is the error defined by ($y - o_i$), α is the learning rate
4:    **end for**
5: **end for**

Algorithm 4 Weight update between input to hidden layer

1: **for** (i in 1 to n) **do**
2:    **for** (j in 1 to m) **do**
3:       $W_{kij} \leftarrow w_{kij} - \alpha * \Delta_{hj}$ ▷ Where $\Delta_{hj} = \delta E/\delta W_{kij}$, E is the error between the activation of a single hidden layer and the actual output generated by all the output layer
4:    **end for**
5: **end for**

---

4 Experimental Setup and Evaluation Matrices

In this paper, for preprocessing purpose, we have used python 3.7 and for classification purpose the most promising machine learning tool **Waikato Environment for Knowledge Analysis (Weka)** [12] has been used. All the experiments have been done on a personal notebook with Intel core i3-6006U CPU @ 2.00GHz configuration and 8GB memory. The experimental section is divided into two parts: binary and multiclass classification. In binary classification, the total UNSW-NB15 dataset has been divided into two classes, namely: normal and attack. In multiclass classification, the dataset is further divided into 9 subclasses based on the individual attack categories. Finally, a critical comparison has been performed with traditional classifier like Simple logistic, Naïve Bayes, SMO, and SVM. Four important performance indicators are used to evaluate the performance of the model.

1. Accuracy(AC): It is basically the percentage ratio of correctly classified instances to the total number of instances.

$$AC = \frac{TP + TN}{TP + TN + FP + FN} \tag{5}$$

2. Precision: It is the ratio of relevant instances to the retrieved instances.

$$Precision = \frac{TP}{TP + FP} \tag{6}$$

3. Recall: It is the ratio of relevant instances that have been retrieved to the total amount of relevant instances.

$$Recall = \frac{TP}{TP + FN} \tag{7}$$

4. F-Measure: It is the weighted average of Precision and Recall.

$$F - Measure = \frac{2 * (Recall * Precision)}{Recall + Precision} \tag{8}$$

where,

– TP (True positive): is the number of positive instances that are corrected classified.
– TN (True negative): is the number of negative instances that are correctly classified.
– FP (False positive): is the number positive instances that are wrongly classified.
– FN (False negative): is the number of negative instances that are wrongly classified.

Hence, a good IDS always have a good detection rate and low false alarm rate.

5 Result Analysis

The result analysis portion is divided into two parts, namely: binary and multiclass classification. The essentiality of various features has been evaluated through various machine learning platforms. The accuracy achieved through various classifiers have been described in the following section.

5.1 Binary Classification

In binary classification, we have directly applied PCA on UNSW-NB15 dataset, which reduces the dimension to 22 features. These reduced feature vectors are fed

Table 1 Accuracy, precision, recall and f-measure obtained by different classifiers

Classifier	Accuracy (in %)	Precision	Recall	F-measure
PCA+Simple Logistic	82.750	0.853	0.827	0.823
PCA+Naive Bayes	85.270	0.827	0.853	0.850
PCA+SMO	86.800	0.882	0.868	0.861
PCA+SVM	90.870	0.909	0.908	0.908
PCA+MLP	92.500	0.925	0.925	0.925

into the said classifier. A 10 fold cross validation technique has been applied. From Table 1, it can be observed that the PCA-MLP model produces an accuracy of 92.5% which is the best score among all the classifiers. At the same time, the Precision, Recall, and F-Measure parameters have also shown that the proposed model performs significantly better than other classifiers.

5.2 Multi Class Classification

In multiclass classification, the record set of UNSW-NB15 dataset is further divided into 9 subcategories according to the type of attack present in the dataset. Table 2 shows each of the individual attack and their respective count.

In multi class classification, PCA has been applied on each of the attack category and the resultant dataset is further fed into the said classifier. From Table 3, it is observed that, except the backdoor attack category, the proposed model outperforms all classifiers.

From the above table, it is observed that although the recognition accuracy falls slightly for backdoor attack, but for the other categories, the success rate of the proposed model is consistent. Table 4 summarizes the success rate for the present work along with some other work mentioned in literature. Work described in [3, 14] have used Decision tree and Reptree method for classification. The records from Table 4 show the high success rate of PCA-MLP model.

Table 2 Different attack categories with their respective count

Attack	Count	Attack	Count
Normal	93,000	Fuzzers	24246
Backdoor	2329	Generic	58817
Dos	16353	Reconnaissance	13987
Analysis	2677	Shellcode	1511
Exploits	44525	Worms	174

Table 3 Accuracy obtained by different classifier in multi categories

Classifier	Backdoor	DOS	Analysis	Exploits	Fuzzers	Generic	Recon	Shellcode	Worm
PCA+Simple Logistic	99.09	93.65	98.98	87.35	90.95	91.24	94.57	99.41	99.93
PCA+Naive Bayes	63.65	89.33	98.48	83.12	58.00	66.74	51.27	61.04	95.09
PCA+ SMO	99.09	93.65	98.96	82.87	90.60	95.93	94.57	99.41	99.41
PCA+ SVM	97.37	92.67	98.98	88.12	91.5	96.71	95.23	99.4	99.52
PCA + MLP	96.37	93.66	98.98	89.18	91.71	98.71	96.07	99.41	99.93

Table 4 Comparison with other work

References	No of features used	Dataset size (Training + Testing)	Accuracy (in %)
[3]	49	257,673	85.56%
[14]	49	257,673	88.95%
Present work	22	257,673	92.50%

6 Conclusion

In this work, we have proposed a novel intrusion detection technique, which is used for identifying modern low footprint attack. The dimension reduction technique used here aided the improvement of the attack identification on the basis of less number of attributes. The basic MLP classifier produced satisfactory results in recognizing intrusion from the above mentioned dataset. We have also implemented the proposed technique for individual categories of attack and the result is quite satisfactory. So it is concluded that different machine learning approaches play a significant role in the construction of IDS. The objective of our future work will be to enhance the recognition accuracy in a higher semantic level. Moreover, we will look forward to apply different deep learning and hybrid approaches for the construction of more intelligent IDS.

References

1. Aburomman, A.A., Reaz, M.B.I.: Survey of learning methods in intrusion detection systems. In: 2016 International Conference on Advances in Electrical, Electronic and Systems Engineering (ICAEES), pp. 362–365. IEEE (2016)
2. Amato, F., Cozzolino, G., Mazzeo, A., Vivenzio, E.: Using multilayer perceptron in computer security to improve intrusion detection. In: International Conference on Intelligent Interactive Multimedia Systems and Services, pp. 210–219. Springer (2018)
3. Belouch, M., El Hadaj, S., Idhammad, M.: A two-stage classifier approach using reptree algorithm for network intrusion detection. Int. J. Adv. Comput. Sci. Appl. **8**(6), 389–394 (2017)

4. Buczak, A.L., Guven, E.: A survey of data mining and machine learning methods for cyber security intrusion detection. IEEE Commun. Surv. Tutor. **18**(2), 1153–1176 (2015)
5. Cannady, J.: Artificial neural networks for misuse detection. In: National Information Systems Security Conference, vol. 26. Baltimore (1998)
6. Enache, A.C., Patriciu, V.V.: Intrusions detection based on support vector machine optimized with swarm intelligence. In: 2014 IEEE 9th IEEE International Symposium on Applied Computational Intelligence And Informatics (SACI), pp. 153–158. IEEE (2014)
7. Jha, J., Ragha, L.: Intrusion detection system using support vector machine. Int. J. Appl. Inf. Syst. (IJAIS) **3**, 25–30 (2013)
8. Li, W.: Using genetic algorithm for network intrusion detection. In: Proceedings of the United States Department of Energy Cyber Security Group, vol. 1, pp. 1–8 (2004)
9. Li, Y., Xia, J., Zhang, S., Yan, J., Ai, X., Dai, K.: An efficient intrusion detection system based on support vector machines and gradually feature removal method. Expert Syst. Appl. **39**(1), 424–430 (2012)
10. Li, Z., Qin, Z., Huang, K., Yang, X., Ye, S.: Intrusion detection using convolutional neural networks for representation learning. In: International Conference on Neural Information Processing, pp. 858–866. Springer (2017)
11. Lippmann, R.P., Cunningham, R.K.: Improving intrusion detection performance using keyword selection and neural networks. Comput. Netw. **34**(4), 597–603 (2000)
12. Markov, Z., Russell, I.: An introduction to the Weka data mining system. In: ACM SIGCSE Bulletin, vol. 38, pp. 367–368. ACM (2006)
13. Moustafa, N., Slay, J.: Unsw-nb15: a comprehensive data set for network intrusion detection systems (unsw-nb15 network data set). In: 2015 Military Communications and Information Systems Conference (MilCIS), pp. 1–6. IEEE (2015)
14. Moustafa, N., Slay, J.: The evaluation of network anomaly detection systems: statistical analysis of the unsw-nb15 data set and the comparison with the kdd99 data set. Inf. Secur. J.: A Glob. Perspect. **25**(1–3), 18–31 (2016)
15. Shaveta, E., Bhandari, A., Saluja, K.K.: Applying genetic algorithm in intrusion detection system: a comprehensive review. In: Association of Computer Electronics and Electrical Engineers (2014)
16. Subba, B., Biswas, S., Karmakar, S.: Enhancing performance of anomaly based intrusion detection systems through dimensionality reduction using principal component analysis. In: 2016 IEEE International Conference on Advanced Networks and Telecommunications Systems (ANTS), pp. 1–6. IEEE (2016)
17. Wang, W., Sheng, Y., Wang, J., Zeng, X., Ye, X., Huang, Y., Zhu, M.: Hast-ids: Learning hierarchical spatial-temporal features using deep neural networks to improve intrusion detection. IEEE Access **6**, 1792–1806 (2017)
18. Wang, W., Zhu, M., Zeng, X., Ye, X., Sheng, Y.: Malware traffic classification using convolutional neural network for representation learning. In: 2017 International Conference on Information Networking (ICOIN), pp. 712–717. IEEE (2017)
19. Yin, C., Zhu, Y., Fei, J., He, X.: A deep learning approach for intrusion detection using recurrent neural networks. IEEE Access **5**, 21954–21961 (2017)

Ion Partitioning Effects on Electroosmotic Flow Through pH Regulated Cylindrical Nanopore

Subrata Bera and S. Bhattacharyya

Abstract The ion partitioning effects on electroosmotic flow have been studied through pH-regulated polyelectrolyte grafted cylindrical nanopore. The permittivity between the polyelectrolyte layer and electrolyte solution are taken differently, which generates ion partitioning. The surface charge density of the nanopore is negative, while the charge density in the polyelectrolyte layer depends on the pH values and concentration in aqueous solution. The present model of the electroosmotic flow through pH-regulated polyelectrolyte layer is considered Nernst–Planck equation for the ionic species, Navier–Stokes equation in modified form for flow and Poisson equation for potential distribution. These governing equations are solved numerically on a staggered grid system by finite volume method. This article investigates the importance of different dielectric permittivity, ionic strength, pH values, softness, and the charged density of the nanopore wall. The permittivity ratio is defined by the permittivity of polyelectrolyte layer and electrolyte solution. The average electroosmotic flow decreases with the permittivity ratio. The increase in the softness parameter decreases the flow rate. The ion partitioning contributes to the penetration effect of counter ions in the polyelectrolyte layer, which decreases the body force, and hence affects the electroosmotic flow rate.

Keywords Ion partition · Electroosmotic flow · Polyelectrolyte layer · Nernst–Planck equation

S. Bera (✉)
Department of Mathematics, National Institute of Technology Silchar, Silchar 788010, India
e-mail: subrata@gmail.com; subrata@math.nits.ac.in

S. Bhattacharyya
Department of Mathematics, Indian Institute of Technology Kharagpur,
Kharagpur 721302, India

1 Introduction

The coating nanochannels with polyelectrolyte layer(PEL) has a new technique in nanofluidic devices for flow control, ion sensing, current rectification, manipulation and detection of biomolecules [1], liquid transport, and many more. Electric double layer (EDL) forms along the channel wall due to contact electrolyte into the channel. The EDL thickness is denoted by Debye length. Electroosmotic flow (EOF) is the bulk motion of the fluid that happens under the applied electric field. Helmholtz–Smoluchowski velocity [2] occurs in a thin EDL approximation with slip velocity condition. Conlisk and McFerran [3] studied the electroosmotic flow in a nanochannel thin electric double layer under an applied electric field. The combined flow of electroosmotic and pressure driven flow have been numerically investigated by Bera and Bhattacharyya [4] using Nernst–Planck distribution for ion transport in micro and nanochannels.

There are some methods for modeling electroosmotic flow. The polyelectrolyte coated nanopores are often used to effectively control the electroosmotic velocity or minimize the interaction between nanopore wall and analyte. Patwary et al. [5] concluded that the polyelectrolyte layer coated nanochannel can be used for highly efficient energy conversion in the pressure-driven flow. Tseng et al. [6] theoretically developed temperature influence on pH-regulated polyelectrolyte layer grafted spherical particle. A simple analytic solution for mobility is developed by Ohshima [7] for low potential. Bera and Bhattacharyya [8] studied the mixing of two different solutes in the surface corrected microchannel with heterogeneous potential. Yeh et al. [9] investigated electroosmotic flow in the multi-ion case and analyzed the charge property of nanopore devices. The effects of fixed charged density in polyelectrolyte layer on ion transport and fluid flow were studied numerically by Bera and Bhattacharyya [10] in a canonical nanopore. The transport phenomena on electrokinetic flow in a pH regulated nanopore were studied by Chen and Das [11] investigated transport phenomena on pH regulated polyelectrolyte layer grafted channel. Ohshima [12] established the analytic solution for the distribution of induced potential in a charged nanopore and compared it with a numerical solution for low surface potential condition.

Most of the previous studies were considered a PEL with fixed charge density and same dielectric permittivity between the electrolyte and polyelectrolyte. In the present model, the medium around the polyelectrolyte layer contains the solvent molecules. Therefore, the PEL ions interact with the neighboring electrolyte ions. The Born formula [13] decides the related free energy which is essential to transfer ion from electrolyte solution to PEL membrane layer. Sima et al. [14] compared the numerical solution of ion partition coefficients of electroosmotic flow using Poission–Boltzmann equation with experimental data. They observed that the numerical solution of the potential distribution gave excellent agreement at low ionic strength, but differ for high ionic strength. Ion partitioning effect on soft particles in the presence of volumetrically charged core were investigated by Ganjizade et al. [15]. Poddar et

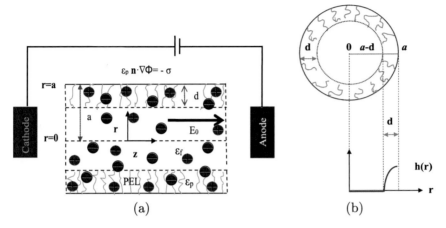

Fig. 1 **a** Schematic representation of PEL in a pH regulated cylindrical nanopore and **b** distribution of soft function in nanopore

al. [16] investigated the effects of Born energy due to the different permittivities of two mediums.

Most of the studies in electrokinetic flow are considered on the Poisson–Boltzmann model, where the ion distribution obeys the Boltzmann equation. This leads the consideration to equilibrium electric double layer, electroneutrality, and nonoverlapping EDL approximation. In this away, it fails to illustrate the fantastic features for overlapping EDL, nonequilibrium EDL distribution, and counterions into the nanopore. The present mathematical model considers the Brinkman extended modified Navier–Stoke equations, Poisson equation, and Nernst–Planck equation.

2 Mathematical Model

We have taken a canonical nanopore with length z and radius a, and an electric field is imposed along z-direction. A PEL of thickness d is coated in the nanopore walls surface. The nanopore is filled with an electrolyte solution of KCl of ionic species concentration n^∞. The charge density in PEL is represented by ρ_{fix} which depends upon the physicochemical properties of the PEL and the solution properties such as pH values and ionic concentration of the background electrolyte. The wall of the nanopore bears negative charge density σ and the electrolyte solution consists of some basic functional groups, which induces negative or positive volumetric charge density to the PEL depending upon pH values. In this away, the volume charge density in PEL varies with pH values. we consider that the polyelectrolyte region distributed into a soft function $h(r)$ and can be expressed

$$h(r) = \begin{cases} 1 - \exp\left[-\frac{r-(a-d)}{\delta}\right], & a - d \leq r \leq a \\ 0 & 0 \leq a \leq a - d \end{cases} \tag{1}$$

where δ is the width of the inhomogeneous length of polyelectrolyte segments near the front edge of PEL (Fig. 1).

We consider the permittivity of the electrolyte to be ϵ_f, while ϵ_p is the permittivity of the polyelectrolyte layer. The medium surrounding the polyelectrolyte ions contains solvent molecules of electrolyte solution with different permittivity and polyelectrolyte layer ions also interact with the surrounding electrolyte ions. The Born formula [13] determines the associated electrostatic free energy required to transfer an ion from the bulk solvent to the bulk of the membrane. Moreover, Δw_i is the Born energy, and the term containing it takes into account the ion partitioning effect. Due to medium to medium interaction, Born energy Δw_i is considered for i^{th} ionic species. It has been proved that $\Delta w_i^* = \frac{(z_i e)^2}{8\pi R_i}(\frac{1}{\epsilon_p} - \frac{1}{\epsilon_f})$, where R_i is the radius of hydrated ionic species. The hydrated radius for all ions are considered to same ($=3.3 \times 10^{-10}$m), which leads to $\Delta w_i^* = \Delta w^*$.

We consider the PEL bears both functional groups acidic (AH) and basic B, the reactions are $AH \Leftrightarrow A^- + H^+$ and $BH^+ \Leftrightarrow B^+ + H^+$. Let K_A and K_B are the equilibrium constants of the reactions, then $K_A = [A^-][H^+]/[AH]$ and $K_B = [B^-][H^+]/[BH^+]$, where $[A^-]$, $[AH]$, $[B]$, $[BH^+]$, and $[H^+]$ are the molar concentrations of ionic species A^-, AH, B, BH^+, and H^+, respectively. Let N_A and N_B be the total concentrations of the acidic and the basic functional groups, respectively, then $N_A = [A^-] + [AH]]$ and $N_B = [B] + [BH^+]$. The charge density of the PEL is

$$\rho_{fix} = -1000F\left([A^-] - [BH^+]\right) \tag{2}$$

where F is the Faraday constant and $\rho_{fix}(pH_0, pK_a, pK_b)$ occurs for the dissociation reaction of functional groups and may be written in nondimensional form as

$$\rho_{fix}(pH_0, pK_a, pK_b) =$$

$$- 1000F\left[\frac{N_A}{1 + \left(\frac{[H^+]\exp(-\frac{\Delta w^*}{k_B T})}{K_A}\right)} - \frac{N_B\left(\frac{[H^+]\exp(-\frac{\Delta w^*}{k_B T})}{K_B}\right)}{1 + \left(\frac{[H^+]\exp(-\frac{\Delta w^*}{k_B T})}{K_B}\right)}\right] \tag{3}$$

Here, the Boltzmann constant is k_B and the temperature is denoted as T.

Total potential E is the combination of induced potential and the polarization of the EDL for applied field. The volumetric charge is generally nonuniform for a pH-dependent PEL. Therefore, electric field is linked to charge density ρ_e and the charge density ρ_{fix} in PEL is connected to the electric field as

$$\nabla \cdot (\epsilon_f \mathbf{E}) = -\epsilon_f \nabla^2 \phi = \rho_e \exp\left(-\frac{\Delta w^*}{k_B T}\right) + h(r)\rho_{fix} \qquad (4)$$

Here, induced electric potential is Φ and the net charge density $\rho_e = \exp(-\frac{\Delta w^*}{k_B T}) F \sum_{i=1}^{4} z_i c_i^*$; z_i is the valance and c_i^* is molar concentration of the i ion. We considered the cylindrical coordinate system for the numerical computation of the mathematical model. The scaled equation for potential is

$$\left[\frac{\partial^2 \phi}{\partial z^2} + \frac{1}{r}\frac{\partial}{\partial r}\left(r\frac{\partial \phi}{\partial r}\right)\right] =$$

$$-\frac{1}{\epsilon_r}\left[\frac{(\kappa a)^2}{2}\exp(-\Delta w)\sum_{i=1}^{4} z_i X_i c_i + h(r)Q_{fix}\right] \qquad (5)$$

Here, Born energy Δw^* is scaled by $k_B T$, and hence $\Delta w = \frac{(z_i e)^2}{8\pi R_i k_B T \epsilon_f}\left(\frac{1}{\epsilon_r} - 1\right)$ and ϵ_r is the ratio of the permittivity of the polyelectrolyte layer to same of the electrolyte solution. We considered that the radius of the nanopore a is the length scale and absolute temperature is T. The potential is non-dimensionalized by $\phi_0 (= RT/F)$, where R is a gas constant and F is a Faraday's constant. Here, $X_i = c_{i0}/I$ and $I = \sum_i^4 z_i^2 c_i$ is the ionic concentration. Electric double layer thickness (λ) is expressed as $\lambda = \left(\epsilon_e RT/F \sum_{i=1}^{4} z_i^2 c_{i0}\right)^{1/2}$ and κ is the inverse of EDL thickness. The scaled charge density $Q_{fix}(r)$ in the diffused layer is

$$Q_{fix} = \frac{Q_A}{1 + \left(10^{pK_A - pH}\exp(-\Delta w)[H^+]\right)} -$$

$$-\frac{Q_B\left(10^{pK_B - pH}\exp(-\Delta w)[H^+]\right)}{1 + \left(10^{pK_B - pH}\exp(-\Delta w)[H^+]\right)} \qquad (6)$$

where the nondimensional parameter $Q_j = -1000 F N_j a^2/\epsilon_e \phi_0$ ($j = A, B$) is the maximum charge density due to acidic and basic functional groups. $[H^+]$ is the scaled concentration of H^+ ion which is normalized by $[H^+]_0$, the bulk molar concentration of H^+ and is related to the pH values of the bulk liquid by $pH = -log[H^+]_0$ and $pK_a = -log K_a$, $pK_b = -log K_b$, where ionization constants K_a is for acid functional group and K_b is for basic functional group.

The ions transport equation is expressed through the Nernst–Planck equation as

$$\frac{\partial c_i}{\partial t} + \nabla \cdot (-D_i \nabla c_i + c_i \omega_i z_i F\mathbf{E} + c_i \mathbf{q}) = 0 \qquad (7)$$

where the net ionic flux $\mathbf{N}_i = -D_i \nabla c_i + c_i \omega_i z_i F\mathbf{E} + c_i \mathbf{q}$. Here, D_i and ω_i are diffusivity and mobility ions, respectively. fluid velocity is $\mathbf{q} = (v, u)$ with component along u and v along radial and axial directions resp. Here, time is dimensionalized by a/U_{HS} and fluid velocity is nondimensionlized by the Helmholtz–Smoluchowski velocity U_{HS} ($= \epsilon_e E_0 \phi_0 / \mu$). Reynolds number is expressed as $Re = U_{HS} a \rho / \mu$, Schmidt number $Sc = \mu / \rho D_i$, Peclet number $Pe_i = ReSc_i$, where μ is the fluid viscosity. The scaled ionic transport are written as

$$Pe_i \frac{\partial c_i}{\partial t} - \left[\frac{\partial^2 c_i}{\partial z^2} + \frac{1}{r} \frac{\partial}{\partial r} \left(r \frac{\partial c_i}{\partial r} \right) \right] + Pe_i \left[\frac{\partial (u c_i)}{\partial z} + \frac{1}{r} \frac{\partial (r v c_i)}{\partial r} \right] + \Lambda \frac{\partial c_i}{\partial z}$$

$$- \left[\frac{\partial c_i}{\partial r} \frac{\partial \phi}{\partial r} + \frac{\partial c_i}{\partial z} \frac{\partial \phi}{\partial z} \right] + \frac{(\kappa a)^2}{2} c_i \left(\sum_{i=1}^{4} z_i c_i \right) \exp(-\Delta w) - h c_i Q_{fix} = 0 \quad (8)$$

The Navier–Stokes equation is written for the flow field in a modified form as

$$\rho \left[\frac{\partial \mathbf{q}}{\partial t} + (\mathbf{q} \cdot \nabla) \mathbf{q} \right] = -\nabla p + \mu \nabla^2 \mathbf{q} + \rho_e \mathbf{E} - \mu \lambda_s^2 \mathbf{q} \tag{9}$$

$$\nabla \cdot \mathbf{q} = 0 \tag{10}$$

Here, fluid density is ρ. Screening length $\lambda_s^2(r)$ is the reciprocal of softness parameter. For diffuse PEL, the softness parameter λ_s was presented by Duval and Ohshima [17] as

$$\lambda_s = \lambda_0 [h(r)]^{1/2} \tag{11}$$

where λ_0 is softness degree in PEL region. Here, hydrodynamics pressure is scaled by $\mu U_{HS} / a$. The scaled equations for fluidflow along cross radial z- and radial r-directions are given as

$$Re \frac{\partial u}{\partial t} + Re \left(u \frac{\partial u}{\partial z} + v \frac{\partial u}{\partial r} \right) = -\frac{\partial p}{\partial z}$$

$$- \frac{(\kappa a)^2}{2\Lambda} \left(-\Lambda + \frac{\partial \phi}{\partial z} \right) \left(\sum_{i=1}^{4} z_i c_i \right) \exp(-\Delta w)$$

$$+ \left[\frac{\partial^2 u}{\partial z^2} + \frac{1}{r} \frac{\partial}{\partial r} \left(r \frac{\partial u}{\partial r} \right) \right] - \beta^2 h u \tag{12}$$

$$Re \frac{\partial v}{\partial t} + Re \left(u \frac{\partial v}{\partial z} + v \frac{\partial v}{\partial r} \right) = -\frac{\partial p}{\partial r}$$

$$- \frac{(\kappa a)^2}{2\Lambda} \frac{\partial \phi}{\partial r} \left(\sum_{i=1}^{4} z_i c_i \right) \exp(-\Delta w)$$

$$+ \left[\frac{\partial^2 v}{\partial z^2} + \frac{1}{r} \frac{\partial}{\partial r} \left(r \frac{\partial v}{\partial r} \right) - \frac{v}{r^2} \right] - \beta^2 h v \tag{13}$$

$$\frac{\partial u}{\partial z} + \frac{\partial v}{\partial r} + \frac{v}{r} = 0 \tag{14}$$

The scaled softness parameter β is related the PEL softness degree (λ_0^{-1}) as $\beta = a/\lambda_0^{-1}$. The softness degree $\lambda_0^{-1} (= \sqrt{\mu/\gamma})$ affects hydrodynamic field in pore where, γ is the hydrodynamics frictional coefficient. For some bio-colloids and polymer gels, it's length varies from 0.1 to 10 nm [18–20] and typically represents penetration length within soft structure.

At the upstream and downstream of the nanopore, we considered the fully developed condition for computation. We imposed no-slip boundary conditions for flow field on nanopore walls. The nanopore surface is considered as ion-impenetrable, $(n \cdot N_i = 0)$ and surface charge density (σ) at the rigid wall surface.

3 Numerical Methods

We divided the whole computational area into certain finite number of rectangular cells and the computation have been performed in each cell. These equations are solved using finite volume techniques [21] considering staggered grid system. The pressure correction base algorithm SIMPLE [22] is used for Navier–Stokes equation using block tri-diagonal system by VARGA [23]. The divergence is towed less than preassigned small number $(\leq 10^{-5})$ for every cell.

To validated our numerical solution and grid independence, we consider (i) 160 $\times 250$, (ii) 400×250, and (iii) 400×500 for electrostatic flow cylindrical nanochannel and compared the axial velocity with Ai et al. [24]. The results shown in Fig. 2

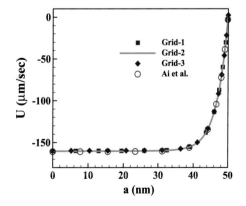

Fig. 2 Comparison of axial velocity Ai et al. [24] and grid size effects of plane EOF flow in a cylindrical channel with $h(r) = 0$ and $\epsilon_r = 1$. The electrolyte concentration is 10 mM and nanopore charge density $(\sigma) = -1$ mC/m^2 with external applied electric filed $E_0 = 10^6$ V/m

indicated that the results obtained from grid (i) and (ii) closely agree for both and the graphs are close with Ai et al. [24]. Thus, we find that grid (iii) can be considered for our computation.

4 Results and Discussions

In this article, we investigated the effects of the electrokinetic flow through polyelectrolyte layer grafted cylindrical nanopore. The polyelectrolyte thickness (d) generally varies from 3–5 nm [25]. We have taken the length of polyelectrolyte coated nanopore as $l = 80$ nm, radius a as 10 nm, and thickness of the polyelectrolyte layer (d) as 1 nm [16]. Here, permittivity of the electrolyte $\epsilon_f = 80.0 \times 8.854 \times 10^{-12}$ C/Vm, density $\rho = 1000$ Kg/m^3, Boltzmann constant ($k_B =$)1.381×10^{-23} J/K, viscosity $\mu = 0.001$ Kg/ms. Solutions are presented here for different values of electrolyte concentration, pH values, softness parameter, applied electric field, polyelectrolyte layer thickness, and charged density of the nanopore.

We have considered the background salt of aqueous solution to be KCl, denoted as C_{KCl}. The main ionic species, H^+, Cl^-, K^+, OH^-. These ions of electrolyte concentration are H^+, Cl^-, K^+, and OH^- are scaled, respectively, by bulk values $[H^+]_0$, $[K^+]_0, [Cl^-]_0, [OH^-]_0$. Let, C_{10}, C_{20}, C_{30}, and C_{40} (in mM) be the bulk ionic concentration of H^+, K^+, Cl^-, and OH^- respectively. The following relations require to satisfy the bulk electroneutrality condition [26]: if $pH \leq 7$ then $C_{10} = 10^{-pH+3}$, $C_{20} = C_{KCl}, C_{30} = C_{KCl} + 10^{-pH+3} - 10^{-(14-pH)+3}, C_{40} = 10^{-(14-pH)+3}$ and if $pH > 7$ then $C_{10} = 10^{-pH+3}$, $C_{20} = C_{KCl} - 10^{-pH+3} + 10^{-(14-pH)+3}$, $C_{30} = C_{KCl}, C_{40} = 10^{-(14-pH)+3}$. The diffusion coefficient (D_i) of these ions are, respectively, 9.31×10^{-9} m^2/s, 1.96×10^{-9} m^2/s, 2.03×10^{-9} m^2/s, and 5.30×10^{-10} m^2/s. In the present study, we considered an aqueous dispersion of amino acid $pK_A = 2.5$ (α-carboxyl), $pK_B = 9.5$ (α-amino) [27].

Figure 3a and b describes the distribution of u-velocity and corresponding induced potential distribution for different pH values. Here, we considered the ionic concentration $C_0 = 10$ mM, charge density of nanopore $\sigma = 1$ mC/m^2, permittivity ratio $\epsilon_r = 0.5$ with $N_A = N_B = 1$ mol/L. The axial velocity increases from negative to positive with the pH values and accordingly reverse case happens in the potential distribution profile. The variation of permittivity ratio are shown in Fig. 4a and b for $pH = 2$ and $pH = 8$, respectively. Here, ϵ_r is the permittivity ratio of polyelectrolyte layer to same of electrolyte solution. The increase of permittivity ratio decreases the fluid velocity for both low and high pH values. Here, $\epsilon_r = 1$ indicates that same permittivity for polyelectrolyte layer and the electrolyte solution. The effects of ionic concentration were shown in Fig. 5a and b for high $pH = 8$ and low $pH = 2$, respectively. Here, we considered the permittivity ratio $\epsilon_r = 0.5$ and softness value $\beta = 1$. The velocity decrease with ionic concentration for low pH but increase for high pH values. The variation of softness parameter are shown in Fig. 6a and b for different $pH = 2$ and $pH = 8$, respectively. The softness parameter β mainly disturb the flow field in the nanopore membrane when the conductance

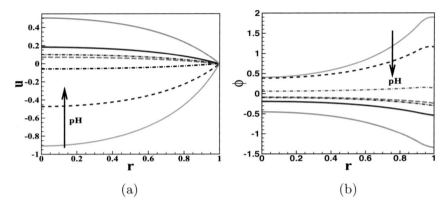

Fig. 3 Profiles of **a** velocity u and **b** induced potential ϕ for different pH values. Here, polyelectrolyte thickness $\delta/d = 0.1$, softness parameter $\beta = 1$, permittivity ratio $\epsilon_r = 0.5$ with $N_A = N_B = 1$ mol/L, and ionic concentration $C_0 = 10$ mM, when the nanopore radius $h = 10\,\mu$m, homogeneous thickness $d = 1$ nm, charge density of the nanopore $\sigma = 1$ mC/m^2. Arrows in the figure indicate the increasing order of $pH = 2, 3, 4, 5, 6, 7$, and 8

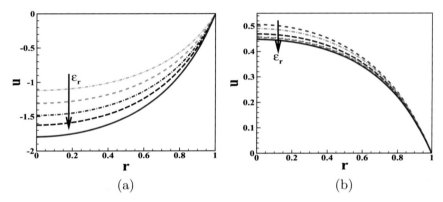

Fig. 4 Profiles of u velocity for different permittivity ratio ϵ_r values. Here, polyelectrolyte thickness $\delta/d = 0.1$, softness parameter $\beta = 1$ with $N_A = N_B = 1$ mol/L and ionic concentration $C_0 = 10$ mM when nanopore radius $h = 10\mu$m, homogeneous thickness $d = 1$ nm, charge density of the nanopore $\sigma = 1$ mC/m^2. Arrows denote increasing of $\epsilon_r = 0.6, 0.7, 0.8, 0.9$ and 1.0. **a** $pH = 2$ and **b** $pH = 8$

is not much more affected by the flow. The increases of softness values decrease the fluid velocity in low and high pH values.

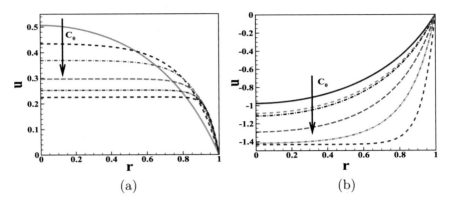

Fig. 5 Distribution u velocity for different values of molar concentration C_0. Here, polyelectrolyte thickness $\delta/d = 0.1$, softness parameter $\beta = 1$ with $N_A = N_B = 1$ mol/L, and permittivity ratio $\epsilon_r = 0.5$ when nanopore radius $h = 10\,\mu$m, homogeneous thickness $d = 1$ nm, charge density of the nanopore $\sigma = 1\,\text{mC/m}^2$. Arrows denote increasing values of ionic concentration $C_0 = 10, 50, 100, 250, 500$, and 1000 mM. **a** $pH = 2$ and **b** $pH = 8$

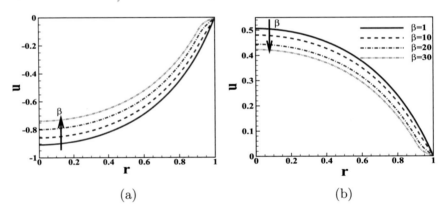

Fig. 6 Distribution u-velocity for different softness values β. Here, polyelectrolyte thickness $\delta/d = 0.1$, ionic concentration $C_0 = 10$ mM with $N_A = N_B = 1$ mol/L and permittivity ratio $\epsilon_r = 0.5$ when nanopore radius $h = 10\,\mu$m, homogeneous thickness $d = 1$ nm, surface charge of the nanopore $\sigma = 1\,\text{mC/m}^2$. Arrows here denote the increasing of softness values $\beta = 1, 10, 20$, and 30. **a** $pH = 2$ and **b** $pH = 8$

5 Conclusions

The effects of ion partitioning in electroosmotic flow have been studied through pH-regulated polyelectrolyte grafted cylindrical nanopore membrane. The equations for the electroosmotic flow are Poisson equation for potential, Nernst–Planck equation for ions transport, and Brinkman extended Navier–Stokes equation for fluid flow. These equations are discretized in computational domain and solved numerically. The crossradial velocity increases with pH values in succinoglycan functional group when the concentration is fixed. The axial velocity decreases with the electrolyte concentration for both high and low pH values. The increase of permittivity ratio decreases the fluid flow for all pH values. The increase of softness values decreases the fluid velocity for all pH values.

Acknowledgements The work is sponsored by the Science and Engineering research Board, Govt. of India, through the project (No: ECR/ 2016/ 000771).

References

1. Squires, A., Hersey, J.S., Grinstaff, M.W., Meller, A.: A nanopore-nanofiber mesh biosensor to control DNA translocation. J. Am. Chem. Soc. **135**, 16304–16307 (2013)
2. Probstein, R.F.: Physicochemical Hydrodynamics: An Introduction, 2nd edn. Wiley Interscience, New York (1994)
3. Conlisk, A.T., McFerran, J.: Mass transfer and flow in electrically charged micro-and nanochannels: Anal. Chem. **74**, 2139–2150 (2002)
4. Bera, S., Bhattacharyya, S.: On mixed electroosmotic-pressure driven flow and mass transport in microchannels. Int. J. Eng. Sci. **62**, 165–176 (2013)
5. Patwary, J., Chen, G., Das, S.: Efficient electrochemomechanical energy conversion in nanochannels grafted with polyelectrolyte layers with pH-dependent charge density. Microfluid. Nanofluid. **20**, 37–51 (2016)
6. Tseng, S., Lin, J.-Y., Hsu, J.-P.: Theoretical study of temperature influence on the electrophoresis of a pH-regulated polyelectrolyte. Anal. Chim. Acta **847**, 80–89 (2014)
7. Ohshima, H.: Electrical phenomena of soft particles. A soft step function model. J. Phys. Chem. A. **116**, 6473–6480 (2012)
8. Bera, S., Bhattacharyya, S.: Effects of geometric modulation and surface potential heterogeneity on electrokinetic flow and solute transport in a microchannel. Theor. Comp. Fluid Dyn. **32**, 201–214 (2018)
9. Yeh, Li-H., Xue, S., Joo, S. W., Qian, S., Hsu, J.P.: Field effect control of surface charge property and electroosmotic flow in nanofluidics. J. Phys. Chem. C **116**, 4209–4216 (2012)
10. Bera, S., Bhattacharyya, S.: Effect of charge density on electrokinetic ions and fluid flow through polyelectrolyte coated nanopore, V01BT10A008-V01BT10A008, ASME, USA (2017). https://doi.org/10.1115/FEDSM201769194
11. Chen, G., Das, S.: Electroosmotic transport in polyelectrolyte-grafted nanochannels with pH-dependent charge density. J. Appl. Phys. **117**, 185304–185313 (2015)
12. Ohshima, H.: A simple algorithm for the calculation of the electric double layer potential distribution in a charged cylindrical narrow pore. Colloid Polym. Sci. **294**, 1871–1875 (2016)
13. Israelachvili, J.N.: Intermolecular and Surface Forces, 3rd edn. Elsevier, Amsterdam (2011)
14. Sims, S.M., Higuchi, W.I., Srinivasan, V., Peck, K.: Ion partitioning coefficient and electroosmotic flow in cylindrical pores: comparison of the prediections of the Poission-Boltzmann equation with experiment. J. Colloid Interface Sci. **155**, 210–220 (1993)
15. Ganjizade, A., Ashrafizadeh, S.N., Sadeghi, A.: Effect of ion partitioning on the electrostatics of soft particles with a volumetrically charged core. Electrochem. Commun. **84**, 19–23 (2017)
16. Poddar, A., Maity, D., Bandopadhyay, A., Chakraborty, S.: Electrokinetics in polyelectrolyte grafted nanofluidic channels modulated by the ion partitioning effect. Soft Matter **12**, 5968–5978 (2016)
17. Duval, J.F.L., Ohshima, H.: Electrophoresis of diffuse soft particle. Langmuir **22**, 3533–3546 (2006)
18. Duval, J.F.L., Gaboriaud, F.: Progress in electrohydrodynamics of soft microbial particle interphases. Curr. Opin. Colloid Interface Sci. **15**, 184–195 (2010)

19. Duval, J.F.L., Slaveykova, V.I., Hosse, M., Buffle, J., Wilkinson, K.J.: Electrohydrodynamic properties of succinoglycan as probed by fluorescence correlation spectroscopy, potentiometric titration and capillary electrophoresis. Biomacromol **7**, 2818–2826 (2016)
20. Yeh, L.H., Hsu, J.P.: Effects of double-layer polarization and counterion condensation on the electrophoresis of polyelectrolytes. Soft Matter **7**, 396–411 (2011)
21. Leonard, B.P.: A stable and accurate convective modelling procedure based on quadratic upstream interpolation. Comput. Methods Appl. Mech. Eng. **19**, 59–98 (1979)
22. Fletcher, C.A.J.: Computational Techniques for Fluid Dynamics, Springer Series in Computational Physics), vol. 1, 2nd edn. Springer, Berlin (1991)
23. Varga, R.S.: Matrix Iterative Numerical Analysis. Wiley, New York (1962)
24. Ai, Y., Zhang, M., Joo, SW., Cheney., M.A., Qian, S.: Effects of electroosmotic flow on ionic current rectification in conical nanopores. J. Phys. Chem. C **114**, 3883–3890 (2010)
25. Yeh, Li-Hsien, Zhang, M., Qian, S., Hsu, Jyh-Ping, Tseng, S.: Ion Concentration Polarization in Polyelectrolyte-Modified Nanopores. J. Phys. Chem. C **116**, 8672–8677 (2012)
26. Yeh, L.H., Hsu, J.P., Qian, S., Tseng, S.: Counterion condensation in pH-regulated polyelectrolytes. Electrochem. Commun. **19**, 97–100 (2012)
27. Switzer, R., Garrity, L.: Experimental Biochemistry. W. H. Freeman Publishing, New York (1999)

Optimal Control of Complementary and Substitute Items in a Production System for Infinite Time Horizon

J. N. Roul⊙, K. Maity, S. Kar, and M. Maiti

Abstract The objective of this paper is to formulate an optimal control problem under infinite time horizon for complementary and substitute items. Here the advertisement rate is unknown which considered as a control variable and the stock level is taken as state variable. The demand is controlled by stock as well as advertisement rate. The unit production cost is a function of production rate and also dependent on raw material cost, development cost due to reliability and wear-tear cost. In the production process, reliability plays an important role to improve the quality of products and to decrease the defective rate. It is formulated to optimize the advertisement rate and maximize the optimal profit. Finally, total profit, which consists of the revenue, production cost, holding cost, and advertisement cost, is formulated as optimal control in the steady state with infinite time horizon. The effect of time value of money is taken into consideration. The model is formulated analytically for substitute and complementary items. The optimum results are illustrated and solved by Lyapunov Stability Theory and Pontryagin's Maximum Principle accordingly. A sensitivity analysis is also presented with respect to time value of money.

Keywords Asymptotically stable · Optimal advertisement policy · Complementary · Substitute items · Infinite time horizon

J. N. Roul (✉)
Department of Patha Bhavana, Visva-Bharati, Santiniketan 731235, West Bengal, India
e-mail: jotin2008@rediffmail.com

K. Maity
Department of Mathematics, M. G. Mahavidyalaya, Bhupatinagar 721425, West Bengal, India

S. Kar
Department of Mathematics, National Institute of Technology,
Durgapur 713209, West Bengal, India

M. Maiti
Department of Applied Mathematics with Oceanology and Computer Programming
Vidyasagar University, Midnapore 711201, West Bengal, India

© The Author(s), under exclusive license to Springer Nature Singapore Pte Ltd. 2023   117
P. Gyei-Kark et al. (eds.), *Engineering Mathematics and Computing*,
Studies in Computational Intelligence 1042,
https://doi.org/10.1007/978-981-19-2300-5_9

1 Introduction

It is well known that stability and convergence are prerequisites for defining the dynamical system under infinite time horizon. Here, a method is developed for the resolution of query regarding the properties of dynamical system and, in particular, of equilibrium, which are associated with asymptotic stability under infinite time horizon. As a consequence, the asymptotically stable point is considered by Lyapunov stability theory. During the course of journey, Lyapunov stability theory (cf. Lyapunov [15]) is used as conditions on the asymptotic stability of equilibrium point for state variable (stock) and control variable (advertisement). Asymptotically stable is defined as if it is stable and locally striking. Essentially, they also approach to the limit condition on the state. Therefore, asymptotic stability is a stronger condition than plain stability because it needs that trajectories satisfy more conditions in sense of restrictions. For the existence of the proposed optimal control model under infinite time horizon, the parametrically connected events like: Boundedness of the system, Condition of equilibrium of the dynamic system, and Analysis of the nature of eigen values of the dynamic system are needed.

Nowadays, an assortment of products can be viewed as a choice strategy that allows consumers to pursue multiple decision objectives. Due to the transaction costs of shopping, it is more efficient for the consumer to buy an assortment of products that will be consumed in the future. One of the most basic strategic decisions a retailer must make involves determining the complement and substitute products to offer. Retailers attempt to offer a balance among number of categories, number of stock-keeping units within a category, and service level. The assortment product items have proposed a field of research domains and common assortment decisions for the relational properties of the items, pricing policies and the variety of items over time. However, these suggest that product assortment plays a key role, not only in satisfying demands, but also in influencing buyer wants and preferences (cf. Simonson [31]). The customers of an item may be influenced to go for a substitute item seeing the stock level of the complementary. Several research works has done due to following authors. Maity and Maiti [16] have developed the optimal production control problem for complementary and substitute items on finite time horizon. Next, Chernev [6] has done an interdisciplinary review on product assortment and consumer choice. Recently, Katsifou et al. [11] have developed the joint product assortment inventory problem to attract the loyal and non-loyal customers. There are some investigations for substitute items in the newsboy setting. Abdel-Malek and Montanari [1] analysed a multi-product newsboy problem with a budget constraint. Das and Maiti [7] studied a single period newsboy type inventory problem for two substitutable deteriorating items with resource constraint involving a wholesaler and several retailers. Stavrulaki [33] modelled the joint effect of demand (stock-dependent) stimulation and product substitution on inventory decisions by considering a single period and stochastic demand. Gurler and Yilmaz [10] assumed substitution of a product when the other one is out of stock and presented a two-level supply chain newsboy problem with two substitutable products. Recently, Zhao et al. [34] developed a two-stage supply chain where two different manufacturers compete to sale substitutable products through common retailer and analysed the problem using game theory. Here the consumer demand function is defined as linear form of the two products' retail prices-downward slopping in its own price. Here the manufacturing costs and the customers demand for each product are characterized as fuzzy variables. Mukhopadhyay et al. [19] considered a duoploy market in which two separate firms offer complementary goods in a leader follower type move and solved, formulating Stackelberg game model. Recently, Zhao and Wang [36] studied pricing and retail service decisions in fuzzy uncertainty environments. They followed a game-theoretical approach. There is a larger manufacturer and two relatively smaller retailers in the market, and the manufacturer is considered as the Stackelberg leader.

Nowadays, the development of global economical market fetches the problem due to the inflation and time value of money. Consequently, its effect on economical review in any dynamic process should be mentioned. Buzacott [5] was the first in this direction who derived expressions for the optimal order quantity, considering inflation and time value of money. In this direction, the great works of Biermann and Thomas [4], Datta and Pal [8], Sana [29], Sarkar et al. [30], Panja and Mondal [20], Panja et al. [21] are worth mentioning, among others.

Production of imperfect units is a natural phenomenon due to different difficulties in a long-run production process. The feedback of current production-inventory optimal control models is that produced all items are not of good quality. The imperfect quality in production process was initially considered by Porteus [23] and later by several researchers such as Salameh and Jaber [27], Maity et al. [17], Khan and Jaber [12], Sivashankari and Panayappan [32], Roul et al. [24–26], and others.

In spite of the above facts developments, there are still following lacunas in the development of dynamic imperfect production-inventory control system.

1. Most of the classical production-inventory models did not take into account the effects of time value of money for infinite time horizon.
2. Here perfect and imperfect production-inventory model are considered for two complementary and substitute items. The demands of complementary items decrease against the prices of its own and complementary item. The demand of a substitute item decreases against its own price but increases against a substitute item. Moreover, in both cases, demand of an item increases with its stock and advertisement.
3. The demands of substitute and complementary items are always influenced by their prices in the advertisements and displayed stocks. Very few researchers have considered this.
4. Another important investigation is the study of the effect of reliability in the dynamical system to determine an optimal profit corresponding to the optimal reliability.

There are many authors like Zoppoli et al. [35], Bagno and Tarasyev [2, 3], and others did not mention the above lacunas.

Taking the above lacunas into account, in this paper, some imperfect and perfect assortment of items production-inventory models are formulated over an infinite time horizon with unknown advertisements policy. Here, imperfect unit production rates depend on the reliability of the system and in the process, unit production cost, directly proportional to the corresponding raw material cost, development cost due to the systems reliability and labour, energy, cost of technology, design, complexity, resources, etc. The development costs and wear-tear costs are inversely and directly proportional to production rate, respectively. The general model is formulated as an optimal control problem over an infinite time horizon for the profit maximization. The asymptotically stable point and stability of the model are considered by Lyapunov stability theory. Then, equivalent problem is solved by the Pontryagins Principle (cf. Pontryagin et al.) [22], Generalized Reduced Gradient (GRG) Technique (Gabriel and Ragsdell) [9], and Mathematica-9.0. The behaviour of time value of money is tabulated along with the total profit. The models are illustrated with numerical examples. A sensitivity analysis is performed due to the change in the time value of money parameter.

2 Preliminaries: Analysis of Stability of a Dynamical System

2.1 Equilibrium Points

Let us suppose a dynamical system

$$\dot{x} = \frac{dx}{dt} = f(x, v), \tag{1}$$

where the function $f_i(.)$ are continuously differentiable. The equilibrium point (x_e, v_e) of the system (1) is defined as

$$f(x_e, v_e) = 0 \tag{2}$$

2.2 General Form of Non-linear System

The general form of non-linear system is $\dot{x} = f(x, v)$ as

$$\frac{dx_1}{dt} = f_1(x_1, x_2, \cdots, x_n, v_1, v_2, \cdots, v_m)$$

$$\frac{dx_2}{dt} = f_2(x_1, x_2, \cdots, x_n, v_1, u_2, \cdots, v_m)$$

$$\cdots = \cdots\cdots$$

$$\frac{dx_n}{dt} = f_n(x_1, x_2, \cdots, x_n, v_1, v_2, \cdots, v_m) \tag{3}$$

By the method of linearization, the above non-linear system can be converted to its linear counter part in a small region about its stable point. By using this method, one can analyse and stabilize the non-linear systems.

2.3　Jacobian Matrix, Eigen Values, and Stability

Let us consider a dynamical system given by the equation

$$\dot{x} = f(x)$$

Let x_e denote the equilibrium point and the Jacobian matrix at the equilibrium point be given by

$$\frac{\partial f}{\partial x}(x_e) = \begin{bmatrix} \frac{\partial f_1}{\partial x_1} & \cdots & \frac{\partial f_1}{\partial x_n} \\ \cdots & \cdots & \cdots \\ \frac{\partial f_n}{\partial x_1} & \cdots & \frac{\partial f_n}{\partial x_n} \end{bmatrix}\Bigg|_{x_e}.$$

The following are the different possibilities and their effect on the system when disturbed from equilibrium points when the eigen values are real:

Nature of eigen values	Effect on dynamic system
Positive real number	Driven away from steady state value
Positive real number and Zero	The origin is unstable
Negative real number	Driven back to steady state value
Negative and positive real number	The exponent is negative, curve's axes as asymptotes but origin is unstable
Negative real number and zero	In this case, the origin is stable, but not asymptotically stable
Zero(0)	Remains at position to which it was disturbed
Identical	Effect can not be determined

2.4　Lyapunov Stability Theory: Notion of Stability

Consider the dynamical system

$$\dot{x} = f(x). \tag{4}$$

Let an equilibrium point of the system be $\bar{x}$, so that

$$f(\bar{x}) = 0. \tag{5}$$

We say that $\bar{x}$ is stable in the sense of Lyapunov stability theory if there exists an $\epsilon > 0$ such that for each $\delta = \delta(\epsilon) > 0$,

$$| x(t_0) - \bar{x} | < \delta \Rightarrow | x(t) - \bar{x} | < \epsilon \quad \forall \, t > t_0. \tag{6}$$

We say that $\bar{x}$ is **asymptotically stable** if it is stable and

$$| x(t) - \bar{x} | \to 0 \text{ as } t \to \infty. \tag{7}$$

We call $\bar{x}$ unstable if it is not stable.

We can determine the stability or in stability of $\bar{x}$ without explicitly solving the dynamic equations as following.

2.5 *Lyapunov's First or Indirect Method*

Taking the non-linear system given in Eq. (3), and Expand in Taylor series around $\bar{x}$ (we also redefine $x \to x - \bar{x}$)

$$\dot{x} = Bx + h(x), \tag{8}$$

where

$$B = \left. \frac{\partial f}{\partial x} \right|_{\bar{x}}$$

is the Jacobian matrix of $f(x)$ evaluated at $\bar{x}$ and $g(x)$ contains the higher order terms, that is

$$\lim_{|x| \to 0} \frac{| h(x) |}{| x |} = 0.$$

Then the non-linear system Eq. (3) is asymptotically stable if and only if the linear system $\dot{x} = Bx$ is stable, that is if all eigen values of B have negative real parts.

3 Proposed Optimal Control Models

Taking substitute and complementary items for the optimal control production system under infinite time horizon, following assumptions and notations are used.

3.1 Assumptions

(i) It is a single period production-inventory model with infinite time horizon $t \epsilon [0, \infty)$;
(ii) Defective rate is reliability dependent;
(iii) Shortages are not allowed;
(iv) There is no repair or replacement of defective units over whole time period;
(v) The development cost of the system increases the reliability of the system;
(vi) Unit production cost depends on produced-quantity, raw material, wear-tear and development costs;
(vii) Demand depends on stock and advertisement rate simultaneously; For substitute items, demand function is defined as a linear form of the two products retail prices-downward slopping in its own price and increasing with respect to its substitute item's selling price. In the case of complementary items, effect of both prices are downwards.
($viii$) The time value of money (i.e., discount rate) is considered on infinite time horizon.

3.2 Notations

γ : Time value of money,
For the i-th ($i = 1, 2$) item,
$v_i(t)$: advertisement rate at time t which is control variable;
$x_i(t)$: the stock level at time which is state variable at time t;
$d_i(t)$: demand rate at time t;
A_i, B_i and k_i : positive constants
$u_i(t) = u_i(1 - \frac{x_i(t)}{k_i})$: stock-dependent production with $\frac{x_i(t)}{k_i} < 1$ and u_i is production parameter; k_i is large quantity;
h_i : holding cost per unit;
L_i : fixed cost like labour, energy, etc;
N_i : cost of technology, design, complexity, resources, etc;
f_i : feasibility of increasing a component reliability and $0 < f_i < 1$;
δ_i : imperfect parameter;
r_i : reliability of the system;
$\delta_i(t) = (1 - r_i)\left(u_i(1 - \frac{x_i(t)}{k_i}) \right)^{\alpha_i}$: rate of defectiveness which is less than 1 and $0 < \alpha_i < 1$;
d_{i0} : constant part of demand function;
c_{i0} : constant material cost;
$c_{di}(r_i)$: development cost to increase the reliability of the system;
c_{di} : reliability independent development cost due to labour utilization of the system;
$r_{i,max}$: maximum reliability for the product item;

$r_{i,min}$: minimum reliability for the product item;

s_i : selling price per unit for i-th item;

β_{ii} : measure of responsiveness of i-th product's consumer demand to its own price;

s_{3-i} : selling price per unit another i-th item;

β_{i3-i} : measure of responsiveness of another i-th product's consumer demand to its own price;

$p_i(t)$: adjoint function treated as shadow price;

β_{i0} : wear-tear cost for the system;

q_i : catch-ability parameter;

c_{v_i} : advertising cost parameter.

3.3 Model-1: Formulation of the Imperfect Production Optimal Control Model for Complementary Items

Model Formulation: Here, the items are produced at a rate $u_i(t)$ with production parameter u_i of which $\delta_i(t)$ (where $\delta_i(t) = (1 - r_i)\left(u_i(1 - \frac{x_i(t)}{k_i})\right)^{\alpha_i} < 1, 0 < \alpha_i < 1$) is reliability dependent defective rate. The increase in reliability r_i decreases the defective rate and improves the quality of the product. Here a two items production-inventory problem of complementary type is considered.

In this model, the demand rate depends on the advertisement rate as well as stock rate simultaneously. Also demand of one item is dependent on the retail prices of its own and complementary item negatively.

$$d_i(t) = d_{i0} + \frac{q_i v_i(t) x_i(t)}{A_i + B_i v_i(t)} - \beta_{ii} s_i - \beta_{i3-i} s_{3-i}, \quad i = 1, 2, \tag{9}$$

where d_{i0} is the constant part of demand, q_i is the catch-ability coefficient to customers by the advertisement rate and stock level, A_i, B_i are two positive constants and β_{ii} are the measure of responsiveness of i−th product's consumer demand to its own price and which are negative for complementary item.

The differential equations for i-th representing above system during an infinite time horizon is

$$\dot{x}_i(t) = \left(1 - (1 - r_i)u_i(t)^{\alpha_i}\right)u_i\left(1 - \frac{x_i(t)}{k_i}\right) - d_i(t), \text{ for } i = 1, 2. \tag{10}$$

It can be written as $\dot{x}_i(t) =$

$$\left(1 - (1 - r_i)\left(u_i(1 - \frac{x_i(t)}{k_i})\right)^{\alpha_i}\right)u_i\left(1 - \frac{x_i(t)}{k_i}\right) - \left(d_{i0} + \frac{q_i v_i(t) x_i(t)}{A_i + B_i v_i(t)} - \beta_{ii} s_i - \beta_{i3-i} s_{3-i}\right) \tag{11}$$

for i = 1, 2; where (.) denotes the differentiation.

Here the advertising cost is dependent upon the catch-ability parameter. The advertising cost is $c_{v_i} e^{q_i} v_i(t)$ per unit advertisement.

According to Roul et al. [24], "In some cases, the cost function for a reliability dependent product system can be derived from basic considerations. It is reasonable to assume that the cost function $c_{di}(r_i)$ would satisfy the following three conditions : (i) $c_{di}(r_i)$ is a positive definite function, (ii) $c_{di}(r_i)$ is non-decreasing, (iii) $c_{di}(r_i)$ increases at a higher rate for higher values of r_i. The third condition suggests that it can be very expensive to achieve the reliability value as 1. Let $r_{i,min}$ and $r_{i,max}$ are the minimum and maximum values of r_i and f_i is a parameter ranging between 0 and 1 that represents the relative difficulty of increasing a component's reliability".

The development cost function is often stated in terms of the reliability (cf. Mettas [18]) which is given by

$$c_{di}(r_i) = L_i + N_i Exp \left[(1 - f_i) \left(\frac{r_i - r_{i,min}}{r_{i,max} - r_i} \right) \right] \qquad (12)$$

The unit production cost is considered as a function of constant produced-quantity, raw material cost, wear-tear and development costs. So unit production cost (cf. Khouja (1995)) is

$$c_{u_i}(r_i, t) = \left(c_{i0} + \frac{c_{di}(r_i)}{u_i(t)} + \beta_{i0} u_i(t) \right). \qquad (13)$$

The total holding cost over the time interval $[0, \infty)$ for the stock $x_i(t)$ is

$$\sum_{i=1}^{2} \int_0^\infty h_i x_i(t) \, dt$$

Then the total profit consisting of selling prices, holding costs, advertising costs and production costs leads to

$$J = \int_0^\infty \sum_{i=1}^{2} \left[s_i d_i(t) - h_i x_i(t) - c_{v_i} e^{q_i} v_i(t) - c_{u_i}(r_i) u_i(t) \right] e^{-\gamma t} dt \qquad (14)$$

Study of Stability of the Model: For the stability of the proposed model under infinite time horizon, validity of the parametrically connected elements like: Boundedness of the system, Condition of equilibrium of the dynamic system, and Analysis of the nature of eigen values the dynamic system are needed. Hence, we proceed as follows:

Boundedness of the System: Let us consider the function:

$$M(x_1, x_2) = x_1 + \frac{1}{l}x_2, \tag{15}$$

where l is a positive constant. The time derivative of equation is

$$\dot{M} = \dot{x}_1 + \frac{1}{l}\dot{x}_2$$

$$= \left(1 - (1 - r_1)u_1^{\alpha_1}\right)u_1\left(1 - \frac{x_1(t)}{k_1}\right) - \left(d_{10} + \frac{q_1 v_1(t)x_1(t)}{A_1 + B_1 v_1(t)} - \beta_{11}s_1 - \beta_{12}s_2\right)$$

$$+ \frac{1}{l}\left[\left(1 - (1 - r_2)u_2^{\alpha_2}\right)u_2\left(1 - \frac{x_2(t)}{k_2}\right) - \left(d_{20} + \frac{q_2 v_2(t)x_2(t)}{A_2 + B_2 v_2(t)} - \beta_{22}s_2 - \beta_{21}s_1\right)\right]$$

For each $s > 0$, we obtain

$$\dot{M} + sM = \left(1 - (1 - r_1)u_1(t)^{\alpha_1}\right)u_1\left(1 - \frac{x_1(t)}{k_1}\right) - \left(d_{10} + \frac{q_1 v_1(t)x_1(t)}{A_1 + B_1 v_1(t)} - \beta_{11}s_1 - \beta_{12}s_2\right)$$

$$+ \frac{1}{l}\left[\left(1 - (1 - r_2)u_2(t)^{\alpha_2}\right)u_2\left(1 - \frac{x_2(t)}{k_2}\right) - \left(d_{20} + \frac{q_2 v_2(t)x_2(t)}{A_2 + B_2 v_2(t)} - \beta_{22}s_2 - \beta_{21}s_1\right)\right]$$

$$+ s\left(x_1 + \frac{1}{l}x_2\right)$$

$$< (u_1 - \delta_1(t)u_1)\left(1 - \frac{x_1(t)}{k_1}\right) - \left(\frac{q_1 v_1(t)x_1(t)}{A_1 + B_1 v_1(t)}\right)$$

$$+ \frac{1}{l}\left[(u_2 - \delta_2(t)u_2)\left(1 - \frac{x_2(t)}{k_2}\right) - \left(\frac{q_2 v_2(t)x_2(t)}{A_2 + B_2 v_2(t)}\right)\right] + s\left(x_1 + \frac{1}{l}x_2\right)$$

$$< (u_1 - \delta_1(t)u_1) + sx_1 - \left(\frac{q_1 v_1(t)x_1(t)}{A_1 + B_1 v_1(t)}\right)$$

$$+ \frac{1}{l}\left[(u_1 - \delta_2(t)u_2) + sx_2 - \left(\frac{q_2 v_2(t)x_2(t)}{A_2 + B_2 v_2(t)}\right)\right]$$

$$= \left[(u_1 - \delta_1(t)u_1) + sx_1 - \frac{q_1 v_1(t)x_1(t)}{A_1 + B_1 v_1(t)}\right]$$

$$+ \frac{1}{l}\left[(u_2 - \delta_2(t)u_2) + sx_2 - \frac{q_2 v_2(t)x_2(t)}{A_2 + B_2 v_2(t)}\right]$$

$$< \left[(u_1 - \delta_1(t)u_1) + sx_1\right] + \frac{1}{l}\left[(u_2 - \delta_2(t)u_2) + sx_2\right]$$

$$= T_{impr}[\text{ considering } T_{impr} = \text{ constant}]$$

since $0 \le x_1 \le k_1$; $0 \le x_2 \le k_2$ with $\frac{x_1}{k_1} < 1, \frac{x_2}{k_2} < 1$

and, $0 < \delta_1 < 1$; $0 < \delta_2 < 1$

Hence,

$$\dot{M} + sM < T_{impr}$$

$$i.e., \quad \frac{dM}{dt} + sM = k_{impr}, [where \ k_{impr} < T_{impr}]$$

$$or, \quad e^{st}\frac{dM}{dt} + e^{st}sM = e^{st}k_{impr}$$

$$or, \quad \frac{d(Me^{st})}{dt} = e^{st}k_{impr}$$

$$or, \quad \int d(Me^{st}) = \int e^{st}k_{impr}dt + A \quad [A = constant]$$

$$So, \quad M = \frac{k_{impr}}{s} + Ae^{-st} < \frac{T_{impr}}{s} + Ae^{-st}$$

$$i.e., \quad M(x_1, x_2) < \frac{T_{impr}}{s} + Ae^{-st}$$

Therefore, we have

$$0 < M(x_1, x_2) < \frac{T_{impr}}{s} + Ae^{-st}, \tag{16}$$

where A is positive constant. When $t \to \infty$ the above yields $0 < M < \frac{T_{impr}}{s}$. Therefore, x_1, x_2 are also bounded, i.e., all solutions that start in R_2^+ are confined to the region R, where

$$R = \{(x_1, x_2) \in R_2^+ : M = \frac{T_{impr}}{s} + \varepsilon, \ for \ any \ \varepsilon > 0\}$$

Condition of Equilibrium of the Dynamic System: For equilibrium,

$$\dot{x}_1(t) = \dot{x}_2(t) = 0. \tag{17}$$

Then $x_i(t)$ gets the constant value for $i = 1, 2$. Taking $x_i(t) = \bar{x}_i$ as $t \to \infty$. Consequently, $v_i(t) = \bar{v}_i$ and $u_i(t) = \bar{u}_i$ for the steady state of an infinite time horizon dynamical system. As a result, the possible non-zero critical points of the dynamical system are $(\bar{x}_1, \bar{x}_2)$ which will be determined by solving the followings:

$$\left[(1 - (1 - r_1)\bar{u}_1^{\alpha_1})u_1(1 - \frac{\bar{x}_1}{k_1}) - (d_{10} + \frac{q_1\bar{v}_1\bar{x}_1}{A_1 + B_1\bar{v}_1} - \beta_{11}s_1 - \beta_{12}s_2)\right] = 0 \tag{18}$$

$$\left[(1 - (1 - r_2)\bar{u}_2^{\alpha_2})u_2(1 - \frac{\bar{x}_2}{k_2}) - (d_{20} + \frac{q_2\bar{v}_2\bar{x}_2}{A_2 + B_2\bar{v}_2} - \beta_{22}s_2 - \beta_{21}s_1)\right] = 0. \tag{19}$$

Solving Eqs. (18) and (19), we obtain the critical points of the dynamical system.

Analysis of the Nature of Eigen Values the Dynamic System: The system is represented as

$$\begin{pmatrix} \dot{x}_1(t) \\ \dot{x}_2(t) \end{pmatrix} = \begin{pmatrix} (1-(1-r_1)u_1(t)^{\alpha_1})u_1(1-\frac{x_1(t)}{k_1}) - (d_{10} + \frac{q_1 v_1(t)x_1(t)}{A_1+B_1 v_1(t)} - \beta_{11}s_1 - \beta_{12}s_2) \\ (1-(1-r_2)u_2(t)^{\alpha_2})u_2(1-\frac{x_2(t)}{k_2}) - (d_{20} + \frac{q_2 v_2(t)x_2(t)}{A_2+B_2 v_2(t)} - \beta_{22}s_2 - \beta_{21}s_1) \end{pmatrix}$$
$$= \begin{pmatrix} F(x_1,x_2) \\ G(x_1,x_2) \end{pmatrix} \tag{20}$$

Now differentiating $F(x_1,x_2)$ and $G(x_1,x_2)$ partially with respect to x_1 and x_2, we have

$$\frac{\partial F(x_1,x_2)}{\partial x_1} = -\frac{\left(1-(1-r_1)(1+\alpha_1)u_1(t)^{\alpha_1}\right)u_1}{k_1^2} - \frac{q_1 v_1(t)}{A_1+B_1 v_1(t)}$$

$$\frac{\partial F(x_1,x_2)}{\partial x_2} = 0$$

$$\frac{\partial G(x_1,x_2)}{\partial x_1} = 0$$

$$\frac{\partial G(x_1,x_2)}{\partial x_2} = -\frac{\left(1-(1-r_2)(1+\alpha_2)u_2(t)^{\alpha_2}\right)u_2}{k_2^2} - \frac{q_2 v_2(t)}{A_2+B_2 v_2(t)}$$

Considering $\tau_1 = (1-r_1)(1+\alpha_1)$ and $\tau_2 = (1-r_2)(1+\alpha_2)$, we investigate the local behaviour of the critical points of the dynamical system. The variational matrix ψ of the system of equations

$$\begin{pmatrix} \frac{\partial F(x_1,x_2)}{\partial x_1} & \frac{\partial F(x_1,x_2)}{\partial x_2} \\ \frac{\partial G(x_1,x_2)}{\partial x_1} & \frac{\partial G(x_1,x_2)}{\partial x_2} \end{pmatrix} \tag{21}$$

It can be written as

$$\begin{pmatrix} \frac{\partial F(x_1,x_2)}{\partial x_1} & 0 \\ 0 & \frac{\partial G(x_1,x_2)}{\partial x_2} \end{pmatrix}$$

The characteristic equation of $\psi(x_1,x_2)$, taken as $\psi(\bar{x}_1,\bar{x}_2)$ for point of equilibrium, is:

$$\left(-\frac{(1-\tau_1 \bar{u}_1^{\alpha_1})u_1}{k_1^2} - \frac{q_1 \bar{v}_1}{A_1+B_1 \bar{v}_1} - \lambda\right) = 0$$

$$\left(-\frac{(1-\tau_2 \bar{u}_2^{\alpha_2})u_2}{k_2^2} - \frac{q_2 \bar{v}_2}{A_2+B_2 \bar{v}_2} - \lambda\right) = 0 \tag{22}$$

By simplifying the above equation, we get

$$\lambda = -\left(\frac{(1-\tau_1 \bar{u}_1^{\alpha_1})u_1}{k_1^2} + \frac{q_1 \bar{v}_1}{A_1+B_1 \bar{v}_1}\right), -\left(\frac{(1-\tau_2 \bar{u}_2^{\alpha_2})u_2}{k_2^2} + \frac{q_2 \bar{v}_2}{A_2+B_2 \bar{v}_2}\right)$$

According to stability theorem, $(\bar{x}_1, \bar{x}_2)$ is a stable node if both the eigenvalues of the above are negative; that is, $\left(\frac{(1-\tau_1 \bar{u}_1^{\alpha_1})u_1}{k_1^2} + \frac{q_1 \bar{v}_1}{A_1 + B_1 \bar{v}_1}\right) > 0$ and $\left(\frac{(1-\tau_2 \bar{u}_2^{\alpha_2})u_2}{k_2^2} + \frac{q_2 \bar{v}_2}{A_2 + B_2 \bar{v}_2}\right) > 0$ are satisfied.

Solution Procedure: Thus the problem reduces to maximize the profit function J subject to the state constraint satisfying the dynamic stock-advertisement relation.

$$\text{Maximize } J = \int_0^\infty \sum_{i=1}^2 [s_i d_i(t) - h_i x_i(t) - c_{v_i} e^{q_i} v_i(t)$$

$$- \left(c_{i0} + \frac{c_{di}(r_i)}{u_i(t)} + \beta_{i0} u_i(t)\right) u_i(t)] e^{-\gamma t} dt$$

subject to

$$\dot{x}_i(t) = \left(1 - (1 - r_i) u_i(t)^{\alpha_i}\right) u_i\left(1 - \frac{x_i(t)}{k_i}\right)$$

$$- \left(d_{i0} + \frac{q_i v_i(t) x_i(t)}{A_i + B_i v_i(t)} - \beta_{ii} s_i - \beta_{i3-i} s_{3-i}\right); \quad i = 1, 2. \tag{23}$$

The above Eq. (23) are defined as an optimal control problem with control variable $v_i(t)$ and state variable $x_i(t)$. It can be deduced to algebraic forms using Pontryagin's Maximum Principle.

Let us consider Hamiltonian H as

$$H(x_i(t), v_i(t), t)$$

$$= \sum_{i=1}^2 \left(s_i d_i(t) - h_i x_i(t) - c_{v_i} e^{q_i} v_i(t) - (c_{i0} + \frac{c_{di}(r_i)}{u_i(t)} + \beta_{i0} u_i(t)) u_i(t)\right) e^{-\gamma t}$$

$$+ \sum_{i=1}^2 p_i(t) \dot{x}_i(t) \tag{24}$$

$$= \sum_{i=1}^2 \left[s_i \left(d_{i0} + \frac{q_i v_i(t) x_i(t)}{A_i + B_i v_i(t)} - \beta_{ii} s_i + \beta_{i3-i} s_{3-i}\right) - h_i x_i(t) - c_{v_i} e^{q_i} v_i(t) \right.$$

$$- \left(c_{i0} + \frac{L_i + N_i Exp(1 - f_i)(\frac{r_i - r_{i,\min}}{r_{i,\max} - r_i})}{u_i(t)} + \beta_{i0} u_i(t)\right) u_i(t)] e^{-\gamma t}$$

$$+ \sum_{i=1}^2 p_i(t) \left[\left(1 - (1 - r_i)\left(u_i(1 - \frac{x_i(t)}{k_i})\right)^{\alpha_i}\right) u_i\left(1 - \frac{x_i(t)}{k_i}\right)\right.$$

$$- \left(d_{i0} + \frac{q_i v_i(t) x_i(t)}{A_i + B_i v_i(t)} - \beta_{ii} s_i - \beta_{i3-i} s_{3-i}\right)]. \tag{25}$$

Using Pontryaginn's Maximum Principle for steady state system with $p_i(t)$ adjoint function treated as shadow price, control variables $v_i(t)$ and state variables $x_i(t)$ which maximize H, we have

$$\frac{\partial H}{\partial v_i(t)} = 0 \tag{26}$$

$$\frac{d(p_i(t))}{dt} = -\frac{\partial H}{\partial x_i(t)}. \tag{27}$$

Now, solving Eqs. (25) and (26), shadow price or adjoint variable is

$$p_i(t) = \left[s_i - \frac{c_{v_i} e^{q_i}(A_i + B_i v_i(t))^2}{q_i A_i x_i(t)}\right] e^{-\gamma t}. \tag{28}$$

Differentiating (28) with respect to t, we have

$$\dot{p}_i(t) = -\gamma\left[s_i - \frac{c_{v_i} e^{q_i}(A_i + B_i v_i(t))^2}{q_i A_i x_i(t)}\right] e^{-\gamma t}. \tag{29}$$

Again from the Eq. (27),

$$\dot{p}_i(t) = -\left[\frac{s_i q_i v_i(t)}{A_i + B_i v_i(t)} - h_i + \frac{c_{i0} u_i}{k_i} + \frac{2\beta_{i0} u_i^2}{k_i} - \frac{2\beta_{i0} x_i(t) u_i^2}{k_i^2}\right] e^{-\gamma t}$$
$$+ \left[\frac{(1 - (1 - r_i)(1 + \alpha_i) u_i(t)^{\alpha_i}) u_i}{k_i} + \frac{q_i v_i(t)}{A_i + B_i v_i(t)}\right] p_i(t). \tag{30}$$

Considering the optimal advertisement $\dot{v}_i(t) = 0$ and the equilibrium condition $\dot{x}_i(t) = 0$. Taking $v_i(t) = \bar{v}_i$, $x_i(t) = \bar{x}_i$ and $u_i(t) = \bar{u}_i$ as $t \to \infty$ for the steady state of an infinite time horizon dynamical system. Here, it is seen from Eq. (30) that $\dot{p}_i(t) = $ constant; i.e., $p_i(t)$ is function of t. Now, from Eq. (26) and comparing (29) and (30), we have

$$- \left[\frac{s_i q_i \bar{v}_i}{A_i + B_i \bar{v}_i} - h_i + \frac{c_{i0} u_i}{k_i} + \frac{2\beta_{i0} u_i^2}{k_i} - \frac{2\beta_{i0} \bar{x}_i u_i^2}{k_i^2}\right] e^{-\gamma t}$$
$$+ \left[\frac{(1 - (1 - r_i)(1 + \alpha_i) \bar{u}_i^{\alpha_i}) u_i}{k_i} + \frac{q_i \bar{v}_i}{A_i + B_i \bar{v}_i)}\right] p_i(t)$$
$$= -\gamma\left[s_i - \frac{c_{v_i} e^{q_i}(A_i + B_i \bar{v}_i)^2}{q_i A_i \bar{x}_i}\right] e^{-\gamma t}. \tag{31}$$

Using the value of shadow price $p_i(t)$, we have

$$- \left[\frac{s_i q_i \bar{v}_i}{A_i + B_i \bar{v}_i} - h_i + \frac{c_{i0} u_i}{k_i} + \frac{2\beta_{i0} u_i^2}{k_i} - \frac{2\beta_{i0} \bar{x}_i) u_i^2}{k_i^2}\right] e^{-\gamma t}$$
$$+ \left[\frac{(1 - (1 - r_i)(1 + \alpha_i) \bar{u}_i^{\alpha_i}) u_i}{k_i} + \frac{q_i \bar{v}_i}{A_i + B_i \bar{v}_i}\right]\left[s_i - \frac{c_{v_i} e^{q_i}(A_i + B_i v_i(t))^2}{q_i A_i \bar{x}_i}\right] e^{-\gamma t}$$
$$= -\gamma\left[s_i - \frac{c_{v_i} e^{q_i}(A_i + B_i \bar{v}_i)^2}{q_i A_i \bar{x}_i}\right] e^{-\gamma t}. \tag{32}$$

Then (32) takes the form

$$
\left(\frac{s_i q_i \bar{v}_i)}{A_i + B_i \bar{v}_i} - h_i + \frac{c_{i0} u_i}{k_i} + \frac{2\beta_{i0} u_i^2}{k_i} - \frac{2\beta_{i0} \bar{x}_i u_i^2}{k_i^2} \right)
$$
$$
+ \left(\frac{(1 - (1 - r_i)(1 + \alpha_i)\bar{u}_i^{\alpha_i}) u_i}{k_i} + \frac{q_i \bar{v}_i}{A_i + B_i \bar{v}_i} + \gamma \right)\left(s_i - \frac{c_{v_i} e^{q_i} (A_i + B_i \bar{v}_i)^2}{q_i A_i \bar{x}_i} \right) = 0.
$$

$$(33)$$

The control variables $v_i(t)$ i.e., $v_1(t)$ and $v_2(t)$ (in steady state $\bar{v}_1$ and $\bar{v}_2$)which maximize H satisfying the Eqs. (18), (19) and (32). Now, the problem reduces to maximize J given by (23) satisfying (18), (19), and (32). To reach the maximum profit for the above optimal control problem, the tools like a gradient-based non-linear optimization method-GRG(LINGO-11.0) and Mathematica-9.0 are used.

3.4 Model-1A: Formulation of an Imperfect Production Optimal Control Model for Substitute Items

Here the demand is associated with substitute items for imperfect production optimal control model. Thus the problem reduces to maximize the profit function J subject to the state constraint satisfying the dynamic stock-advertisement relation. Then the formulation of the imperfect production optimal control model for substitute items can be written as

$$
\text{Maximize } J = \int_0^\infty \sum_{i=1}^2 \left[s_i d_i(t) - h_i x_i(t) - c_{v_i} e^{q_i} v_i(t) \right.
$$
$$
\left. - \left(c_{i0} + \frac{c_{di}(r_i)}{u_i(t)} + \beta_{i0} u_i(t) \right) u_i(t) \right] e^{-\gamma t} dt
$$

subject to
$$
\dot{x}_i(t) = \left(1 - (1 - r_i) u_i(t)^{\alpha_i} \right) u_i(t) \left(1 - \frac{x_i(t)}{k_i} \right)
$$
$$
- \left(d_{i0} + \frac{q_i v_i(t) x_i(t)}{A_i + B_i v_i(t)} - \beta_{ii} s_i + \beta_{i3-i} s_{3-i} \right); \quad i = 1, 2.. \quad (34)
$$

The above model (34) is defined as an optimal control problem with control variable $v_i(t)$ and state variable $x_i(t)$. (in steady state $\bar{v}_i$ and $\bar{x}_i$)It can be deduced and solved to algebraic forms using Pontryagin's Maximum Principle, a gradient-based non-linear optimization method-GRG(Lingo-11.0) and Mathematica-9.0 are used.

3.5 Model-2: Formulation of the Perfect Production Optimal Control Model for Complementary Items

Model Formulation: In this case, putting $r_i = 1$ in (10), we have

$$\dot{x}_i(t) = u_i\left(1 - \frac{x_i(t)}{k_i}\right) - \left(d_{i0} + \frac{q_i v_i(t) x_i(t)}{A_i + B_i v_i(t)} - \beta_{ii} s_i - \beta_{i3-i} s_{3-i}\right) \tag{35}$$

for $i = 1, 2$; where (.) denotes the differentiation.
Similarly, we have

$$c_{u_i}(t) = \left(c_{i0} + \frac{c_{di}}{u_i(t)} + \beta_{i0} u_i(t)\right) \tag{36}$$

and the total holding cost over the infinite time interval $[0, \infty)$ for the stock $x_i(t)$ as $\int_0^\infty \sum_{i=1}^2 h_i x_i(t)\, dt$.
Then the total profit consisting of selling prices, holding costs, advertising costs and production costs leads to

$$J = \int_0^\infty \sum_{i=1}^2 \left[s_i d_i(t) - h_i x_i(t) - c_{v_i} e^{q_i} v_i(t) - c_{u_i}(t) u_i(t)\right] e^{-\gamma t} dt. \tag{37}$$

Study of Stability of the Model: Proceeding as before, we establish the boundedness, derive the condition of equilibrium and analysed. Thus for the perfect production in dynamical system, the required conditions of equilibrium are

$$\dot{x}_1(t) = \dot{x}_2(t) = 0. \tag{38}$$

As a result, the possible non-zero critical points of the dynamical system are $(\bar{x}_1, \bar{x}_2)$ which will be determined by solving the followings:

$$\left[u_1\left(1 - \frac{\bar{x}_1}{k_1}\right) - \left(d_{10} + \frac{q_1 \bar{v}_1 \bar{x}_1}{A_1 + B_1 \bar{v}_1} - \beta_{11} s_1 - \beta_{12} s_2\right)\right] = 0 \tag{39}$$

$$\left[u_2\left(1 - \frac{\bar{x}_2}{k_2}\right) - \left(d_{20} + \frac{q_2 \bar{v}_2 \bar{x}_2}{A_2 + B_2 \bar{v}_2} - \beta_{22} s_2 - \beta_{21} s_1\right)\right] = 0. \tag{40}$$

Solving Eqs. (39) and (40), we have the critical points of the dynamical system.
The characteristic equations and the values are as follows: The characteristic equation of $\psi(x_1, x_2)$, taken as $\psi(\bar{x}_1, \bar{x}_2)$ for point of equilibrium, is:

$$\left(-\frac{u_1}{k_1^2} - \frac{q_1\bar{v}_1}{A_1 + B_1\bar{v}_1} - \lambda\right) = 0$$

$$\left(-\frac{u_2}{k_2^2} - \frac{q_2\bar{v}_2}{A_2 + B_2\bar{v}_2} - \lambda\right) = 0$$

$$(41)$$

By simplifying the above equation, we get

$$\lambda = -\left(\frac{u_1}{k_1^2} + \frac{q_1\bar{v}_1}{A_1 + B_1\bar{v}_1}\right), -\left(\frac{u_2}{k_2^2} + \frac{q_2\bar{v}_2}{A_2 + B_2\bar{v}_2}\right).$$

According to theorem, $(\bar{x}_1, \bar{x}_2)$ is a stable node if both the eigenvalues of the above are negative; that is, $\left(\frac{u_1}{k_1^2} + \frac{q_1\bar{v}_1}{A_1+B_1\bar{v}_1}\right) > 0$ and $\left(\frac{u_2}{k_2^2} + \frac{q_2\bar{v}_2}{A_2+B_2\bar{v}_2}\right) > 0$ are satisfied.

Solution Procedure: Thus the problem reduces to maximize the profit function J subject to the state constraint satisfying the dynamic stock-advertisement relation.

$$\text{Maximize } J = \int_0^\infty \sum_{i=1}^2 \left[s_i d_i(t) - h_i x_i(t) - c_{v_i} e^{q_i} v_i(t)\right.$$

$$\left. - \left(c_{i0} + \frac{c_{di}}{u_i(t)} + \beta_{i0} u_i(t)\right) u_i(t)\right] e^{-\gamma t} dt$$

subject to

$$\dot{x}_i(t) = u_i\left(1 - \frac{x_i(t)}{k_i}\right) - \left(d_{i0} + \frac{q_i v_i(t) x_i(t)}{A_i + B_i v_i(t)}\right.$$

$$\left. - \beta_{ii} s_i - \beta_{i3-i} s_{3-i}\right) \text{ for } i = 1, 2.$$

$$(42)$$

The above formulation (42) is defined as an optimal control problem with control variable $v_i(t)$ and state variable $x_i(t)$ (in steady state $\bar{v}_i$ and $\bar{x}_i$). It can be deduced to algebraic forms using Pontryagin's Maximum Principle.

Let us consider Hamiltonian H as

$$H(x_i(t), v_i(t), t) = \sum_{i=1}^2 \left(s_i d_i(t) - h_i x_i(t) - c_{v_i} e^{q_i} v_i(t) - (c_{i0} + \frac{c_{di}}{u_i(t)}\right.$$

$$\left. + \beta_{i0} u_i(t)) u_i(t)\right) e^{-\gamma t} + \sum_{i=1}^2 p_i(t) \dot{x}_i(t)$$

$$(43)$$

$$= \sum_{i=1}^2 \left[s_i\left(d_{i0} + \frac{q_i v_i(t) x_i(t)}{A_i + B_i v_i(t)} - \beta_{ii} s_i - \beta_{i3-i} s_{3-i}\right) - h_i x_i(t) - c_{v_i} e^{q_i} v_i(t)\right.$$

$$- \left(c_{i0} + \frac{L_i}{u_i(t)} + \beta_{i0} u_i(t)\right) u_i(t)\right] e^{-\gamma t} + \sum_{i=1}^2 p_i(t)\left[u_i\left(1 - \frac{x_i(t)}{k_i}\right)\right.$$

$$\left. - \left(d_{i0} + \frac{q_i v_i(t) x_i(t)}{A_i + B_i v_i(t)} - \beta_{ii} s_i - \beta_{i3-i} s_{3-i}\right)\right].$$

$$(44)$$

Using Pontryagin's Maximum Principle for steady state system with $p_i(t)$ adjoint function treated as shadow price controls variables $v_i(t)$ and state variables $x_i(t)$ which maximize H, we have

$$\frac{\partial H}{\partial v_i(t)} = 0 \tag{45}$$

$$\frac{d(p_i(t))}{dt} = -\frac{\partial H}{\partial x_i(t)}. \tag{46}$$

Similarly, considering the optimal advertisement $\dot{v}_i(t) = 0$ and the equilibrium condition $\dot{x}_i(t) = 0$. Taking $v_i(t) = \bar{v}_i$, $x_i(t) = \bar{x}_i$ and $u_i(t) = \bar{u}_i$ as $t \to \infty$ for the steady state of an infinite time horizon dynamical system. Now, solving Eqs. (45) and (46) we have

$$\left(\frac{s_i q_i \bar{v}_i}{A_i + B_i \bar{v}_i} - h_i + \frac{c_{i0} u_i}{k_i} + \frac{2\beta_{i0} u_i^2}{k_i} - \frac{2\beta_{i0} \bar{x}_i u_i^2}{k_i^2} \right)$$
$$+ \left(\frac{u_i}{k_i} + \frac{q_i \bar{v}_i}{A_i + B_i \bar{v}_i} + \gamma \right) \left(s_i - \frac{c_{v_i} e^{q_i} (A_i + B_i \bar{v}_i)^2}{q_i A_i \bar{x}_i} \right) = 0. \tag{47}$$

The control variables $v_i(t)$ i.e., $v_1(t)$ and $v_2(t)$ (in steady state $\bar{v}_i$ i.e., $\bar{v}_1$ and $\bar{v}_2$)which maximize H satisfying the Eqs. (39), (40), and (42). Now, the problem reduces to maximize J given by (34) satisfying (39), (40), and (42). To reach the maximum profit for the above optimal control problem, the tools like a gradient-based non-linear optimization method-GRG(Lingo-11.0) and Mathematica-9.0 are used.

3.6 Model-2A: Formulation of the Perfect Production Optimal Control Model for Substitute Items

Here the demand is associated with substitute items for perfect production optimal control model. Thus the problem reduces to maximize the profit function J subject to the state constraint satisfying the dynamic stock-advertisement relation.

Then the formulation of the perfect production optimal control model for substitute items can be written as

$$\text{Maximize } J = \int_0^\infty \sum_{i=1}^2 \left[s_i d_i(t) - h_i x_i(t) - c_{v_i} e^{q_i} v_i(t) \right.$$
$$\left. - \left(c_{i0} + \frac{c_{di}}{u_i(t)} + \beta_{i0} u_i(t) \right) u_i(t) \right] e^{-\gamma t} dt$$

subject to

$$\dot{x}_i(t) = u_i\left(1 - \frac{x_i(t)}{k_i}\right) - \left(d_{i0} + \frac{q_i v_i(t) x_i(t)}{A_i + B_i v_i(t)} - \beta_{ii} s_i + \beta_{i3-i} s_{3-i}\right) \text{ for } i = 1, 2.$$

(48)

The above model (48) is defined as an optimal control problem with control variable $v_i(t)$ and state variable $x_i(t)$ (in steady state $\bar{v}_i$ and $\bar{x}_i$). It can be deduced and solved to algebraic forms using Pontryagin's Maximum Principle, a gradient-based non-linear optimization method-GRG(Lingo-11.0) and Mathematica-9.0 are used.

4 Numerical Experiments

4.1 Input Data for Imperfect Items and Perfect Items

To illustrate the models, two items are taken and the optimum available reliability r_i^* for the complementary and substitute items of imperfect production, i.e., for Model-2.1, 2.1A as $(0.718, 0.719)$. Other inputs for the system are given in Table 1.

A_i	B_i	d_{i0}	q_i	u_i	h_i	β_{ii}	β_{i3-i}	k_i
0.316	0.520	12	0.382	14	0.43	0.50	0.48	13
0.318	0.530	14	0.381	15	0.41	0.40	0.38	14
0.316	0.520	12	0.382	14	0.43	0.50	0.48	13
0.318	0.530	14	0.381	15	0.41	0.40	0.38	14

4.2 Output for Perfect and Imperfect Items

Using the above input data, the eigenvalues of the system are derived with the help of Lingo-11.0 and Mathematica-9.0. At the equilibrium state the unknowns $\bar{x}_i$, $\bar{v}_i$ are evaluated. These are given in Table 2.

Table 1 Values of input data or collected data

System	i	N_i	L_i	C_{i0}	f_i	α_i	s_i	s_{3-i}	$r_{i,min}$	$r_{i,max}$	C_{v_i}	β_{i0}	r_i^*
		in $	in $	in $			in $	in $					
Imperfect	1	11	10	0.14	0.51	0.0271	14	16	0.6701	0.747	0.641	0.114	0.718
	2	12	11	0.13	0.52	0.0275	16	14	0.6705	0.748	0.639	0.115	0.719
Perfect	1	–	10	0.14	–	0.0271	14	16	–	–	0.641	0.114	–
	2	–	11	0.13	–	0.0275	16	14	–	–	0.639	0.115	–

Table 2 Values of $\bar{u}_i$, $\bar{d}_i$, $\bar{v}_i$, $\bar{x}_i$, r_i^* and λ_i; i=1,2 for different models

Models	$\bar{u}_i$	$\bar{d}_i$	$\bar{v}_i$	$\bar{x}_i$	λ_i	r_i^*	Total Profit
Model-1 Imperfect production optimal control	4.14	5.26	9.36	3.45	−1.012	0.718	$523.51
Model for complementary items	7.03	6.33	8.42	2.49	−1.113	0.719	
Model-2 Perfect production optimal control	4.829	8.827	9.003	1.561	−1.561	−	$543.59
Model for complementary items	9.371	4.632	8.645	1.238	−1.238	−	
Model-1A Imperfect production optimal control	10.56	11.30	3.30	2.80	−1.113	0.718	$997.60
Model for substitute items	11.11	9.80	3.80	3.20	−1.321	0.719	
Model-2A Perfect production optimal control	12.940	14.70	3.782	1.362	−1.129	−	$1279.42
Model for substitute items	15.828	12.09	1.319	0.219	−1.151	−	

5 Sensitivity Analysis

5.1 Effect of the Time Value of Money on Profit

The economical system in most of the country changes rigorously due to the presence of high inflation. So, it is not possible to avoid the effect of inflation, because inflation declines sharply the purchasing power of money. For the optimal control models, from Table 4, it is seen that when the time value of money gets minimum value, the corresponding profit function gives maximum value. Also it is more clear from Fig. 1 which is depicted for Model-2.2A and for the other models figure will be same. This is as per expectation.

5.2 Effect of Degree of Substitutive (DOS) on Profit

Here the demands (9) are substitutive on the basis of their prices and $-\beta_{ii} + \beta_{i3-i}$; $(i = 1, 2)$ are Degree of Substitutive(DOS). Here, varying the Degree of Substitutive(DOS) of items, we evaluate the profits of the Model-2.2A which are presented in Table 3. From the Table 3, it is seen in the results under group-A that as the Degree of Substitutive(DOS) of both the items remain same with the increase of the absolute values for both items (same for both) profits of the first item decrease, those for second item increase and total profit decreases. This is because the selling price of the first item (14 $) is less than that of the second item (16$). The total profit

Table 3 Values of profit changes with the degree of substitutive in Model-2.2A

System	β_{11}	β_{12}	$\beta_{11} - \beta_{12}$ (DOS)	β_{22}	β_{21}	$\beta_{22} - \beta_{21}$ (DOS)	Profit for item-1 in $	Profit for item-2 in $	Total Profit in $
A	0.40	0.25	0.15	0.40	0.25	0.15	578.52	594.75	1173.27
	0.60	0.45	0.15	0.60	0.45	0.15	565.75	604.57	1170.32
	0.70	0.55	0.15	0.70	0.55	0.15	559.37	609.53	1168.90
	0.80	0.65	0.15	0.80	0.65	0.15	552.98	614.54	1167.52
	0.90	0.75	0.15	0.90	0.75	0.15	546.59	619.60	1166.19
B	0.40	0.30	0.10	0.40	0.25	0.15	600.86	594.75	1195.61
	0.50	0.35	0.15	0.60	0.45	0.15	572.14	604.57	1176.71
	0.60	0.40	0.20	0.70	0.55	0.15	543.39	609.53	1152.82
	0.70	0.45	0.25	0.80	0.65	0.15	514.56	614.54	1129.11
	0.80	0.50	0.30	0.90	0.75	0.15	485.62	630.63	1116.25
C	0.40	0.25	0.15	0.40	0.30	0.10	578.52	735.42	1313.94
	0.60	0.45	0.15	0.50	0.35	0.15	565.75	785.82	1351.57
	0.70	0.55	0.15	0.60	0.40	0.20	559.37	836.22	1395.59
	0.80	0.65	0.15	0.70	0.45	0.25	552.98	886.62	1439.60
	0.90	0.75	0.15	0.80	0.50	0.30	546.59	937.02	1483.61
D	0.40	0.30	0.10	0.40	0.35	0.05	600.86	696.22	1297.08
	0.50	0.35	0.15	0.50	0.40	0.10	572.14	746.62	1318.76
	0.60	0.40	0.20	0.60	0.45	0.15	543.39	797.02	1340.41
	0.70	0.45	0.25	0.70	0.50	0.20	514.56	847.42	1361.98
	0.80	0.50	0.30	0.80	0.55	0.25	485.62	897.82	1383.44

Table 4 Values of profit changes with Time Value of Money (γ)

γ	0.25	0.26	0.27	0.28	0.29	0.30	0.31	0.32	0.33	0.34	0.35
J for Model-1 (in $)	523.51	482.66	398.57	320.55	247.97	220.28	207.01	157.75	102.16	104.51	100.53
J for Model-2 (in $)	543.042	502.66	498.57	420.55	417.97	412.28	407.01	398.75	352.16	319.51	302.53
J for Model-1A (in $)	997.60	982.66	939.57	920.55	914.97	912.28	817.01	757.75	702.16	678.51	662.53
J for Model-2 A (in $)	1279.42	1182.66	1098.57	1020.55	1007.97	1000.28	987.01	957.75	902.16	849.51	832.53

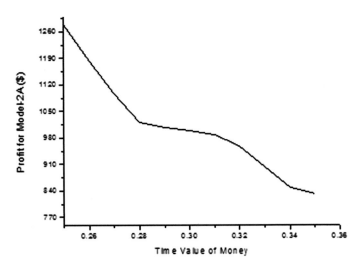

Fig. 1 Values of profit changes with Time Value of Money for Model$-2.2A$

decreases because the effect of first item is marginally more than that of the second item. Under the group-B, the DOSs of the second item remain same with increased values, but the DOSs of the first item gradually are increased. The resultant profits have the same trend as the earlier ones under A. Under group-C, the same procedure as in B is followed, only the positions of the items are interchanged. As DOSs for the second item are increased and its price is higher, the absolute value of the profits are higher than those under group A and B, and the trends of profits are increasing. Under D, the DOSs of both items are increased gradually. In this case also, the trend is same as earlier group-C but the absolute values of profits are less that the earlier values. Attention of the floor manager may be drawn to the fact like C by taking the point of view for maximization of profit of the dynamical system. All these effects are due influence of the expressions: $-\beta_{ii}s_i + \beta_{i3-i}s_{3-i}$; $(i = 1, 2)$.

6 Discussion

From the results in Table 2, it is seen that the total profit for perfect production system of both substitute ($1279.49) and complementary($ 543.59) items is more than the corresponding profits ($997.60 for substitute and $523.51 for complementary items) for imperfect system. It is expected that the profits for an perfect system will be always higher than that of substitute at of imperfect system. Again, in the case of substitute items, the demand of an item decreases against its own price where increases against its competitor's price. On the other hand, in the case of complementary items, demand of an item goes down with its own price and also with its complement's price. Hence, demand of a complementary item decreases more than that of the substitute item. In

Table 2, this is reflected in the total profits. Here, profits for substitute items ($997.6 for imperfect and $1279.42 for perfect system) are more than the corresponding profits of complementary items ($ 523.51 for imperfect and $543.59 for perfect systems). As per expectation, the above-mentioned scenarios are observed in the case of production values and stocks at steady state.

7   Managerial Insights

In many production-inventory systems, demand of an item depends on stock and promotional effort such as advertisement. For substitute items, demand of an item decreases with respect to its own price and increases with respect to substitute item's price. Normally, production-inventory models are of finite time duration, cycle repeats with time. In real- life, some established production-inventory firms such as Tata Metalik firm, ACC cement production firm, etc.; the production is not stopped but continued for years. This situation may be considered as infinite time horizon. Hence, the present investigation may be useful for the established renowned production firms managers.

8   Conclusion

The gate way of the proposed system/model is to establish the consistency of global economic market that will provide to sustain a long-time running system. Here assortment of product items has been considered to make a field of research across various domains, including economics, analytical and empirical modelling. In this section, the key domains are: how consumers perceive the variety of items in an assortment, and how consumers choose an item from a given assortment. Our paper received a customer's viewpoint to examine how product assortment encourages to make decisions for finding the choice. Here we also control the demand by advertisement with time value of money on infinite time horizon. Finally, major contribution is the introduction of effect of Degree of Substitutive (DOS) on profit in which attention of the floor manager may be drawn the fact by taking the point of view for maximization of profit of the dynamical system. The research paper can be extended in future for fuzzy reliability, rework of imperfect items, deteriorating item, etc.

References

1. Abdel-Malek, L., Montanari, R.: An analysis of the multi-product newsboy problem with a budget constraint. Int. J. Prod. Econ. **97**(3), 296–307 (2005)
2. Bagno, A.L., Tarasyev, A.M.: Estimate for the accuracy of a backward procedure for the Hamilton Jacobi equation in an infinite-horizon optimal control problem. Proc. Steklov Inst. Math. **304**, 110–123 (2019)
3. Bagno, A.L., Tarasyev, A.M.: Numerical methods for construction of value functions in optimal control problems with infinite horizon. IFAC-PapersOnLine **53**(2), 6730–6735 (2020)
4. Bierman, H., Thomas, J.: Inventory decisions under inflationary condition. Decis. Sci. **8**, 151–155 (1977)
5. Buzacott, J.A.: Economic order quantities with inflation. Oper. Res. Quat. **26**, 553–558 (1975)
6. Chernev, A.: Product assortment and consumer choice: an interdisciplinary review **6**, 1–61 (2011)
7. Das, B., Maiti, M.: An application of bi-level newsboy problem in two substitutable items under capital cost. Appl. Math. Comput. **190**(1), 410–422 (2007)
8. Datta, T.K., Pal, A.K.: Effects of inflation and time-value of money on inventory model with linear time-dependent demand rate and shortages. Eur. J. Oper. Res. **52**, 1–8 (1991)
9. Gabriel, G.A., Ragsdell, K.M.: The generalized reduced gradient method. AMSE J. Eng. Ind. **99**, 384–400 (1977)
10. Gurler, U., Yilmaz, A.: Inventory and coordination issues with two substitutable products. Appl. Math. Model. **34**, 539–551
11. Katsifou, A., Seifert, R.W., Tancrez, J.: Joint product assortment, inventory and price optimization to attract loyal and non-loyal customers. Omega **46**, 36–50 (2014)
12. Khan, M., Jaber, M.Y.: Optimal inventory cycle in a two-stage supply chain incorporating imperfect items from suppliers. Int. J. Oper. Res. **10**(4), 442–457 (2011)
13. Khouja, M.: The economic production lot-size model under volume flexibility. Comput. Oper. Res. **22**, 515–525 (1995)
14. Krishnamoorthi, C., Panayappan, S.: An EPQ model for an imperfect production system with rework and shortages. Int. J. Oper. Res. **17**(1), 104–124 (2013)
15. Lyapunov, A.M.: The general problem of the stability of motion. Int. J. Control **55**(3), 531–773 (1992)
16. Maity, K., Maiti, M.: Optimal inventory policies for deteriorating complementary and substitute items. Int. J. Syst. Sci. **40**, 267–276 (2009)
17. Maity, K., Maiti, M.: A numerical approach to a multi-objective optimal inventory control problem for deteriorating multi-items under fuzzy inflation and discounting. Comput. Math. Appl. **55**, 1794–1807 (2008)
18. Mettas, A.: Reliability Allocation and Optimization for Complex System. Reliability and Maintainability Symposium, pp. 216–221. Los Angeles, Relia Soft Corporation (2000)
19. Mukhopadhyay, S.K., Yue, X., Zhu, X.: A Stackelberg model of pricing of complementary goods under information asymmetry. Int. J. Prod. Econ. **134**(2), 424–433 (2011)
20. Panja, P., Mondal, S.K.: Stability analysis of coexistence of three species prey predator model. Nonlinear Dyn. **81**(1–2), 373–382 (2015)
21. Panja, P., Mondal, S.K., Chattopadhyay, J.: Stability and bifurcation analysis of Japanese encephalitis model with/without effects of some control parameters. Comp. Appl. Math. **37**, 1330–1351 (2018)
22. Pontryagin, L.S., Boltyanski, V.G., Gamkrelidze, R.V., Mishchenko, E.F.: The Mathematical Theory of Optimal Process (This is the Original Bible of Optimal Control Theory). Inter Science, New York (1962)
23. Porteus, E.L.: Optimal lot sizing, process quality improvement and set up cost reduction. Oper. Res. **34**(1), 137–144 (1986)
24. Roul, J.N., Maity, K., Kar, S., Maiti, M.: Multi-item reliability dependent imperfect production inventory optimal control models with dynamic demand under uncertain resource constraint. Int. J. Prod. Res. **53**(16), 4993–5016 (2015)

25. Roul, J.N., Maity, K., Kar, S., Maiti, M.: Optimal control problem for an imperfect production process using fuzzy variational principle. J. Intell. Fuzzy Syst. **32**(1), 565–577 (2017)
26. Roul, J.N., Maity, K., Kar, S., Maiti, M.: Multi-item optimal control problem with fuzzy costs and constraints using fuzzy variational principle. RAIRO-Oper. Res. **53**, 1061–1082 (2019)
27. Salameh, M.K., Jaber, M.Y.: Economic production quantity model for items with imperfect quality. Int. J. Prod. Econ. **64**(1), 59–64 (2000)
28. Sana, S.S.: A production-inventory model of imperfect quality products in a three-layer supply chain. Decis. Support Syst. **50**(2), 539–547 (2011)
29. Sana, S.S.: The EOQ model-A dynamical system. Appl. Math. Comput. **218**, 8736–8749 (2012)
30. Sarkar, B., Mandal, P., Sarkar, S.: An EMQ model with price and time dependent demand under the effect of reliability and inflation. Appl. Math. Comput. **231**, 414–421 (2014)
31. Simonson, I.: The effect of product assortment on buyer preferences. J. Retail. **75**(3), 347–370 (1999)
32. Sivashankari, C.K., Panayappan, S.: Production inventory model with reworking of imperfect production. Int. J. Manage. Sci. Eng. Manage. **9**, 9–20 (2014)
33. Stavrulaki, E.: Inventory decisions for substitutable products with stock-dependent demand. Int. J. Prod. Econ. **129**(1), 65–78 (2011)
34. Zhao, J., Tang, W., Zhao, R., Wei, J.: Pricing decisions for substitutable products with a common retailer in fuzzy environments. Eur. J. Oper. Res. **216**(2), 409–419 (2012)
35. Zoppoli, R., Sanguineti, M., Gnecco, G., Parisini, T.: Optimal control problems over an infinite horizon. In: Neural Approximations for Optimal Control and Decision. Communications and Control Engineering. Springer, Cham (2020). https://doi.org/10.1007/978-3-030-29693-3_10
36. Zhao, J., Wang, L.: Pricing and retail service decisions in fuzzy uncertainty environments. Appl. Math. Comput. **250**, 580–592 (2015)

On a New Class of Szász-Type Operators Based on Beta Function

Dhawal J. Bhatt⊙, Ranjan Kumar Jana⊙, and Vishnu Narayan Mishra⊙

Abstract In this paper, the Durrmeyer-type modification of Szász-type operators using the Beta function is introduced. First, some fundamental properties of these operators are studied and then approximation properties of a sequence of these operators using the Korovkin theorem are investigated. The rate of approximation by the modulus of continuity is estimated. Further, the Voronovskaja-type asymptotic result for these operators is also established.

Keywords Szász operators · Korovkin theorem · Modulus of continuity · Peetre's K-functional

1 Introduction

Approximation theory has much wider applications in applied and computational mathematics, engineering and allied fields of science. Out of different types of approximation theory such as approximation of functions using polynomials, trigonometric polynomials, spline and rational functions, the present study deals with approximation by positive linear operators. Approximation theory involving positive linear operators mainly deals with the convergence behavior of the sequence

D. J. Bhatt · R. K. Jana (✉)
Applied Mathematics and Humanities Department, Sardar Vallabhbhai National Institute of Technology, Ichchhanath Mahadev Dumas Road, Surat 395 007, Gujarat, India
e-mail: rkjana2003@yahoo.com

D. J. Bhatt
e-mail: dhawal.bhatt@sxca.edu.in

D. J. Bhatt
Department of Mathematics, St. Xavier's College, Navrangpura, Ahmedabad 380 009, Gujarat, India

V. N. Mishra
Department of Mathematics, Indira Gandhi National Tribal University, Lalpur, Amarkantak, Anuppur 484 887, Madhya Pradesh, India

of positive linear operators. In 1953, Korovkin [10] presented a very important and simple criterion to determine whether a given sequence of positive linear operators on $C([0, 1])$ provides desired approximation. Due to this important convergence theorem of Korovkin, many new operators were developed by several researchers. The following are some remarkable works in the area of approximation by positive linear operators. In 1950, Szász generalized the Bernstein polynomials [3] to an unbounded interval $[0, \infty)$ known as the Szász-Mirakyan operators [13] which is defined as

$$S_n(f; x) = \sum_{k=0}^{\infty} s_{n,k}(x) f\left(\frac{k}{n}\right), \tag{1}$$

where the Szász basis function is given by $s_{n,k}(x) = e^{-nx} \dfrac{(nx)^k}{k!}$.

In 1957, Baskakov proposed positive linear operators known as Baskakov operators [2] which is defined as

$$V_n(f; x) = \sum_{k=0}^{\infty} b_{n,k}(x) f\left(\frac{k}{n}\right), \tag{2}$$

where $b_{n,k}(x) = \dbinom{n+k-1}{k} \dfrac{x^k}{(1+x)^{n+k}}$ $(x \in [0, \infty))$.

In 1967, Durrmeyer [4] introduced a generalization of Bernstein operators using summation-integral-type formula as follows.

$$D_n(f; x) = (n+1) \sum_{k=0}^{n} p_{n,k}(x) \int_0^1 p_{n,k}(t) f(t) \, dt, \tag{3}$$

where $p_{n,k}(x) = \dbinom{n}{k} x^k (1-x)^{n-k}$ and $x \in [0, 1]$.

In 1985, Mazhar and Totik [11] introduced a Durrmeyer-type modification of Szász operators as defined below.

$$\overline{S}_n(f; x) = n \sum_{k=0}^{\infty} s_{n,k}(x) \int_0^{\infty} s_{n,k}(t) f(t) \, dt, \tag{4}$$

For $x \in [0, \infty)$, the Szász-Baskakov operators [12] are defined as

$$\mathcal{D}_n(f; x) = (n-1) \sum_{k=0}^{\infty} s_{n,k}(x) \int_0^{\infty} b_{n,k}(t) f(t) \, dt. \tag{5}$$

Lots of work have been done by many researchers on the Szász-type operators. Several generalization and development on the Szász-type operators are studied in [1, 5–9]. The present work is motivated by these developments. For the current discussion, consider a continuous function $f : [0, 1] \to \mathbb{R}$. For the development of the subsequent operators, extend the function f as $f(x) = f(1)$ for $x > 1$. For further development, the function f is considered with the above extended definition. We define Szász-type operators as follows. For $x \in [0, 1]$, $\mathcal{F}_n : C([0, 1]) \to C([0, 1])$ is defined as

$$\mathcal{F}_n(f; x) = n \sum_{k=0}^{n} \mathcal{P}_{n,k}(x) \int_0^{\infty} s_{n,k}(t) f(t)\, dt, \tag{6}$$

where $\mathcal{P}_{n,k}(x) = \binom{n}{k} \dfrac{B(nx + k + 1, 2n - k - nx + 1)}{B(nx + 1, n - nx + 1)}$ and $B(a, b)$ is the beta function given by $B(a, b) = \displaystyle\int_0^1 t^{a-1}(1 - t)^{b-1}\, dt \ (a, \ b > 0)$.

It can be seen that the operators defined in (6) are positive linear operators. In this paper, we study the Korovkin-type theorem, convergence behavior and asymptotic result of the sequence of operators (6).

2 Auxiliary Results

The results given in the following theorem are useful for the further development in the rest of the paper.

Lemma 1 *For a, b > 0 and non-negative integers n, k with $0 \leqslant k \leqslant n$, we have the following expressions.*

(i) $\displaystyle\sum_{k=0}^{n} \binom{n}{k} \frac{B(a + k, b + n - k)}{B(a, b)} = 1.$

(ii) $\displaystyle\sum_{k=0}^{n} \binom{n}{k} \frac{B(a + k, b + n - k)}{B(a, b)} \frac{k}{n} = \frac{a}{a + b}.$

(iii) $\displaystyle\sum_{k=0}^{n} \binom{n}{k} \frac{B(a + k, b + n - k)}{B(a, b)} \frac{k^2}{n^2} = \frac{(n - 1)(a + 1)a}{n(a + b + 1)(a + b)} + \frac{a}{n(a + b)}.$

(iv) $\displaystyle\sum_{k=0}^{n} \binom{n}{k} \frac{B(a + k, b + n - k)}{B(a, b)} \frac{k^3}{n^3} = \frac{(n - 1)(n - 2)(a + 2)(a + 1)a}{n^2(a + b + 2)(a + b + 1)(a + b)}$

$$+ \frac{3(n - 1)(a + 1)a}{n^2(a + b + 1)(a + b)} + \frac{a}{n^2(a + b)}.$$

(v) $\displaystyle\sum_{k=0}^{n}\binom{n}{k}\frac{B(a+k,b+n-k)}{B(a,b)}\frac{k^4}{n^4} = \frac{(n-1)(n-2)(n-3)(a+3)(a+2)(a+1)a}{n^3(a+b+3)(a+b+2)(a+b+1)(a+b)}$

$$+\frac{6(n-1)(n-2)(a+2)(a+1)a}{n^3(a+b+2)(a+b+1)(a+b)}$$
$$+\frac{7(n-1)(a+1)a}{n^3(a+b+1)(a+b)}+\frac{a}{n^3(a+b)}.$$

The proof of the above lemma is omitted. The proof of the above lemma can be obtained using properties of the beta function and binomial coefficient.

We obtain the first four raw moments of the operators defined in (6). The following lemma gives the raw moments of operators defined in (6).

Lemma 2 *For $x \in [0, 1]$ and the operators (6), the following holds:*

(i) $\mathcal{F}_n(1; x) = 1$,

(ii) $\mathcal{F}_n(t; x) = \dfrac{nx+1}{n+2}+\dfrac{1}{n}$,

(iii) $\mathcal{F}_n(t^2; x) = \dfrac{(n-1)(nx+2)(nx+1)}{n(n+3)(n+2)}+\dfrac{4(nx+1)}{n(n+2)}+\dfrac{2}{n^2}$,

(iv) $\mathcal{F}_n(t^3; x) = \dfrac{(n-1)(n-2)(nx+3)(nx+2)(nx+1)}{n^2(n+4)(n+3)(n+2)}+\dfrac{9(n-1)(nx+2)(nx+1)}{n^2(n+3)(n+2)}$

$$+\frac{18(nx+1)}{n^2(n+2)}+\frac{6}{n^3},$$

(v) $\mathcal{F}_n(t^4; x) = \dfrac{(n-1)(n-2)(n-3)(nx+4)(nx+3)(nx+2)(nx+1)}{n^3(n+5)(n+4)(n+3)(n+2)}$

$$+\frac{16(n-1)(n-2)(nx+3)(nx+2)(nx+1)}{n^3(n+4)(n+3)(n+2)}+\frac{72(n-1)(nx+2)(nx+1)}{n^3(n+3)(n+2)}$$
$$+\frac{96(nx+1)}{n^2(n+2)}+\frac{24}{n^4}.$$

Proof (i) From (6) and part (i) of the Lemma 1, we get

$$\mathcal{F}_n(1; x) = n\sum_{k=0}^{n}P_{n,k}(x)\int_0^{\infty}s_{n,k}(t)\,dt$$

$$= \sum_{k=0}^{n}P_{n,k}(x) = 1.$$

(ii) Again, from (6), we obtain

$$\mathcal{F}_n(t; x) = n\sum_{k=0}^{n}P_{n,k}(x)\int_0^{\infty}s_{n,k}(t)\,t\,dt$$

$$= n\sum_{k=0}^{n}P_{n,k}(x)\frac{k+1}{n^2}$$

$$= \sum_{k=0}^{n}P_{n,k}(x)\frac{k}{n}+\frac{1}{n}\sum_{k=0}^{n}P_{n,k}(x).$$

From parts (i) and (ii) of the Lemma 1, we get $\mathcal{F}_n(t; x) = \dfrac{nx + 1}{n + 2} + \dfrac{1}{n}$.

(iii) In a similar manner as above, we obtain

$$\mathcal{F}_n(t^2; x) = n \sum_{k=0}^{n} \mathcal{P}_{n,k}(x) \int_0^{\infty} s_{n,k}(t) \, t^2 \, dt$$

$$= n \sum_{k=0}^{n} \mathcal{P}_{n,k}(x) \frac{(k+1)(k+2)}{n^3}$$

$$= \sum_{k=0}^{n} \mathcal{P}_{n,k}(x) \left(\frac{k^2}{n^2} + \frac{3k}{n^2} + \frac{2}{n^2} \right).$$

From the parts (i), (ii) and (iii) of the Lemma 1, we get

$$\mathcal{F}_n(t^2; x) = \frac{(n-1)(nx+2)(nx+1)}{n(n+3)(n+2)} + \frac{4(nx+1)}{n(n+2)} + \frac{2}{n^2}.$$

(iv) Similarly as above, we obtain

$$\mathcal{F}_n(t^3; x) = n \sum_{k=0}^{n} \mathcal{P}_{n,k}(x) \int_0^{\infty} s_{n,k}(t) \, t^3 \, dt$$

$$= n \sum_{k=0}^{n} \mathcal{P}_{n,k}(x) \frac{(k+1)(k+2)(k+3)}{n^4}$$

$$= \sum_{k=0}^{n} \mathcal{P}_{n,k}(x) \left(\frac{k^3}{n^3} + \frac{6k^2}{n^3} + \frac{11k}{n^3} + \frac{6}{n^3} \right).$$

From the parts (i), (ii), (iii) and (iv) of the Lemma 1, we get

$$\mathcal{F}_n(t^3; x) = \frac{(n-1)(n-2)(nx+3)(nx+2)(nx+1)}{n^2(n+4)(n+3)(n+2)}$$

$$+ \frac{9(n-1)(nx+2)(nx+1)}{n^2(n+3)(n+2)} + \frac{18(nx+1)}{n^2(n+2)} + \frac{6}{n^3}.$$

(v) Following the procedure as above, we obtain

$$\mathcal{F}_n(t^4; x) = n \sum_{k=0}^{n} \mathcal{P}_{n,k}(x) \int_{0}^{\infty} s_{n,k}(t)\, t^4 \, dt$$

$$= n \sum_{k=0}^{n} \mathcal{P}_{n,k}(x) \frac{(k+1)(k+2)(k+3)(k+4)}{n^5}$$

$$= \sum_{k=0}^{n} \mathcal{P}_{n,k}(x) \left(\frac{k^4}{n^4} + \frac{10k^3}{n^4} + \frac{35k^2}{n^4} + \frac{50k}{n^4} + \frac{24}{n^4} \right).$$

Using all results of the Lemma 1, we get

$$\mathcal{F}_n(t^4; x) = \frac{(n-1)(n-2)(n-3)(nx+4)(nx+3)(nx+2)(nx+1)}{n^3(n+5)(n+4)(n+3)(n+2)}$$
$$+ \frac{16(n-1)(n-2)(nx+3)(nx+2)(nx+1)}{n^3(n+4)(n+3)(n+2)}$$
$$+ \frac{72(n-1)(nx+2)(nx+1)}{n^3(n+3)(n+2)} + \frac{96(nx+1)}{n^2(n+2)} + \frac{24}{n^4}.$$

The following lemma gives the moment estimation of operators given in (6) about x which is due to the linearity of the operators (6) and Lemma 2.

Lemma 3 *For $x \in [0, 1]$, the following equalities give central moments for the operators given in (6) about x.*

(i) $\mathcal{F}_n((t-x)^0; x) = 1,$

(ii) $\mathcal{F}_n((t-x); x) = \dfrac{2(n - nx + 1)}{n(n+2)},$

(iii) $\mathcal{F}_n((t-x)^2; x) = \dfrac{(-2x^2 + 3x)n^3 + (6x^2 - 7x + 8)n^2 + (-12x + 20)n + 12}{n^2(n+3)(n+2)},$

(iv)

$$\mathcal{F}_n((t-x)^3; x) = \frac{2nx(12x^2 - 27x + 16)}{(2+n)(3+n)(4+n)} - \frac{6(4x^3 + 3x^2 - 3x - 8)}{(2+n)(3+n)(4+n)}$$
$$+ \frac{2(36x^2 - 73x + 108)}{n(2+n)(3+n)(4+n)} - \frac{24(6x - 13)}{n^2(2+n)(3+n)(4+n)}$$
$$+ \frac{144}{n^3(2+n)(3+n)(4+n)},$$

(v)

$$\mathcal{F}_n((t-x)^4; x) = \frac{3n^2x(4x^3 - 12x^2 + 9x + 32)}{(2+n)(3+n)(4+n)(5+n)}$$
$$- \frac{2n(126x^4 - 336x^3 + 353x^2 - 701x - 48)}{(2+n)(3+n)(4+n)(5+n)}$$
$$+ \frac{120x^4 + 804x^3 - 1959x^2 + 4468x + 1416}{(2+n)(3+n)(4+n)(5+n)}$$
$$- \frac{2(240x^3 + 121x^2 - 415x - 2856)}{n(2+n)(3+n)(4+n)(5+n)}$$
$$+ \frac{4(360x^2 - 2419x + 1590)}{n^2(2+n)(3+n)(4+n)(5+n)} - \frac{48(60x + 43)}{n^3(2+n)(3+n)(4+n)(5+n)}.$$

3 Main Results

The following theorem shows the convergence of the sequence of operators (6) for a function $f \in C([0, 1])$. Here, we consider $C([0, 1])$ equipped with the norm $\|f\| = \sup\limits_{x \in [0,1]} |f(x)|$.

Theorem 1 *For every* $f \in C([0, 1])$, $\|\mathcal{F}_n(f; \cdot) - f(\cdot)\| \to 0$ *uniformly as* $n \to \infty$.

Proof From Lemma 2, we have

$$\mathcal{F}_n(1; x) = 1$$
$$\mathcal{F}_n(t; x) = \frac{nx + 1}{n + 2} + \frac{1}{n}$$
$$\mathcal{F}_n(t^2; x) = \frac{(n-1)(nx+2)(nx+1)}{n(n+3)(n+2)} + \frac{4(nx+1)}{n(n+2)} + \frac{2}{n^2}.$$

It follows that $\mathcal{F}_n(t^m; x)$ converges uniformly to x^m ($m = 0, 1, 2$) on $[0, 1]$.

Hence, the result follows by Korovkin's theorem [10].

For a function $f \in C([a, b])$, the modulus of continuity is defined as

$$\omega_f(\delta) \equiv \omega(f, \delta) = \sup_{\substack{x-\delta \leqslant t \leqslant x+\delta \\ a \leqslant x \leqslant b}} |f(t) - f(x)| \; ; \text{where } \delta > 0.$$

For $x, y \in [a, b]$ with $x \neq y$, $|f(t) - f(x)| \leq \omega(f, |t - x|)$. Also, if $\lambda > 0$ and $\delta > 0$, then $\omega(f, \lambda\delta) \leq (1 + \lambda) \cdot \omega(f, \delta)$. Moreover,

$$|f(t) - f(x)| \leq \left(1 + \frac{(t - x)^2}{\delta^2}\right) \omega(f, \delta). \tag{7}$$

Now, we estimate the rate of approximation of the sequence of operators (6). The following theorem gives the rate of approximation of the sequence of operators (6) in terms of modulus of continuity of a function $f \in C([0, 1])$.

Theorem 2 *For every function $f \in C([0, 1])$,*

$$|\mathcal{F}_n(f; x) - f(x)| \leq 2\,\omega(f, \delta_n),$$

where $\delta_n = \mathcal{F}_n((t - x)^2; x)^{\frac{1}{2}}$.

Proof Using the monotonicity of the operators (6), we have

$$|\mathcal{F}_n(f; x) - f(x)| \leq n \sum_{k=0}^{n} \mathcal{P}_{n,k}(x) \int_{0}^{\infty} s_{n,k}(t)\, |f(t) - f(x)|\, dt$$

$$\leq n \sum_{k=0}^{n} \mathcal{P}_{n,k}(x) \int_{0}^{\infty} s_{n,k}(t) \left(1 + \frac{1}{\delta^2}(t - x)^2\right) \omega(f, \delta)\, dt$$

$$= \left[1 + \frac{n}{\delta^2} \sum_{k=0}^{n} \mathcal{P}_{n,k}(x) \int_{0}^{\infty} s_{n,k}(t)\, (t - x)^2\, dt\right] \omega(f, \delta)$$

$$= \left[1 + \frac{1}{\delta^2} \mathcal{F}_n((t - x)^2; x)\right] \omega(f, \delta).$$

If we take $\delta = \delta_n = (\mathcal{F}_n((t - x)^2; x))^{\frac{1}{2}}$, then we have

$$|\mathcal{F}_n(f; x) - f(x)| \leq 2\,\omega(f, \delta_n).$$

Let $f \in C([0, 1])$, $M > 0$ and $0 < s \leq 1$. A function f is said to be in the (Lipschitz) class $Lip_M(s)$ if the inequality $|f(t) - f(x)| \leq M\,|t - x|^s$ holds for $t, x \in [0, 1]$.

Theorem 3 *Let $f \in Lip_M(s)$. Then*

$$|\mathcal{F}_n(f; x) - f(x)| \leq M\,\delta_n^s(x),$$

where $\delta_n(x) = (\mathcal{F}_n((t - x)^2; x))^{\frac{1}{2}}$.

Proof Using the monotonicity of the operators (6), we obtain

$$|\mathcal{F}_n(f;x) - f(x)| \leqslant n \sum_{k=0}^{n} \mathcal{P}_{n,k}(x) \int_0^{\infty} s_{n,k}(t)\,|f(t) - f(x)|\;dt$$

$$\leqslant Mn \sum_{k=0}^{n} \mathcal{P}_{n,k}(x) \int_0^{\infty} s_{n,k}(t)\,|t - x|^s\,dt.$$

Applying Hölder's inequality, we get

$$|\mathcal{F}_n(f;x) - f(x)| \leqslant M \left(n \sum_{k=0}^{n} \mathcal{P}_{n,k}(x) \int_0^{\infty} s_{n,k}(t)\,(t - x)^2\,dt \right)^{\frac{s}{2}}$$

$$\left(n \sum_{k=0}^{n} \mathcal{P}_{n,k}(x) \cdot \frac{1}{n} \right)^{\frac{2-s}{2}}$$

$$= M \left(n \sum_{k=0}^{n} \mathcal{P}_{n,k}(x) \int_0^{\infty} s_{n,k}(t)\,(t - x)^2\,dt \right)^{\frac{s}{2}}$$

$$= M \left(\mathcal{F}_n((t - x)^2; x) \right)^{\frac{s}{2}} = M\,\delta_n^s(x),$$

where $\delta_n(x) = (\mathcal{F}_n\left((t - x)^2; x\right))^{\frac{1}{2}}$.

Now, we establish a direct result for the operators (6) using Peetre's K-functional and the second-order modulus of continuity.

Let $f \in C([a, b])$. Let $\delta > 0$ and $\mathcal{W}^2 = \{g \in C([a, b]) : g', g'' \in C([a, b])\}$. Peetre's K-functional is defined by

$$K_2(f, \delta) = \inf_{g \in \mathcal{W}^2} \{||f - g|| + \delta||g''||\}.$$

For $f \in C([a, b])$, the second-order modulus of continuity is defined by

$$\omega_2(f, \delta) = \sup_{0 < h < \delta}\ \sup_{a \leqslant x \leqslant b} \{|f(x + 2h) - 2f(x + h) + f(x)|\}\ , \text{ where } \delta > 0.$$

It is important to note that there is a constant $M > 0$ such that

$$K_2(f, \delta^2) \leqslant M\omega_2(f, \delta). \tag{8}$$

Theorem 4 *Let $f \in C([0, 1])$ and $g \in C([0, 1])$ be such that $g', g'' \in C([0, 1])$.*
Then for all $n \in \mathbb{N}$ there is a constant $M > 0$ such that

$$\left| \mathcal{F}_n(f; x) - f(x) - g'(x) \left(\frac{2(n - nx + 1)}{n(n + 2)} \right) \right| \leqslant M\omega_2(f, \delta_n(x)),$$

where $\delta_n(x) = \sqrt{\dfrac{\mathcal{F}_n\left((t - x)^2; x\right)}{4}}$.

Proof Set $\mathcal{W}^2$ for $[0, 1]$. For $g \in \mathcal{W}^2$, employ Taylor's expansion to obtain

$$g(t) = g(x) + g'(x)(t - x) + \int_x^t (t - u)g''(u) \, du.$$

By Lemma (3), we have

$$\mathcal{F}_n(g; x) = g(x) + g'(x) \left(\frac{2(n - nx + 1)}{n(n + 2)} \right) + \mathcal{F}_n \left(\int_x^t (t - u)g''(u) \, du; x \right).$$

$$\therefore \left| \mathcal{F}_n(g; x) - g(x) - g'(x) \left(\frac{2(n - nx + 1)}{n(n + 2)} \right) \right| \leqslant \mathcal{F}_n \left(\int_x^t |t - u||g''(u)| \, du; x \right)$$

$$\leqslant \mathcal{F}_n \left(\int_x^t (t - u) \sup_{u \in [t,x]} |g''(u)| \, du; x \right)$$

$$\leqslant \mathcal{F}_n \left(\int_x^t (t - u) \sup_{u \in [0,1]} |g''(u)| \, du; x \right)$$

$$\leqslant \mathcal{F}_n \left(\int_x^t (t - u)\|g''\| \, du; x \right)$$

$$\leqslant \mathcal{F}_n \left(\int_x^t (t - u) \, du; x \right) \|g''\|$$

$$\leqslant \mathcal{F}_n \left(\frac{(t - x)^2}{2}; x \right) \|g''\|$$

$$\therefore \left| \mathcal{F}_n(g; x) - g(x) - g'(x) \left(\frac{2(n - nx + 1)}{n(n + 2)} \right) \right| \leqslant \mathcal{F}_n \left((t - x)^2; x \right) \frac{\|g''\|}{2}.$$

Now,

$$\left| \mathcal{F}_n(f; x) - f(x) - g'(x) \left(\frac{2(n - nx + 1)}{n(n+2)} \right) \right| \leqslant |\mathcal{F}_n(f - g; x) - (f - g)(x)|$$
$$+ \left| \mathcal{F}_n(g; x) - g(x) - g'(x) \left(\frac{2(n - nx + 1)}{n(n+2)} \right) \right|.$$

Using the relation $|\mathcal{F}_n(f; x)| \leqslant ||f||$, we obtain

$$\left| \mathcal{F}_n(f; x) - f(x) - g'(x) \left(\frac{2(n - nx + 1)}{n(n+2)} \right) \right| \leqslant 2||f - g|| + \mathcal{F}_n\left((t - x)^2; x\right) \frac{||g''||}{2}.$$

Taking infimum on the right side of above inequality over $g \in \mathcal{W}^2$, we get

$$\left| \mathcal{F}_n(f; x) - f(x) - g'(x) \left(\frac{2(n - nx + 1)}{n(n+2)} \right) \right| \leqslant 2K_2(f, \delta_n^2(x)),$$

where $\delta_n(x) = \sqrt{\dfrac{\mathcal{F}_n\left((t - x)^2; x\right)}{4}}$.

In the view of the property of Peetre's K-functional given in (8), we get

$$\left| \mathcal{F}_n(f; x) - f(x) - g'(x) \left(\frac{2(n - nx + 1)}{n(n+2)} \right) \right| \leqslant M\omega_2(f, \delta_n(x)).$$

4 Voronovskaya-Type Theorem

In this section, we establish a Voronovskaya-type asymptotic formula for the operators $\mathcal{F}_n$.

Lemma 4 *For every $x \in [0, 1]$, we have*

$$\lim_{n \to \infty} n\mathcal{F}_n((t - x); x) = 2(1 - x) ,$$

$$\lim_{n \to \infty} n\mathcal{F}_n((t - x)^2; x) = x(3 - 2x).$$

The proof of the above lemma is clear from Lemma 3.

Theorem 5 *If $f \in C([0, 1])$ such that $f', f'' \in C([0, 1])$ and $x \in [0, 1]$, then we have*

$$\lim_{n \to \infty} n(\mathcal{F}_n(f; x) - f(x)) = 2(1 - x)f'(x) + \frac{x(3 - 2x)}{2} f''(x).$$

Proof Let $f, f', f'' \in C([0, 1])$ and $x \in [0, 1]$ be fixed.

By Taylor's expansion, we can write

$$f(t) = f(x) + (t - x)f'(x) + \frac{(t - x)^2}{2} f''(x) + h(t)(t - x)^2, \tag{9}$$

where $h(t) \in C([0, 1])$ is Peano-type remainder and $\lim_{t \to x} h(t) = 0$.

Using linearity of $\mathcal{F}_n$ in (9), we get

$$n(\mathcal{F}_n(f; x) - f(x)) = nf'(x)\mathcal{F}_n((t - x); x) + \frac{n}{2} f''(x)\mathcal{F}_n((t - x)^2; x) + n\mathcal{F}_n(h(t)(t - x)^2; x). \tag{10}$$

Using the Cauchy-Schwarz inequality, we have

$$n\mathcal{F}_n(h(t)(t - x)^2; x) \leqslant \sqrt{\mathcal{F}_n(h^2(t); x)}\sqrt{n^2 \cdot \mathcal{F}_n((t - x)^4; x)}. \tag{11}$$

Now, $0 \leqslant \lim_{n \to \infty} n^2 \mathcal{F}_n((t - x)^4; x) < \infty.$ (From (v) of Lemma 3)

As $h(t) \in C([0, 1])$ and $\lim_{t \to x} h(t) = 0$, using uniform convergence of the operators $\mathcal{F}_n$ for $f(t) \in C([0, 1])$, $\mathcal{F}_n(h^2(t); x) \to h^2(x) = 0$ uniformly.

From (11), we get $\lim_{n \to \infty} n\mathcal{F}_n(h(t)(t - x)^2; x) = 0$.

Hence, from (10), we get

$$\lim_{n \to \infty} n(\mathcal{F}_n(f; x) - f(x)) = 2(1 - x)f'(x) + \frac{x(3 - 2x)}{2} f''(x).$$

5 Graphical Illustration

In this section, we show the approximation of some integrable functions by the operators $\mathcal{F}_n$ graphically using Maple.

We consider the functions $f(x) = \sin 2\pi x$ and $g(x) = 2x^3 - x$ defined on $[0, 1]$. Figures 1 and 2 show the approximation of the functions f and g, respectively, by the operators $\mathcal{F}_n$ for $n = 10, 15, 25, 50$ and 100. Figures 1 and 2 also exhibit the rate of approximation by the operators $\mathcal{F}_n$.

6 Conclusion

In this paper, new Szász-type operators (6) using beta function approximating continuous functions on $[0, 1]$ are proposed. The convergence behavior and rate of convergence of the sequence of operators (6) are studied in different forms. The Voronovskaya-type asymptotic estimate for these operators is also studied. Some graphical illustrations are presented to investigate the convergence of the sequence

Fig. 1 Approximation of f by $\mathcal{F}_n$ for $n = 10, 15, 25, 50, 100$

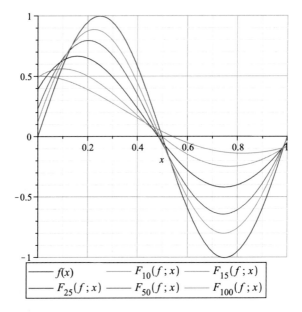

$$\boxed{\begin{array}{lll} \rule[0.5ex]{1.5em}{0.4pt}\ f(x) & \rule[0.5ex]{1.5em}{0.4pt}\ F_{10}(f;x) & \rule[0.5ex]{1.5em}{0.4pt}\ F_{15}(f;x) \\ \rule[0.5ex]{1.5em}{0.4pt}\ F_{25}(f;x) & \rule[0.5ex]{1.5em}{0.4pt}\ F_{50}(f;x) & \rule[0.5ex]{1.5em}{0.4pt}\ F_{100}(f;x) \end{array}}$$

Fig. 2 Approximation of g by $\mathcal{F}_n$ for $n = 10, 15, 25, 50, 100$

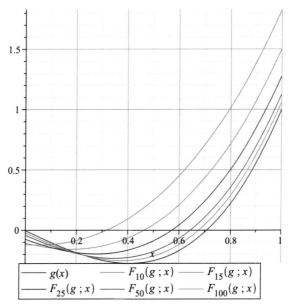

$$\boxed{\begin{array}{lll} \rule[0.5ex]{1.5em}{0.4pt}\ g(x) & \rule[0.5ex]{1.5em}{0.4pt}\ F_{10}(g;x) & \rule[0.5ex]{1.5em}{0.4pt}\ F_{15}(g;x) \\ \rule[0.5ex]{1.5em}{0.4pt}\ F_{25}(g;x) & \rule[0.5ex]{1.5em}{0.4pt}\ F_{50}(g;x) & \rule[0.5ex]{1.5em}{0.4pt}\ F_{100}(g;x) \end{array}}$$

of these types of operators by taking some continuous functions which indicate that operators (6) have a fast rate of convergence for highly oscillating and simple functions as well. These illustrations support theoretical results. Suitable modification in the operators (6) may obtain the desired accuracy in the approximation by these operators.

References

1. Agrawal, P.N., Mohammad, A.J.: On L_p approximation by a linear combination of a new sequence of linear positive operators. Turk. J. Math. **27**, 389–405 (2003)
2. Baskakov, V.A.: Primer posledovatel'nosti lineinyh polozitel'nyh operatorov v prostranstve neprerivnyh funkeil (An example of a sequence of linear positive operators in the space of continuous functons). Dokl. Akad. Nauk. SSSR. **113**, 249–251 (1957)
3. Bernstein, S.N.: Demonstration du Theorem de Weierstrass Fondee sur le Calculu des Probabilites. Comp. Comm. Soc. Mat. Charkow Ser. **13**(2), 1–2 (1912)
4. Durrmeyer J. L.: Une formule d'inversion de la transformée de Laplace: Applications la théorie des moments. Thése de 3e cycle, Faculté des Sciences de l'Université de Paris (1967)
5. Gupta, V., Deo, N., Zeng, X.M.: Simultaneous approximation for Szász-Mirakian-Stancu-Durrmeyer operators. Anal. Theory Appl. **29**(1), 86–96 (2013)
6. Gupta, V., Gupta, P.: Direct theorem in simultaneous approximation for Szász-Mirakyan-Baskakov type operators. Kyungpook Math. J. **41**(2), 243–249 (2001)
7. Gupta, V., Lupas, A.: Direct results for mixed Beta-Szász type operators. Gen. Math. **13**(2), 83–94 (2005)
8. Gupta, V., Srivastava, G.S.: On convergence of derivatives by Szász-Mirakyan-Baskakov type operators. Math. Stud. **64**(1–4), 195–205 (1995)
9. Gupta, V., Tachev, G.: Approximation by Szász-Mirakyan-Baskakov operators. J. Appl. Funct. Anal. **9**(3–4), 308–309 (2014)
10. Korovkin, P.P.: On convergence of linear positive operators in the space of continuous functions. Dokl. Akad. Nauk **90**, 961–964 (1953)
11. Mazhar, S.M., Totik, V.: Approximation by modified Szász operators. Acta Sci. Math. **49**, 257–269 (1985)
12. Prasad, G., Agrawal, P.N., Kasana, H.S.: Approximation of functions on $[0, \infty)$ by a new sequence of modified Szász operators. Math. Forum **6**(2), 1–11 (1983)
13. Szász, O.: Generalizations of S. Bernstein's polynomial to the infinite interval. J. Res. Nat. Bur. Standards, **45**, 239–245 (1950)

Numerical Solution of Intuitionistic Fuzzy Differential Equation by Using the Simpson Method

H. Atti, R. Ettoussi, S. Melliani, M. Oukessou, and L.S. Chadli

Abstract In this paper, we study the numerical solution of intuitionistic fuzzy initial value differential equations $x'(t) = f(t, x(t)), t \in I, \quad x(t_0) = x_0$ by using the Simpson method. Then this method is illustrated by an example.

Keywords Intuitionistic fuzzy differential equation · Intuitionistic fuzzy solution · Simpson method

1 Introduction

In 1965, Zadeh [20] first introduced the fuzzy set theory. Later, many researchers have applied this theory to the well-known results in the classical set theory. The idea of an intuitionistic fuzzy set was first published by Atanassov [1–3] as one of the extensions of the notion of a fuzzy set.

Many authors have paid attention to the IFS theory that has been successfully applied in different areas ([10, 12, 13, 19]). The intuitionistic fuzzy metric space is discussed in [15]. The notions of differential and integral calculus for intuitionistic fuzzy set-valued are given using the Hukuhara difference [16] in intuitionistic Fuzzy theory [9]. Melliani and Chadli [17] introduced intuitionistic fuzzy differential equation and they are introduced in [18] intuitionistic fuzzy partial differential equation. In [7], the authors discussed the existence and uniqueness of the solution of the intuitionistic fuzzy differential equation, using the successive approximation method. Approximate Solution of Intuitionistic Fuzzy Differential Equations by using Picard's method is presented in [8] and other numerical methods for solving intuitionistic fuzzy differential equations are discussed in [4, 11].

H. Atti (✉)
National Higher School of Chemistry, Ibn Tofail University, B.P 133, Kenitra, Morocco
e-mail: hafida.atti@gmail.com

H. Atti · R. Ettoussi · S. Melliani · M. Oukessou · L.S. Chadli
Department of Mathematics, Laboratory of Applied Mathematics and Scientific Computing,
Sultan Moulay Slimane University, PO Box 523, 23000 Beni Mellal, Morocco

© The Author(s), under exclusive license to Springer Nature Singapore Pte Ltd. 2023
P. Gyei-Kark et al. (eds.), *Engineering Mathematics and Computing*,
Studies in Computational Intelligence 1042,
https://doi.org/10.1007/978-981-19-2300-5_11

This paper is organized as follows: in Sect. 2, we give preliminaries which we will use throughout this work. In Sect. 3, we propose the Simpson method to solve the intuitionistic fuzzy differential equation and the convergence of this method is also discussed. Finally, an example is presented in Sect. 4.

2 Preliminairies

An intuitionistic fuzzy set (IFS) $A \in X$ is defined as an object of the following form:

$$A = \{(x, \mu_A(x), \nu_A(x), x \in X)\},$$

where the functions $\mu_A, \nu_A(x) : X \to [0, 1]$ define the degree of membership and the degree of non-membership of the element $x \in X$, respectively, and for every $x \in X$,

$$0 \le \mu_A(x) + \nu_A(x) \le 1.$$

Obviously, each ordinary fuzzy set may be written as

$$\{\langle x, \mu_A(x), 1 - \mu_A(x)\rangle | x \in X\}.$$

We denote by

$$\mathbb{F}_1 = \{\langle u, v \rangle : R \to [0, 1]^2, \ \forall x \in \mathbb{R} \ 0 \le u(x) + v(x) \le 1\}$$

the collection of all intuitionistic fuzzy number by $\mathbb{F}_1$.
An element $\langle u, v \rangle$ of $\mathbb{F}_1$ is said an intuitionistic fuzzy number if it satisfies the following conditions:

(i) is normal, i.e. there exists $x_0, x_1 \in \mathbb{R}$ such that $u(x_0) = 1$ and $v(x_1) = 1$.
(ii) u is fuzzy convex and v is fuzzy concave.
(iii) u is upper semi-continuous and v is lower semi-continuous.
(iv) $supp(u) = cl\{x \in \mathbb{R} : v(x) < 1\}$ is bounded.

Remark 1 If $\langle u, v \rangle$ a fuzzy number, so we can see $[\langle u, v \rangle]_\alpha$ as $[u]^\alpha$ and $[\langle u, v \rangle]^\alpha$ as $[1 - v]^\alpha$.

A Triangular Intuitionistic Fuzzy Number (TIFN) $\langle u, v \rangle$ is an intuitionistic fuzzy set in $\mathbb{R}$ with the following membership function u and non-membership function v:

$$u(x) = \begin{cases} \frac{x-a_1}{a_2-a_1} & \text{if } a_1 \le x \le a_2 \\ \\ \frac{a_3-x}{a_3-a_2} & \text{if } a_2 \le x \le a_3 \\ 0 & \text{otherwise} \end{cases}$$

$$v(x) = \begin{cases} \frac{a_2 - x}{a_2 - a_1'} & \text{if } a_1' \leq x \leq a_2 \\[2mm] \frac{x - a_2}{a_3' - a_2} & \text{if } a_2 \leq x \leq a_3' \\[2mm] 1 & \text{otherwise,} \end{cases}$$

where $a_1' \leq a_1 \leq a_2 \leq a_3 \leq a_3'$ and $u(x), v(x) \leq 0.5$ for $u(x) = v(x)$, $\forall x \in \mathbb{R}$. This TIFN is denoted by $\langle u, v \rangle = \langle a_1, a_2, a_3; a_1', a_2, a_3' \rangle$, where

$$[\langle u, v \rangle]_\alpha = [a_1 + \alpha(a_2 - a_1), a_3 - \alpha(a_3 - a_2)], \quad [\langle u, v \rangle]^\alpha = [a_1' + \alpha(a_2 - a_1'), a_3' - \alpha(a_3' - a_2)].$$

Definition 1 ([15]) Let $\langle u, v \rangle$ an element of IF_1 and $\alpha \in [0, 1]$, we define the following sets:

$$[\langle u, v \rangle]_l^+(\alpha) = \inf\{x \in \mathbb{R} | u(x) \geq \alpha\} \quad [\langle u, v \rangle]_r^+(\alpha) = \sup\{x \in \mathbb{R} | u(x) \geq \alpha\}.$$

$$[\langle u, v \rangle]_l^-(\alpha) = \inf\{x \in \mathbb{R} | v(x) \leq 1 - \alpha\} \quad [\langle u, v \rangle]_r^-(\alpha) = \sup\{x \in \mathbb{R} | v(x) \leq 1 - \alpha\}.$$

Remark 2

$$[\langle u, v \rangle]_\alpha = \left[[\langle u, v \rangle]_l^+(\alpha), [\langle u, v \rangle]_r^+(\alpha)\right], \quad [\langle u, v \rangle]^\alpha = \left[[\langle u, v \rangle]_l^-(\alpha), [\langle u, v \rangle]_r^-(\alpha)\right].$$

We define the following operations by

$$[\langle u, v \rangle \oplus \langle z, w \rangle]^\alpha = [\langle u, v \rangle]^\alpha + [\langle z, w \rangle]^\alpha, \quad [\lambda \langle u, v \rangle]^\alpha = \lambda [\langle u, v \rangle]^\alpha$$

$$[\langle u, v \rangle \oplus \langle z, w \rangle]_\alpha = [\langle u, v \rangle]_\alpha + [\langle z, w \rangle]_\alpha, \quad [\lambda \langle u, v \rangle]_\alpha = \lambda [\langle u, v \rangle]_\alpha,$$

where $\langle u, v \rangle, \langle z, w \rangle \in \mathbb{F}_1$ and $\lambda \in \mathbb{R}$

Definition 2 ([16]) Let $\langle u, v \rangle$ and $\langle u', v' \rangle \in IF_1$, the H-difference is the IFN $\langle z, w \rangle \in \mathbb{F}_1$, if it exists, such that

$$\langle u, v \rangle \ominus \langle u', v' \rangle = \langle z, w \rangle \iff \langle u, v \rangle = \langle u', v' \rangle \oplus \langle z, w \rangle.$$

On the space $\mathbb{F}_1$, we will consider the following metric:

Theorem 1 ([15])

$$d_\infty(\langle u, v \rangle, \langle z, w \rangle) = \frac{1}{4}\left[\sup_{0 < \alpha \leq 1}\left|[\langle u, v \rangle]_r^+(\alpha) - [\langle z, w \rangle]_r^+(\alpha)\right| + \sup_{0 < \alpha \leq 1}\left|[\langle u, v \rangle]_l^+(\alpha) - [\langle z, w \rangle]_l^+(\alpha)\right|\right.$$
$$\left. + \sup_{0 < \alpha \leq 1}\left|[\langle u, v \rangle]_r^-(\alpha) - [\langle z, w \rangle]_r^-(\alpha)\right| + \sup_{0 < \alpha \leq 1}\left|[\langle u, v \rangle]_l^-(\alpha) - [\langle z, w \rangle]_l^-(\alpha)\right|\right]$$

is a metric on $\mathbb{F}_1$.

Definition 3 ([7]) Let $F : [a, b] \to \mathbb{F}_1$ be an intuitionistic fuzzy-valued mapping and $t_0 \in [a, b]$.

Then F is called intuitionistic fuzzy continuous in t_0 iff:

$$\forall(\varepsilon > 0)(\exists \delta > 0)(\forall t \in [a, b] \ tel \ que \ |t - t_0| < \delta) \Rightarrow d_p(F(t), F(t_0)) < \varepsilon.$$

Definition 4 ([7]) F is called intuitionistic fuzzy continuous iff is intuitionistic fuzzy continuous in every point of $[a, b]$.

Definition 5 ([7]) A mapping $F : [a, b] \to \mathbb{F}_1$ is said to be Hukuhara derivable at t_0 if there exists $F'(t_0) \in \mathbb{F}_1$ such that both limits:

$$\lim_{\Delta t \to 0^+} \frac{F(t_0 + \Delta t) \ominus F(t_0)}{\Delta t} \ and \ \lim_{\Delta t \to 0^+} \frac{F(t_0) \ominus F(t_0 - \Delta t)}{\Delta t}$$

exist and they are equal to $F'(t_0) = \langle u'(t_0), v'(t_0) \rangle$, which is called the Hukuhara derivative of F at t_0.

2.1   The Intuitionistic Fuzzy Differential Equation

In this section, we consider the initial value problem for the intuitionistic fuzzy differential equation

$$\begin{cases} x'(t) = f(t, x(t)), \ t \in I \\ x(t_0) = \langle u_{t_0}, v_{t_0} \rangle \in \mathbb{F}_1, \end{cases} \tag{1}$$

where $I = [t_0, T] \subset \mathbb{R}$ and $f : I \times \mathbb{F}_1 \to \mathbb{F}_1$, $x(t_0)$ is intuitionistic fuzzy number.
In this paper, we need the following notations.
Denote the $\alpha-$ level sets

$$[x(t)]_\alpha = \left[[x(t)]_l^+(\alpha), [x(t)]_r^+(\alpha) \right], \ [x(t)]^\alpha = \left[x(t)]_l^-(\alpha), [x(t)]_r^-(\alpha) \right]$$

$$[x(t_0)]_\alpha = \left[[x(t_0)]_l^+(\alpha), [x(t_0)]_r^+(\alpha) \right], \ [x(t_0)]^\alpha = \left[x(t_0)]_l^-(\alpha), [x(t_0)]_r^-(\alpha) \right]$$

$$[f(t, x(t))]_\alpha = \left[f_1^+(t, x(t); \alpha), f_2^+(t, x(t); \alpha) \right], \ [f(t, x(t))]^\alpha = \left[f_3^-(t, x(t); \alpha), f_4^-(t, x(t); \alpha) \right],$$

where

$$f_1^+(t, x(t); \alpha) = \min \left\{ f(t, u) | u \in \left[[x(t)]_l^+(\alpha), [x(t)]_r^+(\alpha) \right] \right\}$$
$$f_2^+(t, x(t); \alpha) = \max \left\{ f(t, u) | u \in \left[[x(t)]_l^+(\alpha), [x(t)]_r^+(\alpha) \right] \right\}$$
$$f_3^-(t, x(t); \alpha) = \min \left\{ f(t, u) | u \in \left[x(t)]_l^-(\alpha), [x(t)]_r^-(\alpha) \right] \right\}$$
$$f_4^-(t, x(t); \alpha) = \max \left\{ f(t, u) | u \in \left[x(t)]_l^-(\alpha), [x(t)]_r^-(\alpha) \right] \right\}.$$

Denote

$$f_1^+(t, x(t); \alpha) = F\left(t, [x(t)]_l^+(\alpha), [x(t)]_r^+(\alpha)\right), \ f_2^+(t, x(t); \alpha) = G\left(t, [x(t)]_l^+(\alpha), [x(t)]_r^+(\alpha)\right)$$
$$f_3^-(t, x(t); \alpha) = L\left(t, [x(t)]_l^-(\alpha), [x(t)]_r^-(\alpha)\right), \ f_4^-(t, x(t); \alpha) = K\left(t, [x(t)]_l^-(\alpha), [x(t)]_r^-(\alpha)\right).$$

2.2 Simpson Method

In this section, we recall the Simpson method. Consider the initial value problem

$$\begin{cases} x'(t) = f(t, x(t)), & t \in I \\ x(t_0) = x_0. \end{cases} \tag{2}$$

We replace the interval $I = [t_0, T]$ by a set of a discrete equally spaced grid points

$$t_0 < t_1 < \ldots < t_N = T, \quad h = \frac{T - t_0}{N}, \quad t_i = t_0 + ih, \quad i = 0, 1, \ldots, N.$$

We integrate the system from t_{i-1} to t_{i+1} using the Simpson method for right-hand side of

$$\int_{t-1}^{t+1} x'(s)ds = \int_{t-1}^{t+1} f(s, x(s))ds.$$

Therefore,

$$x(t_{i+1}) - x(t_{i-1}) = \frac{h}{3}[f(t_{i-1}, x(t_{i-1})) + 4f(t_i, x(t_i)) + f(t_{i+1}, x(t_{i+1}))]$$
$$- \frac{h^5}{90} f^{(4)}(\epsilon_1, x(\epsilon_1)) \quad t_{i-1} \le \epsilon_1 \le t_{i+1}. \tag{3}$$

We substitute $x(t_{i+1})$ by $x(t_i) + hf(t_i, x(t_i)) + \frac{h^2}{2} f'(\epsilon_2, x(\epsilon_2))$ in right-hand side of the equation where $\epsilon \in [t_i, t_i + 1]$.

Therefore,

$$x(t_{i+1}) = x(t_{i-1}) + \frac{h}{3}[f(t_{i-1}, x(t_{i-1})) + 4f(t_i, x(t_i)) + f(t_{i+1}, x(t_i)) + hf(t_i, x(t_i)) + \frac{h^2}{2} f'(\epsilon_2, x(\epsilon_2))]$$
$$- \frac{h^5}{90} f^{(4)}(\epsilon_1, x(\epsilon_1)) \quad t_{i-1} \le \epsilon_1 \le t_{i+1}, \quad t_{i-1} \le \epsilon_2 \le t_{i+1}.$$

But we have

$$f(t_{i+1}, x(t_i) + hf(t_i, x(t_i)) + \frac{h^2}{2} f'(\epsilon_2, x(\epsilon_2))$$
$$= f(t_{i+1}, x(t_i) + hf(t_i, x(t_i)) + \frac{h^2}{2} f'(\epsilon_2, x(\epsilon_2))f_x(t_{i+1}, \epsilon_3), \tag{4}$$

where ϵ_3 is   in   between $x(t_i) + hf(t_i, x(t_i))$ and $x(t_i) + hf(t_i, x(t_i)) + \frac{h^2}{2} f'(\epsilon_2, x(\epsilon_2))$

Then

$$x(t_{i+1}) = x(t_{i-1}) + \frac{h}{3} f(t_{i-1}, x(t_{i-1})) + \frac{4h}{3} f(t_i, x(t_i)) + \frac{h}{3} f(t_{i+1}, x(t_i) + hf(t_i, x(t_i)))$$

$$+ \frac{h^3}{6} f'(\epsilon_2, x(\epsilon_2)) f_x(t_{i+1}, \epsilon_3) - \frac{h^5}{90} f^{(4)}(\epsilon_1, x(\epsilon_1)),$$

where $t_{i-1} \leq \epsilon \leq t_{i+1}$, $t_i \leq \epsilon_2 \leq t_{i+1}$ and ϵ_3 is in between $x(t_i) + hf(t_i, x(t_i))$ and $x(t_i) + hf(t_i, x(t_i))$ $\frac{h^2}{2} f'(\epsilon_2, x(\epsilon_2))$;
thus, we have

$$x_{i+1} = x_{i-1} + \frac{h}{3} f(t_{i-1}, x_{i-1}) + \frac{4h}{3} f(t_{i+1}, x_i + hf(t_i, x_i)) \tag{5}$$

with initial value $x_0 = x(t_0)$ and $x_1 = x_0 + hf(t_0, x_0) + \frac{h^2}{2} f'(t_0, x_0)$

3   Numerical Solution of Intuitionistic Fuzzy Differential Equations

The grid points at which the solution is calculated are

$$h = \frac{T - t_0}{N}, \quad t_n = t_0 + nh, \quad n = 0, 1, \dots N.$$

We denote the lower and upper α-cuts of the exact and approximate solutions at t_n, respectively, by the following expressions:

$$[X(t_n)]_\alpha = \left[[X(t_n)]_l^+(\alpha), [X(t_n)]_r^+(\alpha) \right], \quad [X(t_n)]^\alpha = \left[[X(t_n)]_l^-(\alpha), [X(t_n)]_r^-(\alpha) \right]$$

and

$$[x(t_n)]_\alpha = \left[[x(t_n)]_l^+(\alpha), [x(t_n)]_r^+(\alpha) \right], \quad [x(t_n)]^\alpha = \left[[x(t_n)]_l^-(\alpha), [x(t_n)]_r^-(\alpha) \right].$$

To simplify, we put $x(t_n) = x_n$ and $X(t_n) = X_n$.

$$[x_{n+1}]_l^+(\alpha) = [x_{n-1}]_l^+(\alpha) + \frac{h}{3} F[t_{n-1}, [x_{n-1}]_l^+(\alpha), [x_{n-1}]_r^+(\alpha)] + \frac{4h}{3} F[t_n, [x_n]_l^+(\alpha), [x_n]_r^+(\alpha)] \tag{6}$$

$$+ \frac{h}{3} F\left[t_{n+1}, [x_n]_l^+(\alpha)] + hG[t_n, [x_n]_l^+(\alpha), [x_n]_r^+(\alpha)], [x_n]_l^+(\alpha) \right.$$

$$\left. + hF[t_n, [x_n]_l^+(\alpha), [x_n]_r^+(\alpha)] \right] + h^3[A]_l^+(\alpha).$$

$$[x_{n+1}]_r^+(\alpha) = [x_{n-1}]_r^+(\alpha) + \frac{h}{3} G[t_{n-1}, [x_{n-1}]_l^+(\alpha), [x_{n-1}]_r^+(\alpha)] + \frac{4h}{3} G[t_n, [x_n]_l^+(\alpha), [x_n]_r^+(\alpha)] \tag{7}$$

$$+ \frac{h}{3} G\left[t_{n+1}, [x_n]_l^+(\alpha)] + hF[t_n, [x_n]_l^+(\alpha), [x_n]_r^+(\alpha)], [x_n]_r^+(\alpha) \right.$$

$$\left. + hG[t_n, [x_n]_l^+(\alpha), [x_n]_r^+(\alpha)] \right] + h^3[A]_l^+(\alpha).$$

$$[x_{n+1}]_l^-(\alpha) = [x_{n-1}]_l^-(\alpha) + \frac{h}{3} L[t_{n-1}, [x_{n-1}]_l^-(\alpha), [x_{n-1}]_r^-(\alpha)] + \frac{4h}{3} L[t_n, [x_n]_l^-(\alpha), [x_n]_r^-(\alpha)]$$

$$\tag{8}$$

$$+ \frac{h}{3} L\Big[t_{n+1}, [x_n]_l^-(\alpha)] + hK[t_n, [x_n]_l^-(\alpha), [x_n]_r^-(\alpha)], [x_n]_r^-(\alpha)$$

$$+ hL[t_n, [x_n]_l^-(\alpha), [x_n]_r^-(\alpha)]\Big] + h^3[A]_l^-(\alpha).$$

$$[x_{n+1}]_r^-(\alpha) = [x_{n-1}]_r^-(\alpha) + \frac{h}{3} K[t_{n-1}, [x_{n-1}]_l^-(\alpha), [x_{n-1}]_r^-(\alpha)] + \frac{4h}{3} K[t_n, [x_n]_l^-(\alpha), [x_n]_r^-(\alpha)]$$

$$\tag{9}$$

$$+ \frac{h}{3} K\Big[t_{n+1}, [x_n]_l^-(\alpha)] + hL[t_n, [x_n]_l^-(\alpha), [x_n]_r^-(\alpha)], [x_n]_r^-(\alpha)$$

$$+ hK[t_n, [x_n]_l^-(\alpha), [x_n]_r^-(\alpha)]\Big] + h^3[A]_l^-(\alpha),$$

where $A = [[A]_l^+(\alpha), [A]_r^+(\alpha), [A]_l^-(\alpha), [A]_r^-(\alpha)]$ and $[A]_\alpha = [\frac{1}{6} f'(\epsilon_2, x(\epsilon_2)).f_x(t_{i+1}, \epsilon_3) - \frac{h^2}{90} f^{(4)}(\epsilon_1, x(\epsilon_1))]_\alpha$

Also, we have

$$[X_{n+1}]_l^+(\alpha) = [X_{n-1}]_l^+(\alpha) + \frac{h}{3} F[t_{n-1}, [X_{n-1}]_l^+(\alpha), [X_{n-1}]_r^+(\alpha)] + \frac{4h}{3} F[t_n, [X_n]_l^+(\alpha), [X_n]_r^+(\alpha)]$$

$$\tag{10}$$

$$+ \frac{h}{3} F\Big[t_{n+1}, [X_n]_l^+(\alpha) + hF[t_n, [X_n]_l^+(\alpha), [X_n]_r^+(\alpha)], [X_n]_r^+(\alpha)$$

$$+ hG[t_n, [X_n]_l^+(\alpha), [X_n]_r^+(\alpha)]\Big].$$

$$[X_{n+1}]_r^+(\alpha) = [X_{n-1}]_r^+(\alpha) + \frac{h}{3} G[t_{n-1}, [X_{n-1}]_l^+(\alpha), [X_{n-1}]_r^+(\alpha)] + \frac{4h}{3} G[t_n, [X_n]_l^+(\alpha), [X_n]_r^+(\alpha)]$$

$$\tag{11}$$

$$+ \frac{h}{3} G\Big[t_{n+1}, [X_n]_l^+(\alpha) + hF[t_n, [X_n]_l^+(\alpha), [X_n]_r^+(\alpha)], [X_n]_r^+(\alpha)$$

$$+ hF[t_n, [X_n]_l^+(\alpha), [X_n]_r^+(\alpha)]\Big].$$

$$[X_{n+1}]_l^-(\alpha) = [X_{n-1}]_l^-(\alpha) + \frac{h}{3} L[t_{n-1}, [X_{n-1}]_l^-(\alpha), [X_{n-1}]_r^-(\alpha)] + \frac{4h}{3} L[t_n, [X_n]_l^-(\alpha), [X_n]_r^-(\alpha)]$$

$$\tag{12}$$

$$+ \frac{h}{3} L\Big[t_{n+1}, [X_n]_l^-(\alpha) + hK[t_n, [X_n]_l^-(\alpha), [X_n]_r^-(\alpha)], [X_n]_r^-(\alpha)$$

$$+ hK[t_n, [X_n]_l^-(\alpha), [X_n]_r^-(\alpha)]\Big].$$

$$[X_{n+1}]_r^-(\alpha) = [X_{n-1}]_r^-(\alpha) + \frac{h}{3}K[t_{n-1}, [X_{n-1}]_l^-(\alpha), [X_{n-1}]_r^-(\alpha)] + \frac{4h}{3}K[t_n, [X_n]_l^-(\alpha), [X_n]_r^-(\alpha)]$$

(13)

$$+ \frac{h}{3}K\Big[t_{n+1}, [X_n]_l^-(\alpha) + hL[t_n, [X_n]_l^-(\alpha), [X_n]_r^-(\alpha)], [X_n]_r^-(\alpha)$$

$$+ hK[t_n, [X_n]_l^-(\alpha), [X_n]_r^-(\alpha)]\Big].$$

The following lemma will be applied to show the convergence of our method. For more details, see [5, 6].

Lemma 1 *Let a sequence of non-negative numbers $\{W_n\}_{n=0}^N$ satisfy*

$$|W_{n+1}| \le A|W_n| + B, \ 0 \le n \le N - 1$$

for some given positive constants A and B. Then

$$|W_n| \le A^n|W_0| + B\frac{A^n - 1}{A - 1}, \ 0 \le n \le N.$$

Lemma 2 *[14] Suppose that a sequence of non-negative $\{P_n\}_{n=0}^N$ satisfies*

$$P_n \le AP_n + BP_{n-1} + C, \ 1 \le n \le N - 1$$

for some given positive constants A, B and C. Then

$$P_{n+1} + (\alpha - A)P_n \le \alpha^n[P_1 + (\alpha - A)P_0] + C\frac{\alpha^n - 1}{\alpha - 1},$$

where

$$\alpha = \frac{\sqrt{A^2 + 4B} + A}{2}.$$

Now, we will proof the convergence of our method.

Let $F(t, u^+, v^+)$, $G(t, u^+, v^+)$, $L(t, u^-, v^-)$ and $K(t, u^-, v^-)$ be the functions which are given by the Eqs. (4), (5) where u^+, v^+, u^- and v^- are constants and $u^+ \le v^+$ and $u^- \le v^-$.

The domain of G and H is

$$M_1 = \{(t, u^+, v^+) \backslash \ t_0 \le t \le T, \ \infty < u^+ \le v^+, \ -\infty < v^+ < +\infty\}$$

and the domain of L and K is

$$M_2 = \{(t, u^-, v^-) \backslash \ t_0 \le t \le T, \ \infty < u^- \le v^-, \ -\infty < v^- < +\infty\}$$

with $M_1 \subseteq M_2$.

Theorem 2 *Let $F(t, u^+, v^+)$, $G(t, u^+, v^+)$ belong to $C^1(M_1)$ and $L(t, u^-, v^-)$, $K(t, u^-, v^-)$ belong to $C^1(M_2)$ and suppose that the partial derivatives of G, H, L and K are bounded on M_1 and M_2, respectively. Then, for arbitrarily fixed $0 \le \alpha \le 1$, the Simpson approximations $x(t_n)$ converge to the exact solution $X(t_n)$ uniformly in t. In other words, $\lim\limits_{h \to 0} d_\infty(x(t_n), X(t_n)) = 0$.*

Proof Let

$$W_n^+ = [X(t_n)]_l^+(\alpha) - [x(t_n)]_l^+(\alpha) \quad , \quad V_n^+ = [X(t_n)]_r^+(\alpha) - [x(t_n)]_r^+(\alpha)$$

$$W_n^- = [X(t_n)]_l^-(\alpha) - [x(t_n)]_l^-(\alpha) \quad , \quad V_n^- = [X(t_n)]_r^-(\alpha) - [x(t_n)]_r^-(\alpha).$$

By using the Eqs. (8), (9), (12) and (13), we conclude that

$$|W_{n+1}^+| \le |W_{n-1}^+| + \frac{2hL}{3}\max\{|W_{n-1}^+|, |V_{n-1}^+|\} + \frac{8hL}{3}\max\{|W_n^+|, |V_n^+|\}+$$
$$\frac{2hL}{3}\left[2Lh\max\{|W_n^+|, |V_n^+|\} + \max\{|W_n^+|, |V_n^+|\}\right] + h^3\underline{M}$$

$$|V_{n+1}^+| \le |V_{n-1}^+| + \frac{2hL}{3}\max\{|W_{n-1}^+|, |V_{n-1}^+|\} + \frac{8hL}{3}\max\{|W_n^+|, |V_n^+|\}+$$
$$\frac{2hL}{3}\left[2Lh\max\{|W_n^+|, |V_n^+|\} + \max\{|W_n^+|, |V_n^+|\}\right] + h^3\overline{M},$$

$$(14)$$

where $\underline{M}, \overline{M}$ are upper bound for $\underline{A}^+(r), \overline{A}^+(r)$, respectively, for which
$[A]_\alpha = [\frac{1}{6} f'(\epsilon_2, x(\epsilon_2)).f_x(t_{i+1}, \epsilon_3) - \frac{h^2}{90} f^{(4)}(\epsilon_1, x(\epsilon_1))]_\alpha$.
Consequently,

$$|W_{n+1}^+| \le |W_{n-1}^+| + \frac{2hL}{3}\{|W_{n-1}^+| + |V_{n-1}^+|\} + \frac{8hL}{3}\{|W_n^+| + |V_n^+|\}+$$
$$\frac{2hL}{3}(2Lh+1)\{|W_n^+| + |V_n^+|\} + h^3\underline{M}$$

$$|V_{n+1}^+| \le |V_{n-1}^+| + \frac{2hL}{3}\{|W_{n-1}^+| + |V_{n-1}^+|\} + \frac{8hL}{3}\{|W_n^+| + |V_n^+|\}+$$
$$\frac{2hL}{3}(2Lh+1)\{|W_n^+|, |V_n^+|\} + h^3\overline{M}.$$

$$(15)$$

By adding above two equations and setting $U_n = |W_n^+| + |W_n^+|$. we obtain

$$U_{n+1} \le \frac{4Lh}{3}(5 + 2Lh)U_n + (1 + \frac{4Lh}{3})U_{n-1} + 2h^3 M,$$

where $M = \max\{\underline{M}, \overline{M}\}$. By using Lemma 2. we have

$$U_{n+1} + (\alpha - A)U_n \le \alpha^n[U_1 + (\alpha - A)U_0] + C\frac{\alpha^n - 1}{\alpha - 1},$$

where $\alpha = \frac{\sqrt{A^2+4B}+A}{2}$, $A = \frac{4Lh}{3}(5 + 2Lh)$, $B = 1 + \frac{4Lh}{3}$ and $C = 2h^3 M$.
Because of $U_0 = 0$, for $n = N - 1$, we have

$$\lim_{h \to 0}[\alpha^{N-1}U_1 + +C\frac{\alpha^{N-1} - 1}{\alpha - 1}] = 0.$$

Therefore, we have $\lim_{h \to 0} U_N + (\alpha - A)U_{N-1} = 0$,
 and consequently, $\lim_{h \to 0} U_N = 0$, i.e. $\lim_{h \to 0} |W_n^+| + |W_n^+| = 0$.
Thus, $\lim_{h \to 0}[x(t_n)]_l^+(\alpha) = [X(t_n)]_l^+(\alpha)$ and $\lim_{h \to 0}[x(t_n)]_r^+(\alpha) = [X(t_n)]_r^+(\alpha)$.
We use the same idea to prove that $\lim_{h \to 0}[x(t_n)]_l^-(\alpha) = [X(t_n)]_l^-(\alpha)$ and
$\lim_{h \to 0}[x(t_n)]_r^-(\alpha) = [X(t_n)]_r^-(\alpha)$.

4   Example

Consider the initial value problem

$$\begin{cases} x'(t) = x(t), \ 0 \le t \le 1 \\ x(0) = (0.75 + 0.25\alpha, 1.125 - 0.125\alpha, 0.5 + 0.5\alpha, 1.5 - 0.5\alpha). \end{cases} \quad (16)$$

The exact solution at $t = 1$ is given by

$$X(1, \alpha) = [(0.75 + 0.25\alpha)e, (1.125 - 0.125\alpha)e, (0.5 + 0.5\alpha)e, (1.5 - 0.5\alpha)e].$$

Using the Simpson method approximation, we denote

$$[x_0]_l^+ = 0.75 + 0.25\alpha, \quad [x_0]_r^+ = 1.125 - 0.125\alpha$$

$$[x_0]_l^- = 0.5 + 0.5\alpha, \quad [x_0]_r^- = 1.5 - 0.5\alpha$$

and

$$[x_1]_l^+ = [x_0]_l^+ + h[x_0]_l^+ + \frac{h^2}{2}[x_0]_l^+, \quad [x_1]_r^+ = [x_0]_r^+ + h[x_0]_r^+ + \frac{h^2}{2}[x_0]_r^+$$

$$[x_1]_l^- = [x_0]_l^- + h[x_0]_l^- + \frac{h^2}{2}[x_0]_l^-, \quad [x_1]_r^- = [x_0]_r^- + h[x_0]_r^- + \frac{h^2}{2}[x_0]_r^-$$

as initial values, we have

$$[x_{i+1}]_l^+ = [x_{i-1}]_l^+ + \frac{h}{3}[x_{i-1}]_l^+ + \frac{4h}{3}[x_i]_l^+ + \frac{h}{3}([x_i]_l^+ + h[x_i]_l^+)$$

$$[x_{i+1}]_r^+ = [x_{i-1}]_r^+ + \frac{h}{3}[x_{i-1}]_r^+ + \frac{4h}{3}[x_i]_r^+ + \frac{h}{3}([x_i]_r^+ + h[x_i]_r^+)$$

$$[x_{i+1}]_l^- = [x_{i-1}]_l^- + \frac{h}{3}[x_{i-1}]_l^- + \frac{4h}{3}[x_i]_l^- + \frac{h}{3}([x_i]_l^- + h[x_i]_l^-)$$

$$[x_{i+1}]_r^- = [x_{i-1}]_r^- + \frac{h}{3}[x_{i-1}]_r^- + \frac{4h}{3}[x_i]_r^- + \frac{h}{3}([x_i]_r^- + h[x_i]_r^-).$$

See Tables 1, 2, 3, 4, 5, 6, 7, 8, 9.
The errors can be minimized by taking smaller step size h.

Table 1 The exact solution

α	exact solution
0	2.038711371344284 ; 3.058067057016426 ; 1.359140914229523 ; 4.077422742688569
0.2	2.174625462767236 ; 2.990110011304950 ; 1.630969097075427 ; 3.805594559842664
0.4	2.310539554190189 ; 2.922152965593474 ; 1.902797279921332 ; 3.533766376996759
0.6	2.446453645613141 ; 2.854195919881998 ; 2.174625462767236 ; 3.261938194150855
0.8	2.582367737036093 ; 2.786238874170521 ; 2.446453645613141 ; 2.990110011304950
1	2.718281828459046 ; 2.718281828459046 ; 2.718281828459046 ; 2.718281828459046

Table 2 The approximate solution for h = 0.1

α	h = 0.1
0	2.036927885586508; 3.055391828379762 ; 1.357951923724339 ; 4.073855771173016
0.2	2.172723077958942; 2.987494232193546 ; 1.629542308469207 ; 3.802265386428148
0.4	2.308518270331375 ; 2.919596636007328 ; 1.901132693214074 ; 3.530675001683280
0.6	2.444313462703810 ; 2.851699039821111 ; 2.172723077958942 ; 3.259084616938413
0.8	2.580108655076244 ; 2.783801443634894 ; 2.444313462703810 ; 2.987494232193546
1	2.715903847448677 ; 2.715903847448677 ; 2.715903847448677 ; 2.715903847448677

Table 3 The approximate solution for h = 0.01

α	h = 0.01
0	2.038694286091721 ; 3.058041429137583 ; 1.359129524061149 ; 4.077388572183443
0.2	2.174607238497838 ;2.990084952934527 ; 1.630955428873376 ; 3.805562667371215
0.4	2.310520190903952 ; 2.922128476731469 ; 1.902781333685608 ; 3.533736762558985
0.6	2.446433143310066 ; 2.854172000528412 ; 2.174607238497838 ; 3.261910857746752
0.8	2.582346095716182 ; 2.786215524325356 ; 2.446433143310066 ; 2.990084952934527
1	2.718259048122298 ; 2.718259048122298 ; 2.718259048122298 ; 2.718259048122298

Table 4 The approximate solution for h = 0.001

α	h = 0.001
0	2.038711201354579 ; 3.058066802031856 ; 1.359140800903048 ; 4.077422402709158
0.2	2.174625281444878 ; 2.990109761986705 ; 1.630968961083658 ; 3.805594242528543
0.4	2.310539361535188 ; 2.922152721941560 ; 1.902797121264272 ; 3.533766082347932
0.6	2.446453441625492 ; 2.854195681896404 ; 2.174625281444878 ; 3.261937922167315
0.8	2.582367521715801 ; 2.786238641851246 ; 2.446453441625492 ; 2.990109761986705
1	2.718281601806097 ; 2.718281601806097 ; 2.718281601806097 ; 2.718281601806097

Table 5　The approximate solution for h = 0.0001

α	h = 0.0001
0	2.038711369645266 ; 3.058067054467893 ; 1.359140913096840 ; 4.077422739290531
0.2	2.174625460954925 ; 2.990110008813046 ; 1.630969095716206 ; 3.805594556671146
0.4	2.310539552264619 ; 2.922152963158189 ; 1.902797278335573 ; 3.533766374051778
0.6	2.446453643574327 ; 2.854195917503360 ; 2.174625460954925 ; 3.261938191432412
0.8	2.582367734883982 ; 2.786238871848528 ; 2.446453643574327 ; 2.990110008813046
1	2.718281826193681 ; 2.718281826193681 ; 2.718281826193681 ;2.718281826193681

Table 6　The error for h = 0.1

α	e^+ for h = 0.1	e^- for h = 0.1
0	0.004458714394440	0.004755962020737
0.2	0.004518163919699	0.004755962020737
0.4	0.004577613444959	0.004755962020737
0.6	0.004637062970218	0.004755962020736
0.8	0.004696512495477	0.004755962020736
1	0.004755962020737	0.004755962020737

Table 7　The error for h = 0.01

α	e^+ for h= 0.01	e^- for h = 0.01
0	0.00004271	0.00004556
0.2	0.00004328	0.00004556
0.4	0.00004385	0.00004556
0.6	0.00004442	0.00004556
0.8	0.00004499	0.00004556
1	0.00004556	0.00004556

for $0 \leq \alpha \leq 1$

Table 8　The error for h = 0.0001

α	e^+ for h = 0.0001	e^- for h = 0.0001
0	0.000000004248	0.000000004531
0.2	0.000000004304	0.000000004531
0.4	0.000000004361	0.000000004531
0.6	0.000000004417	0.000000004531
0.8	0.000000004474	0.000000004531
1	0.000000004531	0.000000004531

for $0 \leq \alpha \leq 1$

Table 9 The error for h = 0.001

α	e^+ for h= 0.001	e^- for h= 0.001
0	0.0000004250	0.0000004533
0.2	0.0000004306	0.0000004533
0.4	0.0000004363	0.0000004533
0.6	0.0000004420	0.0000004533
0.8	0.0000004476	0.0000004533
1	0.0000004533	0.0000004533

for $0 \leq \alpha \leq 1$

5 Conclusion

In this paper, we solved intuitionistic fuzzy initial value differential equations numerically by using the Simpson method and we proved that the approximate solution converges uniformly to the exact solution. For future research, one can apply this method to solve intuitionistic fuzzy initial value differential equations of a high order.

References

1. Atanassov, K.: Intuitionistic fuzzy sets. Fuzzy Sets Syst. **20**, 87–96 (1986)
2. Atanassov, K.: Intuitionistic Fuzzy Sets. Springer Physica-Verlag, Berlin (1999)
3. Atanassov K., Intuitionistic Fuzzy Sets Past, Present and Future. CLBME-Bulgarian Academy of Sciences
4. Ben Amma, B., Melliani, S., Chadli, L.S.: Numerical Solution Of Intuitionistic Fuzzy Differential Equations By Euler and Taylor Method Notes on Intuitionistic Fuzzy Sets,**22**(2), 71–86 (2016)
5. Duraisamy, C., Usha, B.: numerical solution of differential equation by Runge-Kutta method of order four. European J. Sci. Res. **67**, 324–337 (2012)
6. Friedman, M., Kandel, A.: Numerical solution of fuzzy differential equations. Fuzzy Sets Syst. **105**, 133–138 (1999)
7. Ettoussi, R., Melliani, S., Elomari, M., Chadli, L.S.: Solution of intuitionistic fuzzy differential equations by successive approximations method. Notes Intuition. Fuzzy Sets **21**(2), 51–62 (2015)
8. Ettoussi, R., Melliani, S., Chadli, L.S.: Approximate solution of intuitionistic fuzzy differential equations by using Picard's method, In: Melliani S., Castillo O. (eds) Recent Advances in Intuitionistic Fuzzy Logic Systems. Stud. Fuzziness Soft Comput. Springer,Cham, **372**, 169–180 (2019)
9. Ettoussi, R., Melliani, S., Chadli, L.S.: Differential equation with intuitionistic fuzzy parameters. Notes Intuition. Fuzzy Sets **23**(4), 46–61 (2017)
10. Kharal, A.: Homeopathic drug selection using intuitionistic fuzzy sets. Homeopathy **98**, 35–39 (2009)
11. Keyanpour, M., Akbarian, T.: Solving intuitionistic fuzzy nonlinear equations. J. Fuzzy Set Valued Anal. **2014**, 1–6 (2014)

12. Li, D.F., Cheng, C.T.: New similarity measures of intuitionistic fuzzy sets and application to pattern recognitions. Pattern Recognit. Lett. **23**, 221–225 (2002)
13. Li, D.F.: Multiattribute decision making models and methods using intuitionistic fuzzy sets. J. Comput. Syst. Sci. **70**, 73–85 (2005)
14. Mahmoud Mohseni Moghadam: Mohammad Shafie Dahaghin, Two-step methos for numerical solution of fuzzy differential equations. Fourth European Congress of Mathematics. Stockholm, Sweden (2014)
15. Melliani, S., Elomari, M., Chadli, L.S., Ettoussi, R.: Intuitionistic fuzzy metric space. Notes Intuition. Fuzzy Sets **21**(1), 43–53 (2015)
16. Melliani, S., Elomari, M., Chadli, L.S., Ettoussi, R.: Extension of hukuhara difference in intuitionistic fuzzy theory. Notes Intuition. Fuzzy Sets **21**(4), 34–47 (2015)
17. Melliani, S., Chadli, L.S.: Introduction to intuitionistic fuzzy differential equations. Notes Intuition. Fuzzy sets **6**(2), 31–41 (2000)
18. Melliani, S., Chadli, L. S.: Introduction to intuitionistic fuzzy partial differential Equations. Notes on Intuition. Fuzzy sets **7**(3), 39–42 (2001)
19. Shu, M.H., Cheng, C.H., Chang, J.R.: Using intuitionistic fuzzy sets for fault-tree analysis on printed circuit board assembly. Microelectron. Reliab. **46**(12), 2139–2148 (2006)
20. Zadeh, L.A.: Fuzzy sets. Inf. Control **8**, 338–353 (1965)

Comparison of AI Techniques in Modeling of Transportation Cost for Persons with Disabilities

Sasalak Tongkaw

Abstract People with disabilities are people with abnormalities or physical disabilities, intellectually or mentally. Usually, persons with disabilities will receive a pension or compensation for living costs from the government. However, the transportation cost is another concern for the handicapped living in a developing country because distinctive disability types may fluctuate the costs. This paper will show the comparison of artificial intelligence techniques, Artificial Neural Network, Decision Tree Classifier, LR, LDA, k-Nearest Neighbor, CART, and Naïve Bayes algorithms to foresight the transportation cost for persons with disabilities, from their home to the government office center, to help the government allocate subsidiary funds to aid persons with disabilities. The details of the cleaning dataset and some better accuracy result configuration will explain in detail. The models can be utilized to predict the expected transportation cost from a bird's-eye view of the government. It also could be adopted for public transport outlining and other amenities for persons with disabilities in Songkhla Province, Thailand.

Keywords ANN · DT · LR · LDA · KNN · CART · NB · Compare

1 Introduction

Per the Constitution of the Kingdom of Thailand B.E. 2550 and the Subordinate Act of Promotion Act and improving the quality of life of persons with disabilities in 2007 and the education management for persons with disabilities 2008, people with disabilities have substantial rights which cover the lifestyle of persons with disabilities since birth until death, including at least nine reasons such as disability premiums—every disabled person who has a book/identity card, persons with disabilities have the right to register to get "Disability premium" per person is 500 baht/month, which was initially only for people with disabilities without an income to have the right to receive "premiums" at 500 baht per month. Also, elderly persons with disabilities

S. Tongkaw (✉)
Faculty of Science and Technology, Songkhla Rajabhat University, Songkhla, Thailand
e-mail: sasalak.to@skru.ac.th

© The Author(s), under exclusive license to Springer Nature Singapore Pte Ltd. 2023
P. Gyei-Kark et al. (eds.), *Engineering Mathematics and Computing*,
Studies in Computational Intelligence 1042,
https://doi.org/10.1007/978-981-19-2300-5_12

or aged 60 years or more are entitled to both "Disability premium" and "Elderly premium" for a total of 1,000 baht per month.

Persons with disabilities have a far higher cost than a normal person. However, there are already certain restrictions on tax deduction. People with disabilities must also pay more expensive charges than ordinary people, such as leasing a car to get about or even employing a professional caretaker if they require assistance.

When it comes to government, the government does not reimburse some fees such as travel expenditures or coordinate with government agencies or even travel to register as a person with disabilities in the province. Some categories of people with disabilities require the use of a car and the assistance of others. This raises the expenditures here; nevertheless, the assessment of the traveling expenditures of people with disabilities relies on a variety of criteria such as distance, type of impairment, age, and so on. This study included elements that may influence travel expenditures, such as gender, age, type, education, and distance from residence to the government office.

While many studies clarify the comparison between functions, minimal research uses data from persons with disabilities to estimate or identify travel costs.

2 Objective

This paper compares artificial intelligence techniques including Artificial Neural Network, Decision Tree Classifier, LR, LDA, k-Nearest Neighbor, CART, and Naïve Bayes algorithms to foresight the transportation cost for persons with disabilities, from their home to the government office center. The distance will only be calculated for the government to allocate funds to people with disabilities more conveniently, putting aside the required auxiliary budget for people with disabilities, which includes transportation and other amenities. The research is limited to providing the best solution for the finances available, without delving into the specific route or distance and considering considerations such as the accommodation for each form of disabilities or a solution tailored to a certain community. The findings include a data model on people with disabilities that may be used to estimate the transportation costs of people with disabilities in Songkhla Province. In addition, the model may be used to forecast future expenses for persons with disabilities in Songkhla Province.

3 Artificial Neural Network

ANN, or Artificial Neural Network, is a computer architecture that is biologically influenced. The network comprises linked coefficients or weights and processing components, or physiologically, neurons [1, p.4]. Neural Networks, according to Deboeck, use mathematical approaches to handle signal processing, forecasting, and clustering problems. Neural networks are used as a regression tool to determine the link between the inputs and outputs: establishing data associations that the users desire [2, p. xxviii]. The capacity of neural networks to be taught by previous decisions paves the way for innovative data regression. This is seen in financial applications such as loan acceptance and credit risk assessment. Previous data may be classified, and decisions are made using previous statistical data and learning from past data [3, p. 7]. Furthermore, the input combinations and weighting of certain data might be linked to liability determinations. A considerably more complicated neural network technique can follow the target system's inner variable interactions, representations, and structures [4, p. 7]. As it is obvious that neural networks can do regression on large amounts of data with many variables and enable decision-making and prediction, a neural network is a strong tool for state developers to use in allocating resources.

3.1 ANN Multi-Layer Perceptron

ANN multi-layer perceptron consists of nodes or units structured in two or more layers, where the input layer is omitted. Real value weights are connected to some nodes, while there are no nodes in the same layer [5, p. 2]. The architecture of MLP could be described mathematically in (1.1).

$$a_{i,q}^l = \sum_{n_j^m \in S_i^l} w_{ij}^{lm} y_{j,q}^m, l > 0 \tag{1}$$

$$y_{i,q}^l = f\left(a_{i,q}^l\right), l > 0$$

where $a_{i,q}^l$ is the activation of node n_i^l for a particular pattern q and layer l. w_{ij}^{lm} is the weight vector that connects node n_i^m and n_i^l, which node n_i^m is the source node for n_i^l. $y_{j,q}^m$ is the source input for pattern q; the set of source nodes is described by the variable S_i^l. Bias nodes denoted by n_0^l. The second notion is that $y_{i,q}^l$ is the output of an activation function, with the input of the node's activation; used most commonly, the sigmoid function in 1.2 [5, pp. 2–3].

$$f(x) = \frac{1}{(1 + e^{-x})} \tag{2}$$

The intermediate layer is sometimes called the hidden layer and within the layers are hidden nodes. The number of hidden layers and hidden nodes impacts the functionality of the MLPs' training. Too few can lead to problems not being solved, while too many can lead to long training periods or inadequate generalization capabilities. The optimal hidden layers and nodes that could solve precision and low minimal approximated error are fantasized for researchers using MLPs. The optimizations could be done by growing and pruning networks depending on node withholding dataset [1, p. 10], [5, p. 4]. MLP researches are present in numerous academic fields, such as forensic sciences, where physical features of the body could determine age; combined with image processing, MLPs served as a powerful tool to dynamically and automatically estimate a person's age [6].

3.2 Logistic Regression

Because the result variable in a logistic regression model is either binary or dichotomous, it differs from a typical linear regression model. However, the same linear regression concepts apply. This research focuses on binomial logistic regression, with the mathematical model established in 2.1 as the probability distribution function [7, p. 297].

$$f(y; p, m) = \prod \exp\left\{ y * \ln\left(\frac{p}{1-p}\right) + m * \ln(1-p) + \ln\binom{m}{y} \right\} \qquad (3)$$

The variable p is the probability of success, while m, n are explanatory variables. The PDF used the application of the logarithmic function to smoothen out the curve of regression. Logistic regression models are proven to be flexible and easy to use and give meaningful estimates of effects [8, p. 7]. The model is considered a valuable tool for GIS and mapping because of its high accuracy and meaningful data interpretation; an example includes validating landslide susceptibility mapping using GIS and remote sensing data. The author concludes that the logistic regression model is accurate at interpreting data [9].

3.3 Linear Discriminant Analysis

The linear discrimination analysis function, known as LDA, is dimensionality reduction, a preliminary step for pattern classification and machine learning applications. The objective of LDA is to sort the dataset onto a lower dimensional space, to reduce the probability of overfitting, and also to minimize computational costs. If there is an n-dimensional sample, LDA would be applied to sort the dataset to a lesser subspace of k which $k \leq n - 1$ while consistently containing the class-discriminatory information. The process of LDA starts by forming l-dimension mean vectors for various classes within the dataset. Next, the scatter matrices are computed, which are the in-between-class and within-class matrices. Scatter matrices are then calculated for eigenvectors $(e_1, e_2, e_3, \ldots, e_l)$ and the eigenvalues $(\lambda_1, \lambda_2, \lambda_3, \ldots, \lambda_l)$. The eigenvectors are then sorted by decreasing eigenvalues; y number of eigenvectors by largest eigenvalues is then used for a $l \times y$ dimensional matrix of W (eigenvector represented at each column). Matrix W is used to transform the samples into a new subspace by matrix multiplication [10]. When trained, LDA is a powerful machine learning tool for classifying multiclass samples as presented in classifying trichomonacidal lead-like compounds by virtual screening. The result proves that 87% accuracy of classification is achieved by applied LDA-QSAR models [11].

3.4 K-Nearest Neighbors Classifier

The K-nearest neighbor classifier or KNN algorithm sorts data according to its similarities in close spacing. KNN classifier uses simple mathematics to compute the distance of data points. While there are multiple ways to compute and define distances between two points on a graph, a popular choice would be using a line or Euclidean distance. To implement the KNN classifier, first, load the data used for sampling. Then a value of k is chosen to index the number of neighbors. For each sample in the dataset, the distance between the query sample and the current sample from the data is computed. The distance and the index of the sample are then appended to an ordered collection. Then indices are ranked from smallest to most considerable distance. Finally, first, k-entries are labeled and returned according to the query: mode for classification and regression for means of k-labels. The right value of k is done by trial and error of sampling unique data and observing the lowest error value [12]. Despite being a simple machine learning algorithm, KNN classification possesses powerful uses in pattern recognition. A study presents using the KNN classifier with RRAM-based parallel computing. The results show that the KNN classifier-based computing enhances pattern recognition capability with less training computation and costs [13].

3.5 Decision Tree Classifier

The decision tree classifier is a non-parametric machine learning method for classification and regression. The decision tree classifier's objective is to predict the target variable's value by learning decision rules governed by the data characteristics. The topmost node of the decision tree is the root node, while the nodes within the tree structure are called internal nodes, the bottom is called leaf nodes, and the connection care called branches. A specific feature defines each internal node of the tree, and branches represent the connection of features that follows classification. Leaf nodes identify samples. The maximum ability of a decision tree classifier is determined by its capability earned from training data. A major advantage of a decision tree classifier is a complex assertion of feature subsets and decision rules at each classification level. From the root, the decision tree splits the source according to feature value and then creates subtrees. The process is done recursively until the leaf nodes are formed. To construct a decision tree, the user needs to apply the gain and gain ratio notion, theoretically measured data. However, the gain computation has disadvantages when dealing with a large number of values. So, the best option would be to employ the gain ratio. The gain ratio equation is shown in 5.1, where x_k is the value of a feature and T is the test set. $split(x_k, T)$ is the acquired information due to splitting the test set on feature x_s [14].

$$gainratio\,(x_k, T) = \frac{gain(x_k, T)}{split(x_k, T)} \qquad (4)$$

$$split\,(x_k, T) = \sum_{i=1}^{n} \frac{|T_i|}{T} \log_2\left(\frac{|T_i|}{|T|}\right) \qquad (5)$$

The application of decision tree models could be seen in analyzing social network metric datasets in [15] and energy efficiency implementation in buildings, including energy-based applications [16]. Moreover, the decision tree application is used, i.e., in real situations, such as predicting medical problems of older people [17].

3.6 Naïve Bayes Algorithm GaussianNB

This research uses the Naïve Bayes algorithm in terms of Gaussian Naïve Bayes, GaussianNB. The Naïve Bayes supervised learning with target, label, and this research used the group of each transportation group as a label. In theory, a Naïve Bayes algorithm is based on Bayes's theorem, which assumes every pair of features. Bayes' theorem is given class variable as y and dependent feature as x_1 to x_n

$$P(y|x_1, \ldots, x_n) = \frac{P(y)P(|x_1, \ldots, x_n|y)}{P(x_1, \ldots, x_n)} \qquad (6)$$

We can use this classification formula for constant as $P(x_1, \ldots, x_n) for the input.$

$$P(y|x_1, \ldots, x_n) \propto P(y) \prod_{i=1}^{n} P(x_i|y) \qquad (7)$$

This research also used the maximum a posteriori (MAP) to estimate $P(y)$ and $P(x_i|y)$. The Naïve Bayes classifiers are used in many aspects since their use for information retrieval in 1998 [18]. It is also suitable for real-world problems such as text classification [19–21]. For the usefulness of Naïve Bayes classification, this research will show the Naïve Bayes with GaussianNB with persons with disabilities dataset for classification the group of transportation cost for those people.

4   Method

The research collected the data from the real situation since 2000. The government needs most persons with disabilities to register for giving many benefits such as expenses each month. Songkhla Government, therefore, let persons with disabilities register and provide some details of their profile. Then, the data need to blind the personal information followed the ethical issue. The researcher needs to clean before process the data. Finally, persons with disabilities data will present only six features: sex, age, type of disability, distance, education, and average cost group.

4.1   Cleansing Data

The government gives datasets from social enterprise organizations with blind personal information. There are including 24609 records and null data was filled with mean.

The distribution of the distance between home and Songkhla Government office show in Fig. 1. For practical use, it is classified into ten groups.

Sex Male = 0 Female = 1

Age : Age

Type : Disabled type

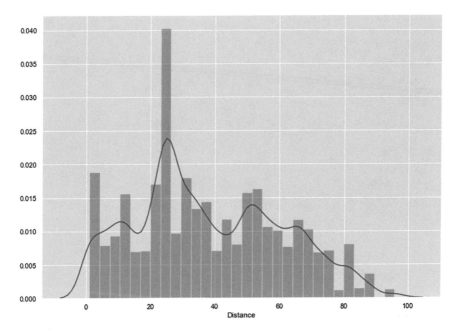

Fig. 1 Distance between persons with disabilities home and Songkhla Government Office

Moving or physical = 1,
   Hearing impaired = 2,
   Redundant = 3,
   Disability notice = 4,
   Intellectual disability = 5,
   Learning disability = 6,
   Mental or behavioral disorders = 7
   Do not known = 8
   Autistic = 9
   Do not specify = 10
   Distance : Distance from home to Songkhla Government Office area (Kilometer)
   0-9.99 = 1
   10-19.99 = 2
   20-29.99 = 3
   30-39.99 = 4
   40-49.99 = 5
   50-59.99 = 6
   60-69.99 = 7
   70-79.99 = 8
   80-89.99 = 9
   >90 = 10
   Edu : Education level

   Prathomsuksa = 0,
   Secondary School = 1,
   Deploma ore equivalent = 2
   Undergraduate = 3
   Graduate = 4
   Uneducated = 5
   Unknown = 6

   CostAvgGroup : Cost of Transportation group (Baht)

0-49.99 = 1
   50-99.99 = 2
   100-199.99 = 3
   200-299.99 = 4
   >300 = 5

4.2 Evaluation

This research chose six attributes sex, age, type of disability, distance, education, and average cost group. Data of persons with disabilities are described in Table 1.

Table 1 Descriptive statistics describe persons with disabilities data

Statistics /Features	SEX	AGE	TYPE	Distance	EDU	Cost average group
Count	24609	24609	24609	24609	24609	24609
Mean	0.553172	52.269466	2.527815	38.046236	3.295827	2.927669
Std	0.497175	23.150345	2.019202	22.540604	2.712823	1.133112
Min	0	0	1	1.005605	0	1
25%	0	35	1	22.407325	0	2
50%	1	54	2	33.901267	5	3
75%	1	71	3	54.777834	6	4
max	1	116	10	95.678406	6	5

The data will split into training and testing set by 75% of the training set and 25% of the testing set.

5 Results

5.1 ANN Multi-Layer Perceptron Classifier Results

Four activation functions for three hidden layers in this research include *identity, logistic, tanh,* and *relu* in MLPClassifier.

'identity', no-op activation, useful to implement linear bottleneck, returns $f(x) = x$

'logistic', the logistic sigmoid function, returns $f(x) = 1 / (1 + \exp(-x))$

'tanh', the hyperbolic tan function, returns $f(x) = \tanh(x)$

'relu', the rectified linear unit function, returns $f(x) = \max(0, x)$

The following paragraph shows some fixed configurations of logistic functions that be adjusted.

MLPRegressor(activation = 'logistic', alpha = 0.0001, batch_size = 'auto', beta_1 = 0.9,

beta_2 = 0.999, early_stopping = False, epsilon = 1e-08,
hidden_layer_sizes = (10, 10, 10), learning_rate = 'constant',
learning_rate_init = 0.001, max_iter = 1000, momentum = 0.9,
n_iter_no_change = 10, nesterovs_momentum = True, power_t = 0.5,
random_state = None, shuffle = True, solver = 'adam', tol = 0.0001,
validation_fraction = 0.1, verbose = True, warm_start = False).

Activation functions use to compare in this paper are *logistics, relu, tanh,* and *identity.*

Table 2 shows the ANN multi-layer perceptron classifier results in four activation

Table 2 ANN Muti-Layer perceptron classifier results

MLP classifier function	Mean squared error	R square score	Accuracy score
Identity	0.1384	0.8918	0.8919
Relu	0.0069	0.9945	0.9947
Tanh	0.0024	0.9980	0.9980
Logistics	0.0882	0.9310	0.9311

functions: logistics, relu, tanh, and identity.

The results show that the '*Tanh*' functions get the 99.80% accuracy results, followed by the '*Relu*' function, 99.41%, '*Logistics*' function, 93.11%, and '*Identity*' function, which is 89.19%.

.

5.2 ANN Model

The ANN model can draw as deep learning with more than one hidden layer. The results in this paper come from three hidden layers. Figure 2 showed the ANN model with six nodes of input layers, ten nodes in three hidden layers, and five nodes in output layers. The model is shown below.

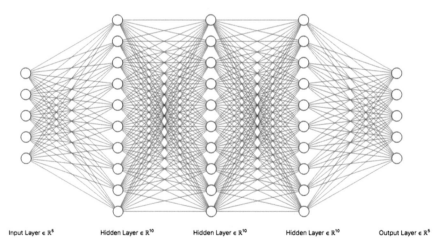

Input Layer ∈ R⁶   Hidden Layer ∈ R¹⁰   Hidden Layer ∈ R¹⁰   Hidden Layer ∈ R¹⁰   Output Layer ∈ R⁵

Fig. 2 ANN model 6-10-10-10-5

5.3   ANN Compare Result

The ANN compares KFold by collecting the results from n_splits: 10, 20, and 30. By using LR, LDA, KNN, CART, and NB algorithm. The boxplot will show in Figs. 3, 4, 5, respectively.

Table 3 shows the ANN Multi-Layer perceptron classifier results, both accuracy (std) in five algorithms: LR, LDA, KNN, CART, and NB. It compares the accuracy result.

The Classification And Regression Trees, CART, has perfect accuracy results, in KFold = 30, equal 1.0000 followed by K-Nearest Neighbor, KNN, has an accuracy of about 98.56%. Naïve Bays Gaussian has an accuracy of about 96.68%. Next, Linear Discriminant Analysis, LDA, has an accuracy of approximately 94.84%. Finally, logistic regression has an accuracy of about 78.89%. Their accuracy scores are the same sequence in all KFold settings, 10, 20, and 30.

Fig. 3 LR, LDA, KNN, CART, and NB accuracy result KFold n_splits = 10

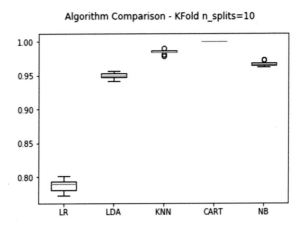

Fig. 4 LR, LDA, KNN, CART, and NB accuracy result KFold n_splits = 20

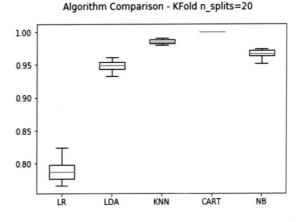

Fig. 5 LR, LDA, KNN, CART, and NB accuracy result KFold n_splits = 30

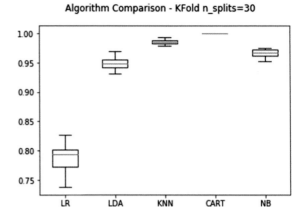

Table 3 ANN Muti-Layer perceptron classifier result (Mean Square Error and Std)

MLP classifier function	KFold = 10	KFold = 20	KFold = 30
LR	LR: 0.788981 (0.018180)	LR: 0.788412 (0.014782)	LR: 0.788981 (0.018180)
LDA	LDA: 0.948474 (0.009016)	LDA: 0.949165 (0.007237)	LDA: 0.948474 (0.009016)
KNN	KNN: 0.985615 (0.003675)	KNN: 0.985615 (0.002930)	KNN: 0.985615 (0.003675)
CART	CART: 1.000000 (0.000000)	CART: 1.000000 (0.000000)	CART: 1.000000 (0.000000)
NB	NB: 0.966843 (0.005975)	NB: 0.966761 (0.005855)	NB: 0.966843 (0.005975)

5.4 Decision Tree Classifier Graph

Figure 6 shows the correlation matrix of all features of persons with disabilities data.

Figure 7 shows the decision tree classifier after plot a decision tree on persons with disabilities data with max_depth = 3. The result showed that the distance (X_3) is the root of the first classification. If the distance value is below or equal to 50.017, entropy is 2.125, samples = 18456, and in-class 2 = 50–99.99 baht. If true, the 18456 samples can be classified 12384 samples, with entropy = 1.461, divided into class 2 = 50–99.99 baht. If false, the 18456 samples can be classified as 6072 samples, with entropy = 0.701, divided into class 3 = 100–199.99 baht.

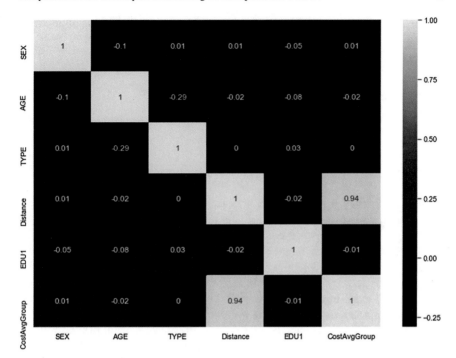

Fig. 6 Heatmap correlation matrix

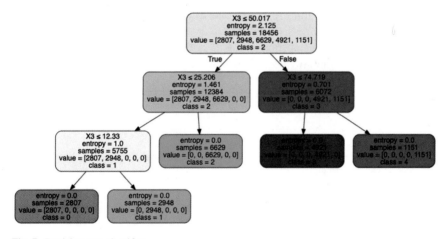

Fig. 7 Decision tree classifier

The accuracy score of decision tree classifier is 1.0 and the confusion matrix of persons with disabilities data shows as below:
array([[938, 0, 0, 0, 0],

[0, 1024, 0, 0, 0],
[0, 0, 2126, 0, 0],
[0, 0, 0, 1671, 0],

6 Conclusion

This research tried to establish the transportation cost model of 24,609 persons with disabilities in Songkhla Province with ANN techniques. The ANN techniques, including ANN, LR, LDA, k-Nearest Neighbor, CART, and Naïve Bayes algorithms. For ANN, the results show that the '*Tanh*' functions get the 99.80% accuracy results, followed by the '*Relu*' function, which is 99.47%, '*Logistics*' function, which is 93.11%, and '*Identity*' function, which is 89.19%. The model of ANN is 6-10-10-10-5, six nodes input layer, ten nodes hidden layers with three hidden layers, and five output layers. The compare accuracy results show that Classification and Regression Trees, CART, has perfect accuracy results equal to 1.0000, followed by a k-nearest neighbor, KNN, 98.56%. Naïve Bays Gaussian 96.68%, Discriminant Analysis, LDA, 94.84%, and logistic regression, 78.89%. The model results can predict persons with disabilities' transportation costs and help the government plan persons with disabilities' expenses and allocate funds accordingly. For future research, the persons with disabilities dataset may be used for other predictions, such as planning public transport outlining and other amenities for persons with disabilities in Songkhla Province, Thailand.

References

1. Shanmuganathan, S., Samarasinghe, S. Eds.: Artificial Neural Network Modelling. Springer International Publishing (2016)
2. Deboeck, G., Kohonen, T. (eds.): Visual Explorations in Finance: with Self-Organizing Maps. Springer-Verlag, London (1998)
3. Surakhman, Y.: Artificial Neural Networks Technology. Accessed 21 Jun 2020. https://www.academia.edu/7152318/Artificial_Neural_Networks_Technology
4. Tosh, C.R., Ruxton, G.D. Eds.: Modelling Perception with Artificial Neural Networks, 1st edn. Cambridge University Press (2011)
5. Shepherd, A.J.: Second-Order Methods for Neural Networks: Fast and Reliable Training Methods for Multi-Layer Perceptrons. Springer-Verlag, London (1997)
6. Avuçlu, E., Fatih, B.: Age estimation by using multi-layer perceptron neural network with image processing techniques. In: Vang-mata, R. (ed.) Multilayer Perceptrons: Theory and Applications, pp. 77–101. Nova Science Pub Inc., Hauppauge, New York (2020)
7. Hilbe, J.M.: Logistic Regression Models. CRC Press (2009)
8. Jr, D.W.H., Lemeshow, S.: Applied Logistic Regression. John Wiley & Sons (2004)

9. Lee, S.: Application of logistic regression model and its validation for landslide susceptibility mapping using GIS and remote sensing data. Int. J. Remote Sens. **26**(7), 1477–1491 (2005). https://doi.org/10.1080/01431160412331331012
10. Raschka, S.: Linear Discriminant Analysis (2014). https://sebastianraschka.com/Articles/2014_python_lda.html. Accessed 21 Jun 2020
11. Meneses-Marcel, A., et al.: A linear discrimination analysis based virtual screening of trichomonacidal lead-like compounds: outcomes of in silico studies supported by experimental results. Bioorg. Med. Chem. Lett. **15**(17), 3838–3843 (2005). https://doi.org/10.1016/j.bmcl.2005.05.124
12. Harrison, O.: Machine Learning Basics with the K-Nearest Neighbors Algorithm. Medium (2019). https://towardsdatascience.com/machine-learning-basics-with-the-k-nearest-neighbors-algorithm-6a6e71d01761. Accessed 22 Jun 2020
13. Jiang, Y., Kang, J., Wang, X.: RRAM-based parallel computing architecture using k-nearest neighbor classification for pattern recognition. Sci. Rep. **7**(1), 45233 (2017). https://doi.org/10.1038/srep45233
14. Du, C.-J., Sun, D.-W.: 4-Object Classification Methods. In: Sun, D.-W. (ed.) Computer Vision Technology for Food Quality Evaluation, pp. 81–107. Academic Press, Amsterdam (2008)
15. Panigrahi, R., Borah, S.: 1-Classification and Analysis of Facebook Metrics Dataset Using Supervised Classifiers. In Dey, N., Borah, S., Babo, R., Ashour, A.S. (eds.) Social Network Analytics, pp. 1–19. Academic Press (2019)
16. Capozzoli, A., Cerquitelli, T., Piscitelli, M.S.: Chapter 11-Enhancing energy efficiency in buildings through innovative data analytics technologiesa. In: Dobre, C., Xhafa, F. (eds.) Pervasive Computing, pp. 353–389. Academic Press, Boston (2016)
17. Tongkaw, A., Tongkaw, S.: Prediction medical problem of elderly people by using machine learning technique. J. Phys. Conf. Ser. **1529**, 032083 (2020). https://doi.org/10.1088/1742-6596/1529/3/032083
18. Lewis, D.D.: Naive (Bayes) at forty: The independence assumption in information retrieval. In Machine Learning: ECML-98, pp. 4–15. Berlin, Heidelberg, (1998). https://doi.org/10.1007/BFb0026666
19. Kim, S.-B., Han, K.-S., Rim, H.-C., Myaeng, S.H.: Some effective techniques for naive bayes text classification. IEEE Trans. Knowl. Data Eng. **18**(11), 1457–1466 (2006)
20. McCallum, A., Nigam, K.: A comparison of event models for naive bayes text classification. In: AAAI-98 Workshop on Learning for Text Categorization, vol. 752, no. 1, pp. 41–48 (1998)
21. Rennie, J.D., Shih, L., Teevan, J., Karger, D.R.: Tackling the poor assumptions of naive bayes text classifiers. In Proceedings of the 20th International Conference on Machine Learning (ICML-03), pp. 616–623 (2003)

Further Results on Weighted Entropy for Doubly Truncated Random Variable

Rajesh Moharana and Suchandan Kayal

Abstract In reliability theory, it is expected that a unit which fails to operate provides different information in two different but equal wide intervals. To deal with such situations, the concept of weighted Shannon's entropy was developed. Misagh and Yari [10] introduced the notion of weighted doubly truncated entropy (WDTE) of random lifetimes and studied various properties of it. Doubly truncated data appear in some applications with survival and astrological data. In the present communication, we obtain further results based on this measure. The effect of monotone transformations and affine transformation on this measure are discussed. It is shown that WDTE characterizes the distribution uniquely. Further, new uncertainty order is obtained. Finally, based on WDTE a new class of lifetime distributions is introduced.

Keywords Weighted entropy · Doubly truncated random variable · Characterization · Uncertainty order · Aging class

1 Introduction

The notion of entropy plays an important role in information theory since its introduction by Shannon [15]. Besides information theory, it has found various applications in different areas of science and technology. For example, molecular biologists use the concept of entropy in the analysis of patterns in gene sequences (see [1]). For a detailed account of importance and applications of entropy in various disciplines we refer to Cover and Thomas [3]. Let X be a non-negative absolutely continuous random variable representing time to failure of a system. Let $F(.)$ be the cumulative

R. Moharana (✉)
Department of Mathematics, School of Advanced Sciences, Vellore Institute of Technology, Vellore, India
e-mail: rajeshmoharana31@gmail.com; rajesh.moharana@vit.ac.in

S. Kayal
Department of Mathematics, National Institute of Technology Rourkela, Rourkela, India
e-mail: suchandan.kayal@gmail.com; kayals@nitrkl.ac.in

P. Gyei-Kark et al. (eds.), *Engineering Mathematics and Computing*,
Studies in Computational Intelligence 1042,
https://doi.org/10.1007/978-981-19-2300-5_13

distribution function (CDF) of X with survival function (SF) $\bar{F}(.)$. We denote the probability density function (PDF), hazard rate function and reversed hazard rate function of X by $f(.)$, $\lambda_F(.) = f(.)/\bar{F}(.)$ and $\eta_F(.) = f(.)/F(.)$, respectively. The Shannon differential entropy of X is defined as

$$S_X = - \int_0^\infty f(x) \ln f(x) dx. \tag{1}$$

Note that S_X in (1) is not a scale invariant measure though it is a translation invariant. For example, S_X remains same when X follows uniform distribution either in (u, v), or in $(u + c, v + c)$, where c is any real number. Besides the importance, it is also important to mention that the Shannon entropy has few drawbacks of being shift independent. There are several situations in reliability and neurobiology, where the shift dependent information measure plays an important role. Because of this, analogous to Guiasu [6], Di Crescenzo and Longobardi [4] proposed the notion of the weighted differential entropy as

$$S_X^w = - \int_0^\infty x f(x) \ln f(x) dx. \tag{2}$$

The notion of weighted entropy is of interest in many areas of applied probability and statistics. For example, it is used to compare the species diversity in two or more habitants (see Guiasu and Guiasu [7]). For some applications of the weighted entropy, we refer to Kelbert et al. [8]. The measure given by (2) is a length-biased shift dependent measure which gives more importance to larger values of X.

Doubly truncated data are often encountered in some applications with astronomical and survival data. In survival studies, one has information about the lifetime only between two time points, that is, individuals whose event time lies within a certain time interval are only observed. We define a random variable $(X|t_1 < X < t_2)$ representing the time to failure of a system which fails in the interval (t_1, t_2), where $(t_1, t_2) \in D = \{(x, y) : F(x) < F(y)\}$. Note that the random variable $(X|t_1 < X < t_2)$ is also dubbed as doubly truncated random variable. A measure of uncertainty for the random variable $(X|t_1 < X < t_2)$, also known as interval entropy of X in the interval (t_1, t_2) is given by (see Sunoj et al. [16])

$$S_X(t_1, t_2) = - \int_{t_1}^{t_2} \frac{f(x)}{F(t_2) - F(t_1)} \ln \left(\frac{f(x)}{F(t_2) - F(t_1)} \right) dx. \tag{3}$$

Equation (3) measures the uncertainty about the lifetime X between t_1 and t_2, given that the system has survived up to time t_1 and has been found to be down at time t_2. Analogous to the notion of (2) and (3), Misagh and Yari [10] proposed weighted doubly truncated (interval) entropy of X in the interval (t_1, t_2) as

$$S_X^w(t_1, t_2) = - \int_{t_1}^{t_2} x \frac{f(x)}{F(t_2) - F(t_1)} \ln \left(\frac{f(x)}{F(t_2) - F(t_1)} \right) dx \tag{4}$$

and obtained various results. The measure given in (4) is a shift-dependent dynamic measure of uncertainty. We call it as weighted doubly truncated entropy. Note that $S_X^w(t_1, t_2)$ respectively reduces to the weighted residual entropy of X at time t_1 and to the weighted past entropy of X at time t_2, when $t_2 \to \infty$ and $t_1 \to 0$. For several properties of the weighted residual entropy and the weighted past entropy, we refer to Di Crescenzo and Longobardi [4] and Yasaei Sekeh et al. [17]. We consider the following example to provide the importance of (4).

Example 1 Let X and Y be two non-negative random variables with joint probability density function

$$f(x, y) = \begin{cases} \frac{1}{2} xe^{-y}, & \text{if } 0 < x < 2, y > 0, \\ 0, & \text{elsewhere.} \end{cases}$$

The marginal densities of X and Y are $f(x) = x/2, 0 < x < 2$ and $g(y) = \exp\{-y\}, y > 0$, respectively. The distributions are not comparable in terms of the weighted entropy as they do not have the same support. In fact, $S_X^w(0.5, 1.9) = 0.247838$ and $S_Y^w(0.5, 1.9) = 0.417902$. Here, the WDTE of X is smaller than that of Y in the interval $(0.5, 1.9)$.

In the following, we recall some preliminary definition, notations and results which will be useful to obtain some results in the rest of this chapter. For an absolutely continuous random variable X with PDF $f(.)$ and CDF $F(.)$, Navarro and Ruiz [12] defined generalized failure rate functions as

$$h_1(t_1, t_2) = \frac{f(t_1)}{F(t_2) - F(t_1)} \quad \text{and} \quad h_2(t_1, t_2) = \frac{f(t_2)}{F(t_2) - F(t_1)} \tag{5}$$

for all $(t_1, t_2) \in D$. Note that when $t_2 \to \infty$, (5) reduces to $\lambda_F(t_1) = f(t_1)/\bar{F}(t_1)$ and when $t_1 \to 0$, (5) reduces to $\eta_F(t_2) = f(t_2)/F(t_2)$. For detail and various properties of $h_1(t_1, t_2)$ and $h_2(t_1, t_2)$, we refer to Navarro and Ruiz [12], Sankaran and Sunoj [13] and Sunoj et al. [16]. The generalized conditional mean function of a doubly truncated random variable is defined by

$$\mu(t_1, t_2) = E(X|t_1 < X < t_2) = \int_{t_1}^{t_2} \frac{xf(x)}{F(t_2) - F(t_1)} dx. \tag{6}$$

Definition 1 Let X and Y be two nonnegative random variables with SFs $\bar{F}(x)$ and $\bar{G}(x)$, respectively. Then, X is said to be smaller than Y in the usual stochastic order, denoted by $X \leq^{ST} Y$, if $\bar{F}(x) \leq \bar{G}(x)$ for all x.

For more results in usual stochastic ordering, we refer to Shaked and Shanthikumar [14].

The purpose of this chaptere is to explore further results of the WDTE given by (4). The rest of the paper is arranged as follows. Section 2 presents few properties and the effect of monotone transformations and affine transformation on WDTE. In

Sect. 3, we obtain some characterization result based on the WDTE. It is proved that WDTE of discrete random variables determines the distribution uniquely. Further, we introduced an uncertainty order which can be used to compare two doubly truncated random variables in terms of uncertainty in Sect. 4. Finally, in Sect. 5, we propose a new class of lifetime distributions based on this measure. Concluding remarks are presented in Sect. 6. Throughout this chapter, we consider that all the random variables are non-negative. Moreover, it is always assumed that all integrals, expectations and derivatives are implicitly assumed to be finite wherever they are given. The terms increasing and decreasing stand for non-decreasing and non-increasing, respectively.

2 Properties of WDTE

In this section, we discuss a few properties of WDTE. The effect of strictly monotone transformations and affine transforation on WDTE are presented here.

The measure given by (4) can be re-written as

$$S_X^w(t_1, t_2) = - \int_{t_1}^{t_2} \int_0^x \frac{f(x)}{F(t_2) - F(t_1)} \ln\left(\frac{f(x)}{F(t_2) - F(t_1)}\right) dy dx. \quad (7)$$

The expression in (7) can also be written in the following form

$$S_X^w(t_1, t_2) = t_1 S_X(t_1, t_2) - \int_{t_1}^{t_2} \int_{t_1}^x \frac{f(x)}{F(t_2) - F(t_1)} \ln\left(\frac{f(x)}{F(t_2) - F(t_1)}\right) dy dx. \quad (8)$$

Differentiating (8) with respect to t_1 partially and simplifying, we obtain

$$\frac{\partial S_X^w(t_1, t_2)}{\partial t_1} = t_1 \frac{\partial S_X(t_1, t_2)}{\partial t_1} + h_1(t_1, t_2)[S_X^w(t_1, t_2) - t_1(S_X(t_1, t_2) - 1) - \mu(t_1, t_2)].$$

Again (4) can be written as

$$S_X^w(t_1, t_2) = t_2 S_X(t_1, t_2) - \int_{t_1}^{t_2} \int_x^{t_2} \frac{f(x)}{F(t_2) - F(t_1)} \ln\left(\frac{f(x)}{F(t_2) - F(t_1)}\right) dy dx. \quad (9)$$

Differentiating (9) with respect to t_2 partially and after simplification, we get

$$\frac{\partial S_X^w(t_1, t_2)}{\partial t_2} = t_2 \frac{\partial S_X(t_1, t_2)}{\partial t_2} - h_2(t_1, t_2)[S_X^w(t_1, t_2) - t_2(S_X(t_1, t_2) - 1) - \mu(t_1, t_2)].$$

Remark 1 It is worthwhile to mention that $S_X^w(t_1, t_2)$ can be smaller or larger than $S_X(t_1, t_2)$. For example, let X be a non-negative random variable with failure distribution $F(x|a, b) = \frac{x-a}{b-a}$, $0 < a < x < b < \infty$. Then it is easy to show that

$$S_X^w(t_1, t_2) = \mu(t_1, t_2)S_X(t_1, t_2),$$

where $\mu(t_1, t_2) = \frac{t_1+t_2}{2}$. Hence, for $\mu(t_1, t_2) \leq 1$, we get $S_X^w(t_1, t_2) \leq S_X(t_1, t_2)$ and for $\mu(t_1, t_2) \geq 1$, we get $S_X^w(t_1, t_2) \geq S_X(t_1, t_2)$.

Remark 2 Note that there exist non-negative random variables whose weighted doubly truncated entropy can be $-\infty$. For example, for a random variable with probability density function $f(x|a, b) = \frac{1}{b-a}$, $0 < a < x < b < \infty$, we have

$$\lim_{t_1 \to t_2^-} S_X^w(t_1, t_2) = \lim_{t_1 \to t_2^-} \left(\frac{t_1 + t_2}{2}\right) \ln(t_2 - t_1) = -\infty.$$

The effect of strictly monotone transformations on the weighted residual entropy and the weighted past entropy has been considered by Di Crescenzo and Longobardi [4]. In the following theorem we present the effect of the WDTE given by (4) in case of the strictly monotone transformations. The proof of the following theorem is directly follows from Theorem 4.1 of Kundu [9], hence omitted.

Theorem 1 *Let X be an absolutely continuous non-negative random variable and $Y = \phi(X)$ be a strictly monotone and differentiable function. Then for all $0 \leq t_1 < t_2 < \infty$, we have*

$$S_{\phi(X)}^w(t_1, t_2) = \begin{cases} S_X^{w,\phi}(\phi^{-1}(t_1), \phi^{-1}(t_2)) \\ \quad + E(\phi(X)\ln(\phi'(X))|\phi^{-1}(t_1) < X < \phi^{-1}(t_2)), \\ \qquad\qquad\qquad\qquad\qquad \textit{if } \phi(x) \textit{ is strictly increasing,} \\ S_X^{w,\phi}(\phi^{-1}(t_2), \phi^{-1}(t_1)) \\ \quad + E(\phi(X)\ln(-\phi'(X))|\phi^{-1}(t_2) < X < \phi^{-1}(t_1)), \\ \qquad\qquad\qquad\qquad\qquad \textit{if } \phi(x) \textit{ is strictly decreasing.} \end{cases}$$

Remark 3 Let $\phi_1(x) = F(x)$ and $\phi_2(x) = \bar{F}(x)$, where both $\phi_1(X)$ and $\phi_2(X)$ follow uniform distribution in $(0, 1)$. Clearly, $\phi_1(x)$ and $\phi_2(x)$ satisfy the assumptions given in Theorem 1. Therefore, as an application of Theorem 1, we get

$$S_{F(X)}^w(t_1, t_2) = S_X^{w,F}(F^{-1}(t_1), F^{-1}(t_2)) \\ + E((F(X)\ln(f(X))|F^{-1}(t_1) < X < F^{-1}(t_2))$$

and

$$S_{\bar{F}(X)}^w(t_1, t_2) = S_X^{w,\bar{F}}(\bar{F}^{-1}(t_2), \bar{F}^{-1}(t_1)) \\ + E(\bar{F}(X)\ln(f(X))|\bar{F}^{-1}(t_2) < X < \bar{F}^{-1}(t_1)).$$

The following lemma provides the expression of WDTE under affine transformation. The proof is straightforward, hence omitted.

Lemma 1 *Let X be a non-negative absolutely continuous random variable. Define $\phi(X) = aX + b$, where $a > 0$ and $b \geq 0$ are constants. Then for $t_1 > b$,*

$$S^w_{\phi(X)}(t_1, t_2) = a S^w_X\left(\frac{t_1 - b}{a}, \frac{t_2 - b}{a}\right) + b S_X\left(\frac{t_1 - b}{a}, \frac{t_2 - b}{a}\right)$$
$$+ (a \ln a)\mu\left(\frac{t_1 - b}{a}, \frac{t_2 - b}{a}\right) + b \ln a. \tag{10}$$

Proposition 1 *Consider an absolutely continuous random variable X with support $[0, a]$. Assume that it is symmetric with respect to $\frac{a}{2}$. Then*

$$S^w_X(t_1, t_2) = a S_X(a - t_2, a - t_1) - S^w_X(a - t_2, a - t_1).$$

Proof Under the given assumptions we have $F(x) = \bar{F}(a - x)$ when $0 \leq x \leq a$, that is, $f(x) = f(a - x)$. Thus for $0 \leq t_1 < t_2 \leq a$, we obtain from (4)

$$S^w_X(t_1, t_2) = -\int_{t_1}^{t_2} x \, \frac{f(a - x)}{\bar{F}(a - t_2) - \bar{F}(a - t_1)} \ln\left(\frac{f(a - x)}{\bar{F}(a - t_2) - \bar{F}(a - t_1)}\right) dx$$

$$= -\int_{a - t_2}^{a - t_1} a \, \frac{f(u)}{\bar{F}(a - t_2) - \bar{F}(a - t_1)} \ln\left(\frac{f(u)}{\bar{F}(a - t_2) - \bar{F}(a - t_1)}\right) du$$

$$+ \int_{a - t_2}^{a - t_1} u \, \frac{f(u)}{\bar{F}(a - t_2) - \bar{F}(a - t_1)} \ln\left(\frac{f(u)}{\bar{F}(a - t_2) - \bar{F}(a - t_1)}\right) du$$

$$= a S_X(a - t_2, a - t_1) - S^w_X(a - t_2, a - t_1).$$

The second equality is due to the transformation $u = a - x$. This completes the proof. ☐

We consider the following example as an application of Proposition 1.

Example 2 Suppose X follows uniform distribution in the interval $(0, 1)$. Clearly, it is symmetric at $1/2$. Here $S^w_X(t_1, t_2) = \frac{t_1 + t_2}{2} \ln(t_2 - t_1)$, $S^w_X(1 - t_2, 1 - t_1) = \frac{2 - t_1 - t_2}{2} \ln(t_2 - t_1)$ and $S_X(1 - t_2, 1 - t_1) = \ln(t_2 - t_1)$. Hence we conclude that Proposition 1 holds for a symmetric distribution.

In the following Table 1, we obtain expressions of the WDTE for some distributions in terms of the generalized conditional mean function $\mu(t_1, t_2)$ and the generalized failure rate functions $h_1(t_1, t_2)$ and $h_2(t_1, t_2)$.

3   Characterization

In the study of a general characterization problem, our aim is to determine conditions under which the WDTE uniquely determines the underlying distribution. Di Crescenzo and Longobardi [4] and Yasaei Sekeh et al. [17] respectively showed that

Table 1 Cumulative distribution functions and the expressions of $S_X^w(t_1, t_2)$

Distribution	CDF	$S_X^w(t_1, t_2)$
Uniform	$F(x\mid a, b) = \frac{x-a}{b-a}$, $0 < a < x < b < \infty$.	$-\mu(t_1, t_2)\ln h_1(t_1, t_2)$
Exponential	$F(x\mid\lambda) = 1 - e^{-\lambda x}$, $x > 0, \lambda > 0$.	$\mu(t_1, t_2)(2 - \ln h_1(t_1, t_2) - \lambda t_1)$ $+t_1^2 h_1(t_1, t_2) - t_2^2 h_2(t_1, t_2)$
Pareto-I	$F(x\mid a, b) = 1 - (\frac{a}{x})^b$, $x > a > 0, b > 0$.	$-\mu(t_1, t_2)(\ln h_2(t_1, t_2) + (b+1)\ln t_2 - \left(\frac{1+b}{1-b}\right)$ $\left(h_2(t_1, t_2)t_2^2 \ln t_2 - h_1(t_1, t_2)t_1^2 \ln t_1 - \mu(t_1, t_2)\right)$
Pareto-II	$F(x\mid p, q) =$ $1 - (1 + px)^{-q}$, $x > 0, p > 0, q > 0$	$-\mu(t_1, t_2)[\ln h_1(t_1, t_2) + (q+1)\ln(1 + pt_1)]$ $+\left(\frac{q+1}{p^2 q}\right)[(1 + pt_1)h_1(t_1, t_2)\ln(1 + pt_1)$ $-(1 + pt_2)h_2(t_1, t_2)\ln(1 + pt_2) + p]$ $+\left(\frac{q+1}{p^2(1-q)}\right)[(1 + pt_2)^2 h_2(t_1, t_2)\ln(1 + pt_2)$ $-(1 + pt_1)^2 h_1(t_1, t_2)\ln(1 + pt_1)$ $-\left(\frac{1}{1-q}\right)\{(1 + pt_2)^2 h_2(t_1, t_2)$ $-(1 + pt_1)^2 h_1(t_1, t_2)\}]$
Power	$F(x\mid\alpha, \beta) = (\frac{x}{\alpha})^\beta$, $0 < x < \alpha, \beta > 0$.	$\mu(t_1, t_2)(\beta\ln\alpha - \ln h_1(t_1, t_2) + (\beta - 1)\ln t_1)$ $-\left(\frac{\beta-1}{\beta+1}\right)\left(h_2(t_1, t_2)t_2^2 \ln t_2 - h_1(t_1, t_2)t_1^2 \ln t_1$ $-\mu(t_1, t_2)\right)$

the weighted residual entropy and the weighted past entropy determine the underlying distribution uniquely. After that Kundu [9] obtained characterization results of weighted doubly truncated inaccuracy measure for continuous random variables. Also, in his study he derived several characterization results for some well-know continuous distributions. In similar way, the characterizations of WDTE for continuous random variable can be obtained. Hence, the characterizations of continuous random variables are not considered in this study. In this section, we obtain a characterization result for discrete distribution which shows that under some assumptions the WDTE determines the survival function uniquely.

Let X be a discrete random variable taking values $t_1, t_2, \ldots, t_k, \ldots$ with respective probabilities $p_1, p_2, \ldots, p_k, \ldots$. Let T be the support of X, where $T = \{t_k : t_k < t_{k+1}, k \in \mathbf{N}\}$. Here $\mathbf{N}$ represents the set of natural numbers. Denote by $p(t_k) = P(X = t_k)$ and $P(t_k) = P(X \leq t_k)$ as the probability mass function and discrete distribution function of X, respectively. The probability mass function of the doubly truncated random variable X with left truncation at t_i and right truncation at t_j, where $i < j$; $i, j \in \mathbf{N}$ is given by

$$P(X = t_k \mid t_i \leq X \leq t_j) = \frac{p(t_k)}{P(t_j) - P(t_{i-1})}, \quad (11)$$

where $1 \leq i < k < j \leq \infty$. Note that for $i = 1$, (11) reduces to the probability mass function of discrete right truncated random variable and for $j = \infty$, it reduces to the probability mass function of discrete left truncated random variable. The generalized failure rates are given by (see Navarro and Ruiz [12])

$$h_i(t_i, t_j) = \frac{p(t_i)}{P(t_j) - P(t_{i-1})} \quad \text{and} \quad h_j(t_i, t_j) = \frac{p(t_j)}{P(t_j) - P(t_{i-1})}. \tag{12}$$

Thus the discrete doubly truncated entropy of $[X | t_i \le X \le t_j]$ is given by

$$S(t_i, t_j) = -\sum_{k=i}^{j} \frac{p(t_k)}{P(t_j) - P(t_{i-1})} \ln \left(\frac{p(t_k)}{P(t_j) - P(t_{i-1})} \right). \tag{13}$$

Note that when $j = \infty$, then (13) reduces to discrete residual entropy (see Belzunce et al. [2]) and when $i = 1$, then it reduces to discrete past entropy (see Nanda and Paul [11]). Analogous to the weighted residual and past entropy due to Di Cresenzo and Longobardi [4], Yasaei Sekeh et al. [17] proposed discrete weighted residual and past entropy and showed that under some conditions these measures determine the distribution uniquely. In this part of the chapter, we consider discrete weighted doubly truncated entropy, which is given by

$$S_X^w(t_i, t_j) = -\sum_{k=i}^{j} t_k \frac{p(t_k)}{P(t_j) - P(t_{i-1})} \ln \left(\frac{p(t_k)}{P(t_j) - P(t_{i-1})} \right), \tag{14}$$

where $w_k = t_k$; $k \in \mathbf{N}$ is the weight function. In the following we obtain a characterization based on the measure given by 14.

Theorem 2 *Let X be a discrete random variable with discrete distribution function $P(t)$. Assume that X has support T, such that $T = \{t_k : t_k < t_{k+1}, k \in \mathbf{N}\}$. If $S_X^w(t_i, t_j)$ is increasing in t_i (for fixed t_j) and decreasing in t_j (for fixed t_i), then $S_X^w(t_i, t_j)$ uniquely determine the distribution function.*

Proof From 14 we have

$$\sum_{k=i}^{j} t_k p(t_k) \ln p(t_k) = \ln[P(t_j) - P(t_{i-1})] \sum_{k=i}^{j} t_k p(t_k)$$
$$- [P(t_j) - P(t_{i-1})] S_X^w(t_i, t_j). \tag{15}$$

Now replacing i by $i + 1$ in (15) and then subtracting from (15), we obtain

$$t_i p(t_i) \ln p(t_i) = t_i p(t_i) \ln[P(t_j) - P(t_{i-1})] + \ln \left[\frac{P(t_j) - P(t_{i-1})}{P(t_j) - P(t_i)} \right] \sum_{k=i+1}^{j} t_k p(t_k)$$
$$- [P(t_j) - P(t_{i-1})] S_X^w(t_i, t_j) + [P(t_j) - P(t_i)] S_X^w(t_{i+1}, t_j). \tag{16}$$

Again, replacing j by $j+1$ in (15) and then subtract (15), we obtain

$$t_{j+1}p(t_{j+1})\ln p(t_{j+1}) = t_{j+1}p(t_{j+1})\ln[P(t_{j+1}) - P(t_{i-1})] + [P(t_j) - P(t_{i-1})]$$

$$\times S_X^w(t_i, t_j) + \ln\left[\frac{P(t_{j+1}) - P(t_{i-1})}{P(t_j) - P(t_{i-1})}\right]\sum_{k=i}^{j}t_k p(t_k)$$

$$- [P(t_{j+1}) - P(t_{i-1})]S_X^w(t_i, t_{j+1}). \qquad (17)$$

Keeping t_j fixed and substituting $p(t_i) = [P(t_j) - P(t_{i-1})] - [P(t_j) - P(t_i)]$ and $\lambda_i = (P(t_j) - P(t_i))/(P(t_j) - P(t_{i-1}))$ in (17) we get

$$t_i[P(t_j) - P(t_{i-1})](1 - \lambda_i)\ln(1 - \lambda_i) = -\ln\lambda_i\sum_{k=i+1}^{j}t_k p(t_k)$$

$$- [P(t_j) - P(t_{i-1})]S_X^w(t_i, t_j) + [P(t_j) - P(t_i)]S_X^w(t_{i+1}, t_j). \qquad (18)$$

Moreover, using the notation

$$M(t_i, t_j) = E(X|t_i \leq X \leq t_j) = \sum_{k=i}^{j}t_k\frac{p(t_k)}{P(t_j) - P(t_{i-1})},$$

Equation(18) can be re-written as

$$g_1(x) = t_j(1 - x)\ln(1 - x) + M(t_{i+1}, t_j)x\ln x + S_X^w(t_i, t_j) - xS_X^w(t_{i+1}, t_j) = 0, \qquad (19)$$

where $x = \lambda_i$, $\lambda_i \in (0, 1)$. Again keeping t_i fixed and substituting $p(t_{j+1}) = [P(t_{j+1}) - P(t_{i-1})] - [P(t_j) - P(t_{i-1})]$ and $\theta_j = (P(t_j) - P(t_{i-1}))/(P(t_{j+1}) - P(t_{i-1}))$ in (18) we get

$$g_2(x) = t_{j+1}(1 - x)\ln(1 - x) + M(t_i, t_j)x\ln x + S_X^w(t_i, t_{j+1}) - xS_X^w(t_i, t_j) = 0, \qquad (20)$$

where $x = \theta_j$, $\theta_j \in (0, 1)$. It can be noted that for fixed t_j, $g_1(x) = 0$ has a unique positive root $x = \lambda_i$, for all t_i, and for fixed t_i, $g_2(x) = 0$ has a unique positive root $x = \theta_j$, for all t_j. We give the detail for $g_1(x) = 0$. The other case when $g_2(x) = 0$ can be shown analogously. Under the given assumption, we have

$$g_1(0) = S_X^w(t_i, t_j) \geq 0 \text{ and } g_1(1) = S_X^w(t_i, t_j) - S_X^w(t_{i+1}, t_j) \leq 0,$$

and hence there exists at least one root of $g_1(x) = 0$ in $(0, 1)$. Now differentiating (19) with respect to x we get

$$g_1'(x) = M(t_{i+1}, t_j)\log x - t_j\log(1 - x) + M(t_{i+1}, t_j) - S_X^w(t_{i+1}, t_j) - t_j. \qquad (21)$$

Again differentiating (21) with respect to x we obtain

$$g_1''(x) = \frac{M(t_{i+1}, t_j)}{x} + \frac{t_j}{1 - x}. \tag{22}$$

From (21) we conclude that $x = \frac{-M(t_{i+1}, t_j)}{t_j - M(t_{i+1}, t_j)}$ can be the only solution of $g_1''(x) = 0$. Note that if $t_j < M(t_{i+1}, t_j)$ then $x > 0$ and if $t_j > M(t_{i+1}, t_j)$ then $x < 0$. Hence $g_1''(x) = 0$ has a unique positive solution. Therefore, $g_1'(x) = 0$ has at most one positive solution in $(0, 1)$. Thus $g_1(x) = 0$ has a unique positive solution in $(0, 1)$ for all t_i implies λ_i is the unique solution for $g_1(x) = 0$. Using similar arguments, it is easy to show that θ_j is the unique solution for $g_2(x) = 0$. Note that $1 - \lambda_i = h_i(t_i, t_j)$, $1 - \theta_j = h_{j+1}(t_i, t_{j+1})$ and $h_i(t_i, t_j)$ uniquely characterizes the distribution. This completes the proof. □

4 Uncertainty Order Based on WDTE

Yasaei Sekeh et al. [17] defined orders based on weighted dynamic measures. Similar to that, here we define a new order based on the WDTE given by (4).

Definition 2 Let X and Y be two non-negative absolutely continuous random variables. Then X is said to be smaller than Y in weighted doubly truncated entropy order, denoted by $X \leq^{WDTE} Y$, if $S_X^w(t_1, t_2) \leq S_Y^w(t_1, t_2)$, for all $0 \leq t_1 < t_2 < \infty$.

From Definition 2, we state that the expected weighted uncertainty contained in the conditional density of X given that $t_1 < X < t_2$ about the predictability of the failure time is less than that of Y given that $t_1 < Y < t_2$ implies $X \leq^{WDTE} Y$. To illustrate the statement in Definition 2, we consider the following example.

Example 3 Let X and Y be two random variables with probability density functions $f(x) = 2x$ and $g(x) = 2(1 - x)$, respectively, where $x \in (0, 1)$. Here, $S_X = \frac{1}{2} - \ln 2 = S_Y$. But, from Fig. 1, we observe that $S_X^w(t_1, t_2) \leq S_Y^w(t_1, t_2)$ implying $X \leq^{WDTE} Y$.

In the following we consider an example which shows that $X \geq^{ST} Y$ does not imply $X \geq^{WDTE} Y$.

Example 4 Let X and Y be two non-negative absolutely continuous random variables with probability density functions $f(x) = \frac{1}{(1+x)^2}$, $x > 0$ and $g(x) = \frac{2}{(1+x)^3}$, $x > 0$, respectively. Now it can be easily shown that $F(x) \leq G(x)$, for all $x > 0$ implying $X \geq^{ST} Y$. Also, at $(t_1 = 1, t_2 = 2)$, we have $S_X^w(t_1, t_2) - S_Y^w(t_1, t_2) = 0.0158$ and at $(t_1 = 1.4, t_2 = 1.6)$ $S_X^w(t_1, t_2) - S_Y^w(t_1, t_2) = -0.0015$. Hence, we conclude that $X \ngeq^{WDTE} Y$.

The following example shows that $X \ngeq^{ST} Y$ but $X \geq^{WDTE} Y$.

Fig. 1 It represents the surface plot of the weighted doubly truncated entropy of the random variables as described in Example 3

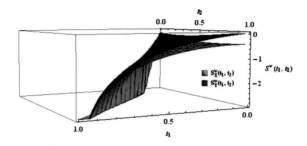

Fig. 2 Surface plot of $S_X^w(t_1, t_2) - S_Y^w(t_1, t_2)$, where X and Y are described in Example 5

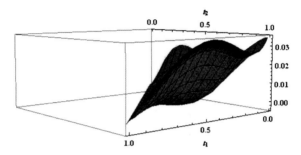

Example 5 Let X and Y be two non-negative absolutely continuous random variables with probability density functions $f(x) = 2(1 - x) \exp(-x) + (1 - x)^2 \exp(-x)$, for $x \in (0, 1)$ and $g(x) = \frac{2(1-x)}{1+4x^2} + \frac{8x(1-x)^2}{(1+4x^2)^2}$, for $x \in (0, 1)$, respectively. Now $F(x) - G(x) = 0.0277$ for $x = 0.2$ and $F(x) - G(x) = -0.0143$ for $x = 0.7$, hence $X \not\geq^{ST} Y$. But $X \geq^{WDTE} Y$ as shown in Fig. 2.

5 Class of Lifetime Distributions

Various aging classes based on residual and past entropies and their weighted versions have been proposed by several authors. For some aging class results, we refer to Ebrahimi [5], Nanda and Paul [11] and Di Crescenzo and Longobardi [4]. In this section, we propose a new class of life distributions based on the WDTE defined by (4).

Definition 3 The non-negative random variable X is said to have decreasing WDTE in t_1, denoted by DWDTE(t_1) if and only if for fixed t_2, $S_X^w(t_1, t_2)$ is decreasing in t_1.

We notice that there exist several distributions which belong to this class. For example, the uniform random variable X with failure distribution $F(x) = x$, where $x \in (0, 1)$ belongs to DWDTE(t_1).

Example 6 Let X_1 and X_2 be the lifetimes of two components of a parallel system. Further, assume that these components are independently working. Then,

$Y = \max\{X_1, X_2\}$ denotes the lifetime of this system. Let the components of the systems are uniformly distributed on $(0, 1)$. Therefore, the distribution function of Y is $G(x) = x^2, 0 < x < 1$. Clearly, X_i's belongs to DWDTE(t_1). Also, one can easily check the system Y is belonging to DWDTE(t_1).

The following theorem provides a partial answer in finding whether the DWDTE(t_1) property of X are inherited by a transformation of X.

Theorem 3 *Let $\phi(x)$ be non-negative, strictly increasing and concave function such that $\phi(0) = 0$ and $\phi'(0) < 1$. Also let $S_X^{w,\phi}(\phi^{-1}(t_1), \phi^{-1}(t_2))$ be decreasing in t_1. Then $\phi(X)$ is also DWDTE (t_1).*

Proof From Theorem 1, we have

$$S_{\phi(X)}^w(t_1, t_2) = S_X^{w,\phi}(\phi^{-1}(t_1), \phi^{-1}(t_2))$$
$$+ E[(\phi(X) \ln \phi'(X))|\phi^{-1}(t_1) < X < \phi^{-1}(t_2)], \qquad (23)$$

when $\phi(x)$ is strictly increasing. Under the given assumptions it can be showed that the second term in the right hand side expression of (23) is decreasing in t_1. This completes the proof. □

In the following proposition, we obtain an upper bound of WDTE given by (4) for the distributions which belong to DWDTE(t_1) class.

Proposition 2 *Let X be an absolutely continuous non-negative random variable. Then for fixed t_2, if $S_X^w(t_1, t_2)$ is decreasing in t_1, we have*

$$S_X^w(t_1, t_2) \le \ln(h_1(t_1, t_2)^{-t_1} \exp\{\mu(t_1, t_2)\}).$$

Proof It is given that $S_X^w(t_1, t_2)$ is decreasing in t_1, that is, $\frac{\partial S_X^w(t_1, t_2)}{\partial t_1} \le 0$. Therefore, the result directly follows from after some simplifications. □

As an application of Proposition 2 we consider the following example.

Example 7 Let X be a random variable with probability density function $f(x|c) = cx^{c-1}, 0 < x < 1, c > 0$. Note that for $c = 2$, it belongs to DWDTE(t_1) class. Using Proposition 2, an upper bound of WDTE of the given distribution can be obtained as $\ln[(\frac{t_2^2 - t_1^2}{2t_1})^{t_1} \exp\{\frac{2}{3}(\frac{t_2^3 - t_1^3}{t_2^2 - t_1^2})\}].$

Our next theorem shows that under some conditions, the DWDTE (t_1) class is closed under affine transformation.

Theorem 4 *Let $X \in DWDTE(t_1)$. Also let $S_X(t_1, t_2)$ is decreasing in t_1. Define $Z = aX + b$, where $0 < a < 1$ and $b \ge 0$. Then $Z \in DWDTE(t_1)$.*

Proof The proof follows from Definition 3 and Lemma 1. Hence omitted. □

6 Conclusion

Sometimes, the random variables are truncated from both left and right sides. For the case of this type of random variables, the mathematical equation of the weighted Shannon entropy gets modified. In this chapter, we have considered and studied various properties of the weighted doubly truncated Shannon entropy. New order as well as new class of lifetime distributions are introduced. The characterization result of weighted doubly truncated Shannon's entropy is obtained. It is shown that this measure determine the distribution function uniquely. To illustrate the proposed results, several examples are considered.

References

1. Adami, C.: Information theory in molecular biology. Phys. Life Rev. **1**, 3–22 (2004)
2. Belzunce, F., Navarro, J., Ruiz, J.M., Aguila, Y.: Some results on residual entropy function. Metrika **59**, 147–161 (2004)
3. Cover, T.M., Thomas, J.A.: Elements of Information Theory. Wiley, New York (2006)
4. Di Crescenzo, A., Longobardi, M.: On weighted residual and past entropies. Scientiae Math. Jpn. **64**, 255–266 (2006)
5. Ebrahimi, N.: How to measure uncertainty about residual life time. Sankhya **58**, 48–57 (1996)
6. Guiasu, S.: Grouping data by using the weighted entropy. J. Stat. Plann. Inference **15**, 63–69 (1986)
7. Guiasu, R.C., Guiasu, S.: The Rich-Gini-Simpson quadratic index of biodiversity. Nat. Sci. **2**, 1130–1137 (2010)
8. Kelbert, M., Stuhl, I., Suhov, Y.: Weighted entropy and its use in computer science and beyond. In: Rykov, V., Singpurwalla, N., Zubkov, A. (eds) Analytical and Computational Methods in Probability Theory. ACMPT 2017. Lecture Notes in Computer Science, vol. 10684. Springer, Cham (2017)
9. Kundu, C.: On weighted measure of inaccuracy for doubly truncated random variables. Commun. Stat. Theory Methods **46**, 3135–3147 (2016)
10. Misagh, F., Yari, G.H.: On weighted interval entropy. Stat. Probab. Lett. **81**, 188–194 (2011)
11. Nanda, A.K., Paul, P.: Some properties of past entropy and their applications. Metrika **64**, 47–61 (2006)
12. Navarro, J., Ruiz, J.M.: Failure rate functios for doubly truncated random variables. IEEE Trans. Reliab. **45**, 685–690 (1996)
13. Sankaran, P.G., Sunoj, S.M.: Identification of models using failure rate and mean residual life of doubly truncated random variables. Stat. Papers **45**, 97–109 (2004)
14. Shaked, M., Shanthikumar, J.G.: Stochastic Orders. Springer, New York (2007)
15. Shannon, C.: The mathematical theory of communication. Bell Syst. Techn. J. **27**, 379–423 (1948)
16. Sunoj, S.M., Sankaran, P.G., Maya, S.S.: Characterizations of life distributions using conditional expectations of doubly (interval) truncated random variables. Commun. Stat. Theory Methods **38**, 1441–1452 (2009)
17. Yasaei Sekeh, S., Borzadaran, G.R.M., Roknabadi, A.H.R.: Some results based on weighted dynamic entropies. Rendiconti del Seminario Matematico **70**, 369–382 (2012)

Manning-Rosen Potential with Position Dependent Mass in Quantum Mechanics via LTM

S. Sur, B. Biswas, and S. Debnath

Abstract Schrödinger equation within the framework of position-dependent mass for Manning-Rosen potential is studied with the help of Laplace transform method combining with Point Canonical transformation. The general solutions are obtained via Pekeris approximation appropriate for potential analogues to Manning-Rosen potential. The bound state solutions are obtained in an analytical form.

Keywords Manning-Rosen potential · Pekeris approximation · Point canonical transformation · Laplace transform method

1 Introduction

In recent years explanation of Quantum Mechanical system with position dependent mass (PDM) is attracting more attention of researchers. Particles with PDM constitute useful models for the study of many physical problems such as quantum dots [1, 2], He clusters [3], metal clusters [4], the properties of heterojuncions, quantum wells [1], quantum liquids [5], semi-conductors [6] and the study of condensed matter physics of impurities in crystals [7].

It is important to create a model which contain potential concepts i.e. to describe the behaviour and interaction between atoms and particles. Potentials play an important role to describe the interaction between nuclei, nuclear particle and the structures of the diatomic molecules. Various potentials are used to analyse the nature of vibration of Quantum System such as pseudo-harmonic [8], modified Eckart plus Hylleraas [9], morse type [10], Wood-Saxon [11], Rosen-Morse [12], harmonic oscillator [13] specially on lower dimensions. The solutions are also crucial in quantum soluble systems. Methods involve in literature are Nikiforov-Uvarov method

S. Sur · S. Debnath
Department of Mathematics, Jadavpur University, Kolkata 700032, India

B. Biswas (✉)
Department of Mathematics, P.K.H.N. Mahavidyalaya, Howrah 711410, India
e-mail: bbiswas.math@gmail.com

© The Author(s), under exclusive license to Springer Nature Singapore Pte Ltd. 2023
P. Gyei-Kark et al. (eds.), *Engineering Mathematics and Computing*,
Studies in Computational Intelligence 1042,
https://doi.org/10.1007/978-981-19-2300-5_14

[14, 15], asymptotic iteration method [16], Point-Cannonical transformation [17], Lie algebraic method [18], super symmetry approach [19], factorization method [20] etc.

Here we use Manning Rosen potential (MRP) [21] to solve the Schrödinger equation with PDM. These type of potentials are used to describe the quark interactions [22] in particle and high energy physics, spectroscopy [23] in nuclear physics, binding energy and inclusive momentum distributions [24] in atomic physics, the inter and intra molecular interactions and atomic pair correlations in molecular physics/chemistry [25, 26].

Ikdhair [27] has considered a mass function $m = \frac{m_0}{(1-\delta e^{\frac{-\alpha(r-r_0)}{r_0}})^2}$, where m_0 is the rest mass and δ is a free parameter and $0 \leq \delta < 1$ to deal with the q-deformed morse potential. Here we use a similar mass function $m = \frac{m_0}{(1-qe^{-\alpha r})^2}$, where m_0 is the rest mass and α determines the inverse range of potential. To investigate the behaviour of MRP within the frame work of Schrödinger equation we use Pekeris approximation [28] and applying some simple constraints we can construct mass function such that the equation can be solved by LTA.

One of the most effective and different method to solve Schrödinger equation with PDM for a hyperbolic potential is Laplace transform method (LTM). LTM is an integral transform method which has been used by many authors [29–33]. Schrödinger had made the first attempt to solve the eigen function of hydrogen atom [29] by this method. It is a powerful method, and it helps us to solve second order differential equation by converting them into a simpler form whose solutions may be obtained easily. Thus LTM is a very effective method to solve the radial equations.

Our work is organized as follows:- In Sect. 2 we have converted the Schrödinger equation for Manning-Rosen Potential with position-dependent mass by Point Cannonical Transformation and Pekeris Approximation in the solvable form related to LTM. In Sect. 3 we discuss the solutions of Schrödinger Equation by using LTM and we get the bound state solution in terms of confluent hypergeometric function. In Sect. 4 we get the energy spectrum for the equation. Section 5 contains the concluding remark.

2 Schrödinger Equation with Position Dependent Mass for Manning-Rosen Potential

The most general form of Hamiltonian for the position dependent mass $m = m(r)$ is given by

$$H = \frac{1}{4(a+1)} \left\{ a \left[\frac{1}{m} \mathbf{P}^2 + \mathbf{P}^2 \frac{1}{m} \right] + m^\alpha \mathbf{P} m^\beta \mathbf{P} m^\gamma + m^\gamma \mathbf{P} m^\beta \mathbf{P} m^\alpha \right\} + \mathbf{V}(\mathbf{r}) \quad (1)$$

where $\mathbf{P}$ denotes the momentum operator and $\mathbf{V}(\mathbf{r})$ is an arbitrary potential. Also α, β, γ and a are the ambiguity parameters satisfying the constrain $\alpha + \beta + \gamma = -1$ and

r is the radial coordinate. The commutation relation by the differentiating properties of the momentum operator $\mathbf{P}$ is

$$[\mathbf{P}, f(r)] = \mathbf{P}f - f\mathbf{P} = -i\hbar\frac{df}{dr}\hat{r} \tag{2}$$

where $f(r)$ is the arbitrary function of the radial coordinate r. Using Eqs. (2), (1) turns into:

$$H = \frac{1}{2m}\mathbf{P}^2 + \frac{i\hbar}{2}\frac{1}{m^2}\frac{dm}{dr}\mathbf{P_r} + U_{\alpha,\beta,\gamma,a}(r) \tag{3}$$

where

$$U_{\alpha,\beta,\gamma,a}(r) = -\frac{\hbar^2}{4m^3(a+1)}\left[(\alpha+\gamma-a)m\frac{d^2m}{dr^2} + 2(a-\alpha-\gamma-\alpha\gamma)(\frac{dm}{dr})^2\right]$$
$$+ \mathbf{V(r)} \quad (4)$$

Imposing some conventional constrain on ambiguity parameters like $(\alpha+\gamma-a) = 0$ and $(a-\alpha-\gamma-\alpha\gamma) = 0$ with two possible solutions: (i) $\alpha = 0$ and $a = \gamma$ (ii) $a = \alpha$ and $\gamma = 0$, the effective potential can be reduced to $U_{\alpha,\beta,\gamma,a}(r) = V(r)$. Here, we are interested in this case where the Schrödinger equation yeilds:

$$-\frac{\hbar^2}{2m}\left[\nabla^2 - \frac{1}{m}\frac{dm}{dr}\nabla\right]\varphi(r) = [E - \mathbf{V(r)}]\varphi(r) \tag{5}$$

The wave function can be separated to the following form:

$$\varphi(r) = \frac{1}{r}\psi(r)Y(\theta,\phi). \tag{6}$$

Using Eq. (6) into Eq. (5), one can easily obtain the radial wave equation as:

$$\left[\frac{d^2}{dr^2} - \frac{1}{m}\frac{dm}{dr}\left(\frac{d}{dr} - \frac{1}{r}\right) - \frac{l(l+1)}{r^2}\right]\psi(r) = -\frac{2m}{\hbar^2}[E - \mathbf{V(r)}]\psi(r) \tag{7}$$

To eliminate first derivative term we use the transformation based on point canonical transformation method [34]:

$$\psi(r) = \sqrt{m(r)}\phi(r). \tag{8}$$

Substituting Eq. (8) into Eq. (7) one obtains

$$\frac{d^2\phi(r)}{dr^2} + \left[\frac{1}{2m}\frac{d^2m}{dr^2} - \frac{3}{4}\left(\frac{1}{m}\frac{dm}{dr}\right)^2 + \frac{1}{rm}\frac{dm}{dr} - \frac{l(l+1)}{r^2}\right]\phi(r)$$

$$= -\frac{2m}{\hbar^2}[E - \mathbf{V(r)}]\phi(r) \qquad (9)$$

The Manning-Rosen potential is

$$\mathbf{V(r)} = \frac{-V_1(1 + qe^{-\alpha r})}{(1 - qe^{-\alpha r})} + \frac{V_2 e^{-2\alpha r}}{(1 - qe^{-\alpha r})^2} \qquad (10)$$

The range of potential is determined by the dimensionless parameter α, q is the deformation parameter, and V_1 and V_2 are two general potential parameters.

The solution is mainly depending on replacing the orbital centrifugal term of singularity with the help of a suitable transformation for Pekeris approximation as

$$r \to \frac{r - r_0}{r} \qquad (11)$$

Here r_0 is the equilibrium position of molecules.

The centrifugal potential barrier term of Eq. (9) can be written using (11) as:

$$\frac{l(l+1)}{r^2} = \frac{l(l+1)}{r_0^2}\frac{1}{(1+r)^2}$$

$$\cong \frac{l(l+1)}{r_0^2}\left[C_0 + C_1\frac{e^{-\alpha r}}{(1 - qe^{\alpha r})} + C_2\frac{e^{-2\alpha r}}{(1 - qe^{\alpha r})^2}\right] \qquad (12)$$

where

$$C_0 = 1 - \frac{1}{\alpha}(1 - q)(3 + q) + \frac{3}{\alpha^2}(1 - q)^2; \quad C_1 = \frac{2}{\alpha}(1 - q)^2(2 + q) - \frac{6}{\alpha^2}(1 - q)^3;$$

$$C_2 = -\frac{1}{\alpha}(1 - q)^3(1 + q) + \frac{3}{\alpha^2}(1 - q)^4 \qquad (13)$$

taking $q \to 0$,

$$C_0 = 1 - \frac{3}{\alpha} + \frac{3}{\alpha^2}; \quad C_1 = \frac{4}{\alpha} - \frac{6}{\alpha^2}; \quad C_2 = \frac{3}{\alpha^2} - \frac{1}{\alpha} \qquad (14)$$

In the same way, we expand the term $\frac{1}{rm}\frac{dm}{dr}$ in Eq. (9) as follows:

$$\frac{1}{rm}\frac{dm}{dr} = \frac{1}{r_0 m}\frac{dm}{dr}\frac{1}{(1+r)} \cong \frac{1}{r_0 m}\left[B_0 + B_1\frac{e^{-\alpha r}}{(1 - qe^{\alpha r})} + B_2\frac{e^{-2\alpha r}}{(1 - qe^{\alpha r})^2}\right] \qquad (15)$$

$$B_0 = 1 - \frac{2(1+q)}{\alpha(1+2q)} + \frac{(1+q)}{\alpha^2(1+4q)} - \frac{(1+q)}{2\alpha(1+2q)};$$

$$B_1 = \frac{2}{\alpha(1+2q)} - \frac{2}{\alpha^2(1+4q)}; \ B_2 = -\frac{1}{2\alpha(1+2q)} + \frac{1}{\alpha^2(1+4q)} \quad (16)$$

taking $q \to 0$,

$$B_0 = 1 + \frac{1}{\alpha^2} - \frac{3}{2\alpha}; \ B_1 = \frac{2}{\alpha} - \frac{2}{\alpha^2}; \ B_2 = -\frac{1}{2\alpha} + \frac{1}{\alpha^2} \quad (17)$$

By substituting Eqs. (10)–(12) and Eq. (15) to Eq. (9) and taking,

$$y = e^{-\alpha r} \quad (18)$$

we have

$$\left[y^2 \frac{d^2}{dy^2} + y \frac{d}{dy} + \Lambda \right] \phi = \frac{l(l+1)}{\alpha^2 r_0^2} \left(C_0 + C_1 y + C_2 y^2 \right) \phi \quad (19)$$

where Λ is as follows:

$$\Lambda = \frac{y^2}{2m} \frac{d^2m}{dy^2} - \frac{3}{4} \frac{y^2}{m^2} (\frac{dm}{dy})^2$$

$$+ \frac{y}{m} \frac{dm}{dy} \left[\frac{1}{2} - \frac{1}{\alpha r_0} (B_0 + B_1 y + B_2 y^2) \right] \frac{m}{m_0} (P^2 + Qy + Ry^2) \quad (20)$$

where

$$P^2 = -\frac{2(E+V_1)m_0}{\alpha^2 \hbar^2}; \ Q = -\frac{2(2qV_1 - V_2)m_0}{\alpha^2 \hbar^2}; \ R = -\frac{4(-qV_2 + V_1 q^2)m_0}{\alpha^2 \hbar^2} \quad (21)$$

Using the following effective mass distribution:

$$m = \frac{m_0}{(1-qy)^2} \quad (22)$$

where $m \to m_0$ when $q \to 0$ and m_0 is the rest mass.

Considering mass distribution given in Eq. (22) we can convert Λ as follows:

$$\Lambda = D_0 + D_1 y + D_2 y^2 \quad (23)$$

where

$$D_0 = -P^2 \ ; \ D_1 = -Q \ ; \ D_2 = -R \quad (24)$$

Substituting Eq. (23) in Eq. (19) we have

$$y^2 \frac{d^2\phi}{dy^2} + y \frac{d\phi}{dy} - [\mu^2 - \nu^2 + \eta^2 y^2]\phi = 0 \tag{25}$$

where

$$-\mu^2 = D_0 - \frac{l(l+1)}{\alpha^2 r_0^2} C_0; \ \nu^2 = D_1 - \frac{l(l+1)}{\alpha^2 r_0^2} C_1; \ -\eta^2 = D_2 - \frac{l(l+1)}{\alpha^2 r_0^2} C_2 \tag{26}$$

3 Bound State Solution

The Schrödinger equation with position dependent mass for Manning Rosen Potential turns into Eq. (25) by using the mass distribution given in (22). We consider the function ϕ as follows to get finite solutions for large values of y:

$$\phi(y) = y^{-\mu} f(y) \tag{27}$$

then Eq. (25) turns into,

$$y^2 \frac{d^2 f}{dy^2} - (2\mu - 1) y \frac{df}{dy} + y(\nu^2 - \eta^2 y^2) f(y) = 0 \tag{28}$$

By using Laplace Transform [35] Eq. (28) can be transformed into,

$$(s^2 - \eta^2) \frac{dF}{ds} + [(2\mu + 1)s - \nu^2] F(s) = 0 \tag{29}$$

Therefore the equation given in Eq. (28) transform into a first order differential equation given in (29) and the solutions are in the form

$$F(s) = N(y + \eta)^{-(2\mu+1)} \left(\frac{y - \eta}{y + \eta} \right)^{\frac{\nu^2}{2\eta} - \frac{(2\mu+1)}{2}} \tag{30}$$

where N is a constant. To have a well-behaved wave function we must impose the condition,

$$\frac{\nu^2}{\eta} - (2\mu + 1) = 2n, \quad \text{where,} \ n = 0, \pm 1, \pm 2, \ldots \tag{31}$$

Here n is positive or negative according as the magnitude of $\frac{\nu^2}{\eta}$ is greater or smaller than the magnitude of $(2\mu + 1)$. To get the finite solution for large y the parameter μ needs to be large enough as per Eq. (27). We can get positive value of μ for sufficiently large values of $\frac{\nu^2}{\eta}$ and for positive or negative values of n. Again we can get positive value of μ for smaller values of $\frac{\nu^2}{\eta}$, which is only possible for negative values of n.

To apply inverse Laplace Transform into (30) we expand it in power series as:

$$F(s) =$$
$$\sum_{m=0}^{\infty} \frac{(2\eta)^m}{m!} \left\{ \begin{array}{l} N_+ \frac{(-1)^m n!}{(n-m)!} (s+\eta)^{-(2\mu^+ + 1 + m)}, n > 0 \\ N_- \frac{(-1)^{2m}(n+m-1)!}{(n-1)!} (s+\eta)^{-(2\mu^- + 1 + m)}, n < 0 \end{array} \right. \tag{32}$$

where $N_\pm$ are two integrating constants and $\mu^\pm$ corresponds to positive or negative values of n as given in Eq. (31). Now applying inverse Laplace Transform to Eq. (32) we have,

$$f(y) = \sum_{m=0}^{\infty} \frac{(2\eta)^m y^m e^{-\eta y}}{m!} \left\{ \begin{array}{l} N_+ \frac{(-1)^m n! y^{2\mu^+}}{(n-m)! \Gamma(2\mu^+ + 1 + m)}, n > 0 \\ N_- \frac{(n+m-1)! y^{2\mu^-}}{(n-1)! \Gamma(2\mu^- + 1 + m)}, n < 0 \end{array} \right. \tag{33}$$

By the series expansion of the confluent hypergeometric function Eq. (33) becomes

$$f(y) = \left\{ \begin{array}{l} N_+ y^{2\mu^+} e^{-\eta y} {}_1F_1(-\eta; 2\mu^+ + 1, 2\eta y), n > 0 \\ N_- y^{2\mu^-} e^{-\eta y} {}_1F_1(-\eta; 2\mu^- + 1, 2\eta y), n < 0 \end{array} \right. \tag{34}$$

Thus using Eqs. (27) and (34) in Eq. (8) the solution becomes

$$\psi(y) = N\sqrt{m(y)} y^{2\mu} e^{-\eta y} {}_1F_1(-\eta; 2\mu + 1, 2\eta y) \tag{35}$$

where N is normalizing constant and the mass function $m(y)$ is given in Eq. (22). The parameter μ is obtained from Eq. (31) as,

$$\mu = \frac{\nu^2}{2\eta} - n - \frac{1}{2}. \tag{36}$$

where ν and η are given in Eq. (26).

4 Energy Spectrum

We get the bound state solution for position dependent Schrödinger equation with the Manning Rosen Potential via Laplace Transform Approach. We get the energy eigen function and now we will find out the energy eigen value for that function.

Comparing Eqs. (24) and (21), we get value of D_0. Using this value of D_0 and the value of μ (given by Eq. (36)) into Eq. (26), we get

$$E = \frac{\alpha^2 \hbar^2}{2m_0}\left[\frac{l(l+1)}{\alpha^2}C_0 - \left\{n + \frac{1}{2} - \frac{1}{2\eta}(D_1 - \frac{l(l+1)}{\alpha^2 r_0^2}C_1)\right\}^2\right] - V_1 \quad (37)$$

where

$$\eta = [\frac{l(l+1)}{\alpha^2 r_0^2}C_2 - D_2]^{\frac{1}{2}} \quad (38)$$

5   Conclusion

In this article, we have obtained the bound state solutions of the Schrödinger equation for the Manning-Rosen potential with position dependent mass. The energy equation have been obtained with the help of LTM, a powerful method to solve second order differential equation via conversion of it into a more simpler one. The eigen function have been obtained in terms of confluent hypergeometric function. The outcome of this article is applicable for any kind of mass function for which one can set the condition $\Lambda = D_0 + D_1 y + D_2 y^2$. One schematic graphical representations for potential function, mass distribution and energy spectrum are presented in Fig. 1, Fig. 2 and Fig. 3 respectively.

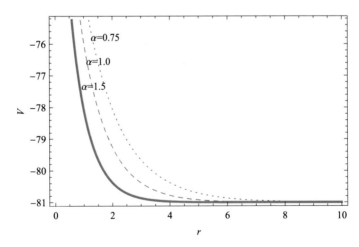

Fig. 1 Manning-Rosen potential versus r with $V_2 = \frac{1}{3}V_1$ and $V_1 = 81\ eV$ for different values of α

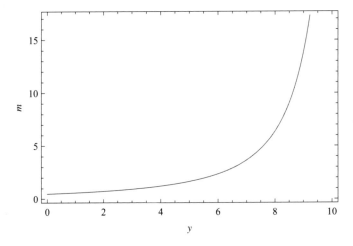

Fig. 2 Graphical representation of mass distribution for $q = 0.9$

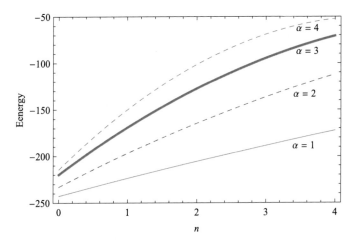

Fig. 3 Energy spectrum for different values of α

Acknowledgements The authors would like to thank the kind referee for positive and invaluable suggestions which have improved the manuscript greatly. And a special thank to Dr. Dipak Kumar Jana for his continuous support to qualitative enrichment of this article.

References

1. Harison, P.: Quantum Wells Wires and Dots. Wiley, New York (2000)
2. Peter, A.J.: The effect of position dependent effective mass of hydrogenic impurities in parabolic GaAs/GaA/As quantum dots in a strong magnetic field. Int. J. Mod. Phys. B **23**(26), 5109 (2009)

3. Barranco, M., Pi, M., Gatica, S.M., Hernandez, E.S., Navarro, J.: Structure and energetics of mixed ^{4}He-^{3}He drops. Phys. Rev. B **56**, 8997 (1997)

4. Puente, A., Serra, L., Casas, M.: Dipole excitation of Na clusters with a non-local energy density functional. Z. Phys. D **31**, 283 (1994)

5. Arias de Saavedra, F., Boronat, J., Polls, A., Fabrocini, A.: Effective mass of one ^{4}He atom in liquid ^{3}He. Phys. Rev. B **50**, 4248 (1994)

6. Young, K.: Position-dependent effective mass for inhomogeneous semiconductors. Phys. Rev. B **39**, 13434 (1989)

7. Geller, M.R., Kohn, W.: Quantum Mechanics of electrons in crystals with graded composition. Phys. Rev. Lett. **70**(20), 3103 (1993)

8. Aydogdu, O., Server, R.: Solution of the Dirac equation for pseudoharmonic potential by using the Nikiforov-Uvarov method. Phys. Scr. **80**, 015001 (2009)

9. Sur, S., Debnath, S.: Relativistic Klein-Gordan Equation with Position Dependent Mass for q-deformed Modifed Eckart plus Hylleraas potential. EJTP **14**, No. 37, 79–90 (2018)

10. Berkdemir, C., Han, J.: Any l-state solutions of the Morse potential through the Pekeris approximation and Nikiforov-Uvarov method. Chem. Phys. Lett. **409**, 203 (2005)

11. Biswas, B., Debnath, S.: Exact solutions of the Klein-Gordon equation for the mass-dependent generalized Woods-Saxon potential. Bull. Cal. Math. Soc. **104**(5), 481–490 (2012)

12. Debnath, S., Biswas, B.: Analytical solutions of the Klein-Gordon equation for Rosen-Morse potential via asymptotic iteration method. EJTP **9**(26), 191–198 (2012)

13. Guo, J.Y., Fang, X.Z., Xu, F.X.: Pseudospin symmetry in the relativistic harmonic oscillator. Nucl. Phys. A **757**, 411 (2005)

14. Nikiforov, A.F., Uvarov, V.B.: Special Functions of Mathematical Physics. Birkhäuser, Basel (1988)

15. Ikot, A.N., Udoimuk, A.B., Akpabio, L.E.: Bound states solution of Klein-Gordon equation with type—I equal vector and Scalar Poschl-Teller potential for Arbitray—State. Ameri. J. Scientific Indus. Res. **2**(2), 179 (2011)

16. Zhang, L.H., Li, X.P., Jia, C.S.: Analytical approximation to the solution of the Dirac equation with the Eckart Potential including the Spin-Orbit coupling term. Phys. Lett. A **372**, 2201 (2008)

17. Alhaidari, A.D.: Nonrelativistic Green's function for systems with Position-Dependent mass. Int. J. Theor. Phys. **42**, 2999 (2003)

18. Dong, S.H.: Wave Equations in Higher Dimensions. Springer, New York (2011)

19. Plastino, A.R., Rigo, A., Casas, M., Garcias, F., PLastine, A.: Super symmertric approach to quantum systems with position dependent effective mass. Phys. Rev. A **60**(6), 4318 (1999)

20. Dong, S.H.: Factorization Method in Quantum Mechanics. Springer Science and Business Media (2007)

21. Manning, M.F., Rosen, N.: A potential function for the vibration of diatomic molecules. Phys. Rev. **44**, 953 (1932)

22. Khelashvili, A.A.: Radial quasipotential equation for a fermion and antifermion and infinitely rising central potentials. Theor. Math. Phys. **51**, 447 (1982)

23. Ginocchio, J.N., Leviatan, A.: On the relativistic foundations of pseudospin symmetry in nuclei. Phys. Lett. B **425**, 1 (1998)

24. Zhukov, M.V., Danilin, B.V., Fedorov, D.V., Vaagen, J.S., Gareev, F.A., Bang, J.: Calculation of ^{11}Li in the framework of a three-body model with simple central potentials. Phys. Lett. B **265**, 19 (1991)

25. Arda, A., Sever, R.: Approximate analytical solutions of a two-term diatomic molecular potential with centrifugal barrier. J. Math. Chem. (2012). https://doi.org/10.1007/s10910-012-0011-0

26. Rahan, A., Stillinger, F.H., Lemberg, H.L.: Study of a central force model for liquid water by molecular dynamics. J. Chem. Phys. **63**, 5223 (1975)

27. IKdhair, S.M.: Rotation and vibration of diatomic molecule in the spatially dependent mass Schrödinger equation with generalised q-deformed Morse Potential. Chem. Phys. **361**, 9 (2009)

28. Ferreira, F.J.S., Prudente, F.V.: Pekeris approximation—another perspective. Phys. Lett. A **377**, 3027–3032 (2013)

29. Schrödinger, E.: Quantisierung als Eigenwertproblem. Ann. Physik **384**, 361 (1926)

30. Swainson, R.A., Drake, G.W.F.: A unified treatment of the non-relativistic and relativistic hydrogen atom I: the wavefunctions. J. Phys. A: Math. Gen. **24**, 79 (1991)

31. Chen, G.: The exact solutions of the Schrödinger equation with the Morse potential via Laplace transforms. Phys. Lett. A **326**, 55 (2004)

32. Chen, G.: Exact solutions of N-dimensional harmonic oscillator via Laplace transformation. Chin. Phys. **14**, 1075 (2005)

33. Arda, A., Sever, R.: Exact solutions of the Schrödinger equation via Laplace transform approach: pseudoharmonic potential and Mie-type potentials. J. Math. Chem. **50**, 971 (2012)

34. Bagchi, B., Gorain, P.S., Quense, C.: Morse potential and its relationship with the Coulomb in a PDM background. Mod. Phys. Lett. A **60**(6), 4318 (1999)

35. Spiegel, M.R.: Schaum's Outline of Theory and Problems of Laplace Transforms. Schaum Publishing Co., NY (1965)

Physics Simulation Based Approach to Node Clustering

Kapil Kalra, K. N. Nikhila, and Sujit Kumar Chakrabarti

Abstract We present a novel method of node clustering that is based on carrying out a physical simulation. We treat nodes of a graph as point-sized unit mass particles that interact with each other as well as the space (multi-dimensional) that they are present in through certain defined physical forces. As the configuration of the system evolves during the simulation, *similar* nodes coalesce while the *dissimilar* nodes separate out, thus allowing node clusters to emerge. We have experimented with this idea on graphs with up to 300 nodes and have found it to work well. Doing so also allowed us to solve problems of network community detection by utilizing existing density based clustering algorithms, which otherwise would not be possible.

Keywords Node clustering · Network community detection · Physics simulation · Experimentation

The authors thank Dr. Manish Gupta (Professor, IIIT-B and Director, Google Research India) for his technical inputs as well as for providing partial financial support to the project through his Infosys Foundation chair professorship fund. The authors also acknowledge the financial support from Machine Intelligence and Robotics (MINRO) Center at IIIT Bangalore through a grant from the Department of ITBT&ST, Government of Karnataka.

K. Kalra (✉)
Birla Institute of Technology and Science, Pilani, K K Birla Goa Campus, Goa 403726, India
e-mail: kapilkalra04@gmail.com; f20140707g@alumni.bits-pilani.ac.in

K. N. Nikhila · S. K. Chakrabarti
International Institute of Information Technology, Bangalore 560100, India
e-mail: nikhila.kn@iiitb.org

S. K. Chakrabarti
e-mail: sujitkc@iiitb.ac.in

1 Introduction

Node clustering [1] or Network Community Detection [7] is a well-known computing problem and has been studied in a variety of contexts. Several software engineering problems can be modelled as a node clustering problem [1, 8]. Given an undirected simple graph $G < V, E >$ [5], the problem of node clustering or network community detection, involves the assignment of each node $n \in V$ of G, to a set $C \in \mathcal{C}$, such that all the nodes in a given set (or cluster) C are 'similar' to every other node in C. This formulation assumes a prior understanding of the notion of similarity measure that any node clustering algorithm would employ [7, 15]. This similarity measure would typically be reflected in the form of edge weights between the nodes.

Even though numerous node clustering approaches exist in literature [1, 7], most of these approaches find node clusters/communities from a given undirected graph/network in the following way. First, a scoring function, which quantifies the relation between communities and densely (in terms of either the number of edges or the value of edge weights) linked sets of nodes, is selected. Thereafter, various sets of nodes are found, such that they all result in a high or a low value for the chosen scoring function [15]. This has known to be a daunting task for two main reasons. Firstly, there exist numerous ways to define a scoring function, as it is sensitive to the structural definition of a densely linked sets of nodes [2, 9]. And secondly, finding clusters through maximization or minimization of the score function for each cluster is a NP-Hard Optimization Problem [12]. Therefore, the fundamental problem of node clustering (and all its variants) is intractable and consequently, all proposed methods are heuristics of some form [15] with their effectiveness depending on the scenarios in which they are used.

In this paper, we propose a novel physics simulation based approach to node clustering, in an effort to overcome the above-mentioned problems by adopting a radically different approach. The basic idea is to model the nodes in the graph as point particles in a multi-dimensional space. Attractive, repulsive and dissipative forces are effective on these particles, which consequentially define and control their movement in this space. Our hypothesis is that the forces acting on these particles will eventually cause similar particles to cluster together, while pushing individual clusters away from each other. We have conducted some preliminary tests to validate the hypothesis. We have gotten strong evidences that the physics based simulation can be an effective approach to node clustering. Furthermore, on applying existing density based clustering algorithms [10] on the nodes' final positional values; obtained after subjecting them to a simulation of finite length, we observed that definite node clusters or network communities can be easily obtained.

Section 2 presents our model universe and the basic simulation algorithm. Section 3 presents the experimental results that provide evidence that this approach could be an effective alternative to existing node clustering algorithms. In Sect. 4, the various open problems associated with our approach have been presented while the steps we plan to take with reference to them are listed in Sect. 5. Section 6 concludes the paper.

2 Simulation Model

Our aim is to completely rethink the approach to node clustering. Instead of thinking it as an optimization problem [7] we approach the problem as an *N-body simulation* [13].

In our model, we consider each data point present in the data set, as a node of a graph that is present in a *D* dimensional space. We consider the edges of the graph, which connect these nodes to one another, as a measure of similarity between the corresponding nodes (data points). This means that any data set can be mapped out onto a weighted undirected simple graph. The nodes of such a graph can now be re-imagined as point-sized unit mass particles. Doing so allows us to reduce the problem of running an undefined physical simulation over a data set to just running a standard *N-body simulation*. Therefore, we draw inspiration from *N-body simulations* to build our universe. Our universe contains 2 kinds of interactions that have been explained below.

2.1 Particle-Particle Interaction

Interactions between the nodes are modelled with two ranged forces, namely attraction and repulsion.

2.1.1 Attraction

Each and every node attracts all the other nodes present in the *D* dimensional space, but the value of this attractive force between any two nodes is dependent on the edge weight present between them. All edge weights (E_w) in the graph $\in [0, 1]$, where $E_w = 0$ indicates the absence of an edge between the two concerned nodes and $E_w = 1$ indicates that the two concerned nodes attract each other with the maximum possible force. The force of attraction experienced by node i due to node j is analogous to the gravitational force [14] and the electrostatic force of attraction [14] and hence it has been defined as

$$\frac{k_a * E_{w(i,j)}}{(|\vec{x}_i - \vec{x}_j|^m + z)} * \frac{\vec{x}_j - \vec{x}_i}{|\vec{x}_i - \vec{x}_j|} \tag{1}$$

where k_a is the attractive force constant and $E_{w(i,j)}$ is the edge weight between the nodes i and j. Variables $\vec{x}_i, \vec{x}_j$ are the position vectors of the concerned nodes. A constant z has been added in the denominator to soften the singularity caused when $\vec{x}_i$ and $\vec{x}_j$ are nearly coinciding. While m here is any positive real value number representing the order of this ranged force.

2.1.2 Repulsion

Each and every node repels all the other nodes present in the D dimensional space. However, the force of repulsion between any two nodes is independent of the edge weight present between them. The force of repulsion experienced by node i due to node j is analogous to the electrostatic force of repulsion [14] and hence it has been defined as

$$\frac{k_r}{(|\vec{x}_i - \vec{x}_j|^m + z)} * \frac{\vec{x}_i - \vec{x}_j}{|\vec{x}_i - \vec{x}_j|} \tag{2}$$

where k_r is the repulsive force constant and the variables $\vec{x}_i$, $\vec{x}_j$, z and m are same as the ones defined in the Force of Attraction.

2.2 Particle-Medium Interaction

Through the virtue of friction, each and every node interacts with the medium it is present in. This includes *kinetic* and *static* friction [14]. We have considered *kinetic friction* in our universe for two reasons. Firstly, to counter the problem of fast moving nodes that gain very high velocities due to singularities caused by the ranged forces. And secondly, to help in achieving clustering faster by dissipating a part of the nodes' energies. The *kinetic frictional force* experienced by node i is inspired from the *Stoke's law* [6] and has hence been defined as

$$- k_f * \vec{v}_i \tag{3}$$

Here, k_f is the frictional force constant and $\vec{v}_i$ is the velocity vector of node i.

Whereas, the *static frictional force* provides a sense of inertia to the nodes at rest and also allows for a situation where the nodes can come to rest momentarily. The static frictional force has a fixed value that is pre-determined before the simulation runs.

2.3 Simulation Algorithm

The pseudo-code for the simulation algorithm is given in Algorithm 3 which utilizes Algorithm 1 and 2 and can be described as follows—First, we initialize the position, velocity and acceleration vectors of each of the nodes present in the graph, using the INITIALIZE procedure that is called within the procedure, NBODYSIMULATION. Then, the NBODYSIMULATION procedure uses the SIMULATE procedure to update position, velocity and acceleration vectors of each of the nodes for a given time step in the simulation. The SIMULATE procedure formulates the two single order differential equations

$$\frac{d\vec{v}}{dt} = \vec{a} \tag{4}$$

$$\frac{d\vec{x}}{dt} = \vec{v} \tag{5}$$

that alone govern the movements of the nodes. Therefore, NBODYSIMULATION calls SIMULATE at every time step to carry out the simulation movements of all the N nodes in G through the D dimensional space. It is important to note that we are dealing with unit mass particles, therefore the net force experienced by a node is also its net acceleration. Moreover, we do not account for any collisions, as such an event is very rare given the size of the nodes and for the fact that collisions only add to the complexity of the algorithm and nothing to help aid the obtainment of node clusters.

Algorithm 1 Acceleration Algorithm

1: **procedure** ACCELERATION(G, n_i)

2:     $\mathbf{f_{a_i}} = \sum_{j \neq i}^{N} \frac{k_a * E_{w(i,j)}}{(|\mathbf{x_j} - \mathbf{x_i}|^m + z)} * \frac{\mathbf{x_i} - \mathbf{x_j}}{|\mathbf{x_i} - \mathbf{x_j}|}$

3:     $\mathbf{f_{r_i}} = \sum_{j \neq i}^{N} \frac{k_r}{(|\mathbf{x_i} - \mathbf{x_j}|^m + z)} * \frac{\mathbf{x_i} - \mathbf{x_j}}{|\mathbf{x_i} - \mathbf{x_j}|}$

4:     $\mathbf{f_{k_i}} = -k_f * \mathbf{v_i}$

5:     **if** $|\mathbf{v_i}| \neq 0$ **then**

6:        $\mathbf{a_i} = \mathbf{f_{a_i}} + \mathbf{f_{r_i}} + \mathbf{f_{k_i}}$

7:     **else**

8:        $\mathbf{a_i} = \mathbf{f_{a_i}} + \mathbf{f_{r_i}} - \mathbf{f_s}$

9:     **end if**

10:    **return** $\mathbf{a_i}$

11: **end procedure**

Algorithm 2 Initialization Algorithm

1: **procedure** INITIALIZE(G)

2:     $\mathbf{x_{i,0}}$ = D-dimensional random vector, whose projection onto any of its axis $\in [-b, b]$

3:     $\mathbf{v_{i,0}}$ = D-dimensional zero vector

4:     Repeat Steps 2 and 3 for each node $n_i \in V(G)$.

5:     $\mathbf{a_{i,0}} \leftarrow$ ACCELERATION(G, i)

6:     Repeat Step 5 for each node $n_i \in V(G)$.

7: **end procedure**

Algorithm 3 Simulation Algorithm

1: **procedure** SIMULATE(G, t)
2:     $x_{i,t} = x_{i,t-1} + (v_{i,t-1} * \delta t)$
3:     **if** $|v_{i,t-1} + (a_{i,t-1} * \delta t)| > v_c$ **then**
4:         $v_{i,t} = v_{i,t-1} + (a_{i,t-1} * \delta t)$
5:     **else**
6:         $v_{i,t} = D$ dimensional zero vector
7:     **end if**
8:     Repeat Steps 2 to 7 for each node $n_i \in V(G)$
9:     $a_{i,t} \leftarrow$ ACCELERATION(G, n_i)
10:     Repeat Step 9 for each node $n_i \in V(G)$
11: **end procedure**
12: **procedure** NBODYSIMULATION(G)
13:     For each node, $n_i \in V(G) \leftarrow$ INTIALIZE(G)
14:     SIMULATE(G, t)
15:     Repeat Step 15 for each time step t in T.
16: **end procedure**

Variables and notations used in the Algorithm 1, 2 and 3:

G: Weighted undirected simple graph

D: Dimensionality of the Space

m: Order of the ranged forces

z: Softening constant

δt: Size of the Time Step

T: Total number of time steps in a simulation run

n_i: ith node

$\vec{f}_{a_i}/\vec{f}_{r_i}/\vec{f}_{k_i}$: Net force of attraction/repulsion/kinetic-friction experienced by the ith node

$k_a/k_r/k_f$: Attractive/Repulsive/Frictional Force Constants

$\vec{x}_i/\vec{v}_i/\vec{a}_i$: Position/Velocity/Acceleration vector of the ith node

$\vec{x}_{i,0}/\vec{v}_{i,0}/\vec{a}_{i,0}$: Initial position/velocity/acceleration vector of the ith node

$\vec{x}_{i,t}/\vec{v}_{i,t}/\vec{a}_{i,t}$: Position/Velocity/Acceleration vector of the ith node at time t

$E_{w(i,j)}$: Edge weight between nodes i and j

b: Initialization Boundary

f_s: Static Frictional Force

v_c: Cut-Off Velocity.

3 Experimental Observations

We conducted our experiments with the following questions in mind about our simulation based approach for clustering:

1. Does it lead to high quality clustering?
2. Is it a viable pre-processing step to density based clustering algorithms for the task of network community detection?

In order to answer these two questions, we performed a set of experiments. In these experiments we dealt with small-sized graphs but nonetheless, successfully achieved the desired high-quality clustering. However, these experiments though small in size were crucial in understanding the motion of each data point before and after clustering. In cases of both the small-scale and large-scale graphs, we essentially aim to carry out an N-body simulation, with the number of data points (N) being the only difference. Therefore, we can directly apply the insights gained from these experiments with small graphs to experiments with large and complex graphs.

In all the experiments, we ran an N-body simulation over a pre-defined graph for a fixed number of time steps and for a unique set of values of $\{k_a, k_r, k_f\}$. At the end of each simulation run—for every unique value of $\{k_a, k_r, k_f\}$—the resultant clusters were obtained through a density-based clustering algorithm like HDBSCAN [4]. Across all of the experiments the following parameters were kept constant:

1. D: The dimensionality of the system = 2D
2. b: The initialization boundary = 5 SI Units
3. δt: The size of the time step used in solving the differential equations of motion for each node = 0.1 SI Units.
4. T: Total number of time-steps = 6000
5. Range of values for k_a, k_r and $k_f = k_a, k_r, k_f \in \{0, 1, 2....10\}$ while $k_a > k_r$
6. f_s: The force of static friction = 0.001 SI units
7. v_c: The cutoff velocity = 0.0001 SI units
8. z: The value of softening constants = 0.1 SI Units
9. m: The order of the ranged forces = 2.

3.1 Experiment 1

The predefined graph and the expected outcome:

– Number of nodes: 2
– Starting positions of N_1 and N_2: (-5,0) and (5,0) respectively
– Edge-Weight: $E_{w(1,2)} = 1$
– Desired Outcome: N_1 and N_2 should form a cluster

The results observed for this experiment point out a critical piece of information regarding the possible values of k_f. It is observed that for simulation runs with $k_f = 0$ the desired clustering was never achieved. This is due to the fact that when any two nodes start approaching each other, we start edging towards the singularity caused by the continuously decreasing value of the distance separating the two nodes. Hence, the force of attraction becomes quite large—approaches infinity—as the nodes N_1 and N_2 move closer and closer to one another. It is important to note that this singularity also causes the repulsive force between them to be quite large, but as we consider $k_a > k_r$, it eventually results in nothing but a reduced force of attraction between them, hence we can consider the force of attraction as the only force that acts on N_1 and N_2. Therefore, the increasing large attraction between nodes N_1 and N_2 causes them to possess a very large but also increasing approach velocity. As a result, the nodes end-up crossing each other after a certain time. They do so with a large enough velocity that the deceleration due to net attraction after crossing, is not enough to stop the nodes from moving apart until a very long time. This causes the nodes to move quite far away from each other, before they begin to approach each other again. Though the nodes do start moving towards each other again, the situation is nothing but a state similar to the start of the simulation. Hereby, once again the nodes end up crossing each other with very high velocities after some time. This comes down to the fact that the reduction of the possessed velocities—after the two nodes cross each other—is quite slow in the absence of any friction. This process continues to take place as there is no dissipative force in the system and therefore the nodes continue to execute a periodic oscillatory motion about the origin along the x-axis, which is their line of approach, until the simulation is stopped. We can conclude that the two nodes perform an oscillatory motion with a very large magnitude of periodic displacement about their barycenter (location of the system's center of mass), which is the origin in this case. This system of N_1 and N_2 executing an oscillatory motion about the origin cannot be termed as a cluster like system because at certain time steps the nodes might be very close to each other but at time steps shortly after these, they would be observed to be quite far apart. Hence, we can conclude that the desired clustering is not achieved.

But if we have $k_f \neq 0$, then the nodes would decelerate at a faster rate than before, as the kinetic frictional force experienced by a node is directly proportional to its velocity. So when the nodes N_1 and N_2 do cross each other, the resistance offered by the kinetic frictional force would then allow the net force of attraction between the two to overcome the problem of fast escaping nodes. The effect of this would be that the nodes now will approach each other much sooner after the crossing, leading to periodic systems with much smaller oscillations. This is exactly what is obtained for simulations runs with $k_f \neq 0$. Many sets of values of $\{k_f, k_r, k_a\}$ lead to clustering. It is important to note that we can only conclude that the nodes have actually formed a cluster by studying their movement in the 2D space and not by observing where N_1 and N_2 end up in the space after the simulation has ended. Due to the symmetry of our considered system, we just need to observe the movement of the nodes along the x-axis. As there is kinetic friction present, the reduction in the velocities of the nodes after they have crossed each other is much faster than before. This results in the nodes approaching each other again, much faster after crossing each other. Hence, now N_1 and N_2 oscillate about their barycenter with a smaller periodic displacement as compared to cases with $k_f = 0$. The smaller the periodic displacement, the more close the nodes remain to each other, the better is the clustering picture and hence better is the quality of clustering.

All this can be observed in Fig. 1 where the top plot shows the starting and ending positions of N_1 and N_2. The green 'cross' symbol represents the starting position of N_2(-5,0) and the blue 'cross' symbol represents the starting position of N_1(5,0) while the green and blue 'diamond' symbols represent the ending positions of N_2 and N_1 respectively. The middle plot tracks the motion of N_1 along the x-axis with respect to time while the bottom plot tracks the motion of N_2 along the x-axis with respect to time. We can see that about 1800 time steps into the simulation, the time when the two nodes get quite close to each other, they enter the above described oscillatory system of periodic motion about their cluster's barycenter.

To summarize, each of the simulation run resulted in a periodic motion of the two nodes about their barycenter i.e. the origin. The only difference between these runs was the magnitude of the periodic displacements along the x-axis. Interestingly, a pattern was observed in all of the simulation runs. It was observed that the later the two nodes entered the periodic system of performing oscillations, the smaller the periodic movements about the barycenter were, resulting in a better and richer clustering. In Figs. 1, 2 and 3 we can see this occurring. The average magnitude of periodic displacements is the least for Fig. 3 where the nodes enter the stable periodic system in $t > 5000$ time steps, then for Fig. 2 where the nodes enter the stable periodic system in $t < 3500$ time steps and lastly for Fig. 1 where the nodes enter the stable periodic system in $t < 2000$ time steps. This pattern observed is also observed in standard control-system problems, and hence strengthens our claim of achieving stable clusters through a physical simulation for a given data set.

3.2 Experiment 2

The predefined graph and expected outcome:

– Number of nodes: 3

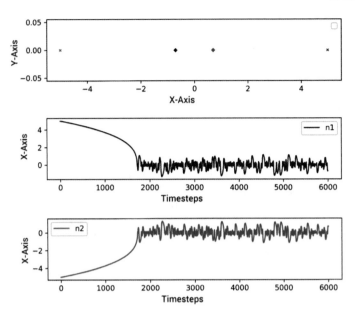

Fig. 1 Experiment 1 run with $k_f = 1, k_r = 1, k_a = 2$

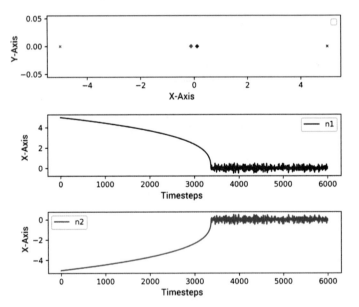

Fig. 2 Experiment 1 run with $k_f = 2, k_r = 0, k_a = 1$

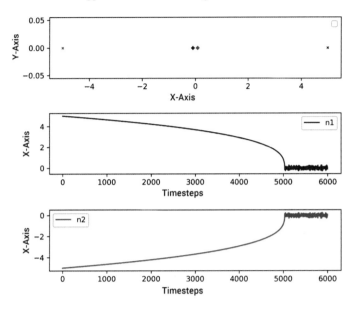

Fig. 3 Experiment 1 run with $k_f = 3, k_r = 4, k_a = 5$

- Starting positions of N_1, N_2 and N_3: (5,0), $(-5\sin(30),5\cos(30))$ and $(-5\sin(30), 5\cos(30))$ respectively
- Edge-Weights: $E_{w(1,2)} = 1$, $E_{w(2,3)} = 1$, $E_{w(3,1)} = 1$
- Desired Outcome: N_1, N_2 and N_3 should form a cluster.

We obtained good clustering results for simulation runs with $k_f \neq 0$ while oscillations of large magnitudes were observed for simulation runs with $k_f = 0$. As was observed in Experiment 1, the nodes performed an oscillatory motion about their barycenter—the origin—along their direction of approach to the barycenter. For N_1, the approach direction to the barycenter lies only along the x-axis hence the periodic motion of N_1 was limited in the x-axis only. As the approach direction of N_2 and N_3 lied along both x-axis and y-axis, we observed the two nodes to have a periodic motion also along both the axial directions, as can be seen in Fig. 4. The pattern of obtaining a better quality clustering picture for systems with later occurrence of periodic motions of its nodes, holds true here as well. This can also be seen in runs with $k_f = 0$. As the nodes enter into periodic motion the earliest in simulations with $k_f = 0$ due to zero resistance offered by the medium, the nodes in such cases have the largest magnitude of average periodic displacements about their system's barycenter. The top most plot in Fig. 4 shows the initial and the final positions of the nodes. The 'cross' symbol represents the starting positions, while the 'diamond' symbol represents the ending positions of the nodes. The colour scheme is blue for N_1, green for N_2 and red for N_3. The next part of Fig. 4 is divided into two halves, where the LHS contains the plots capturing the movement of the nodes along the x-axis, while the RHS contains the plots capturing the movement of the nodes along

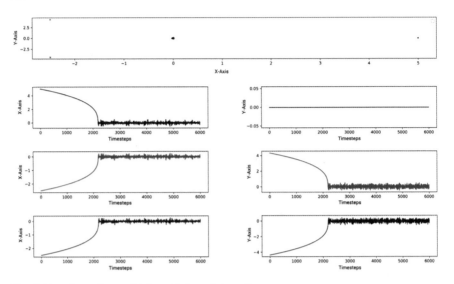

Fig. 4 Experiment 2 run with $k_f = 3$, $k_r = 4$, $k_a = 5$

the y-axis. We observe that the nodes clearly oscillate about the origin also with a very small value of average periodic displacement confirming that a good clustering picture was obtained.

3.3 Experiment 3

The predefined graph and expected outcome:

– Number of nodes: 4
– Starting positions of N_1, N_2, N_3 and N_4: (5,0), (-5,0), (-5,0) and (5,0) respectively
– Edge-Weights: $E_{w(1,2)} = 1$, $E_{w(1,3)} = 1$, $E_{w(1,4)} = 1$, $E_{w(2,3)} = 1$, $E_{w(2,4)} = 1$, $E_{w(3,4)} = 1$
– Desired Outcome: N_1, N_2, N_3 and N_4 should form a cluster

We again obtained good clustering results for simulation runs with $k_f \neq 0$ while very large oscillations were observed in simulation runs with $k_f = 0$. As was observed in Experiment 1 and 2, the nodes performed an oscillatory motion about their barycenter, in this case again the origin, along their direction of approach to the barycenter. For N_1 and N_3, the approach direction to the barycenter lies only along the x-axis; hence, the periodic motion of N_1 and N_3 was limited in the x-axis only. As the approach direction of N_2 and N_4 lies along only y-axis, we observed the two nodes to have a periodic motion along only the y-axis. The pattern of obtaining better quality clustering for systems with a later occurrence of periodic motions of its nodes, holds true here as well.

3.4   Experiment 4

The predefined graph and expected outcome:

- Number of nodes: 7
- Starting positions of nodes: Randomized
- Edge-Weights: $E_{w(1,2)} = 1$, $E_{w(1,3)} = 1$, $E_{w(1,4)} = 1$, $E_{w(2,3)} = 1$, $E_{w(2,4)} = 1$, $E_{w(3,4)} = 1$, $E_{w(5,6)} = 1$, $E_{w(5,7)} = 1$, $E_{w(6,7)} = 1$
- Desired Outcome = N_1, N_2, N_3, N_4 should form one cluster and N_5, N_6, N_7 should form an another cluster

In this experiment the nodes experience in addition to a net force of attraction from nodes similar to it, a net force of repulsion from nodes dissimilar to it. In all the simulation runs, both N_1, N_2, N_3, N_4 and N_5, N_6, N_7 sub-systems when considered in isolation, showed similar behaviour to what was observed in Experiment 4 and 3 respectively. This means that N_1, N_2, N_3, N_4 showed periodic oscillations about their barycenter or their cluster center, while N_5, N_6, N_7 showed periodic oscillations about their barycenter or their cluster center. But due to a net repulsion due to presence of dissimilar nodes in the system, the clusters as a whole, and also the cluster centers moved away from each other. This is a useful feature in clustering. The greater the inter-cluster separations are, the better the clustering picture we obtain. Therefore we can conclude that in each cluster, the intra-cluster interactions of particles are net attractive while the inter-cluster interactions are net repulsive.

For cases where we observed poor intra-cluster interactions (systems that had large values of average periodic displacement) in the previous experiments, the clustering picture obtained for this experiment is incorrect as well. In these negative cases, due to large oscillations and now additional repulsions in the system, the particles move unpredictability with no obvious periodic movement about a moving point, which is in contrast to cases where a rich intra-cluster interaction was observed in the previous experiments (the positive cases, ex: $k_f = 3$, $k_r = 4$, $k_a = 5$). For these positive cases we observed the correct cluster formation in this experiment as well. The nodes belonging to a particular cluster showed periodic movement about their moving cluster center while the clusters as a whole moved away from each other. This is a desired result as the simulation has caused similar particles to coalesce while dissimilar particles to separate out.

The plot in Fig. 5 shows the initial and the final positions of the nodes. The 'cross' symbol represents the starting positions and the 'diamond' symbol represents the ending positions of the nodes, whereas the colour scheme for each node has been mentioned in the plot itself. By observing Fig. 5 we can see that after the simulation was stopped, clear clusters with large inter-cluster distances are obtained. The difference between each positive simulation run lies in the total sum of inter-cluster and intra-cluster distances and not the final clusters obtained. The smaller the intra-cluster distances and the larger the inter-cluster distances, the better the quality of clusters obtained.

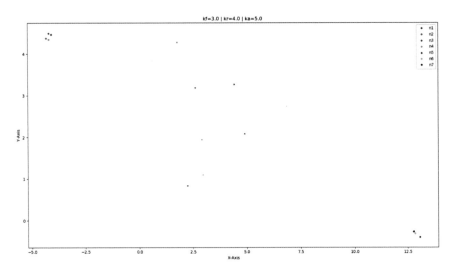

Fig. 5 Experiment 4 run with $k_f = 3, k_r = 4, k_a = 5$

3.5 Experiment 5

The predefined graph and expected outcome:

- Number of nodes: 10
- Starting positions of nodes: Randomized
- Edge-Weights: $E_{w(1,2)} = 1$, $E_{w(1,3)} = 1$, $E_{w(1,4)} = 1$, $E_{w(2,3)} = 1$, $E_{w(2,4)} = 1$, $E_{w(3,4)} = 1$, $E_{w(5,6)} = 1$, $E_{w(5,7)} = 1$, $E_{w(6,7)} = 1$, $E_{w(8,9)} = 1$, $E_{w(8,10)} = 1$, $E_{w(9,10)} = 1$
- Desired Outcome = N_1, N_2, N_3, N_4 should form one cluster, N_5, N_6, N_7 should form a second cluster and N_8, N_9, N_{10} should form the third cluster.

This experiment is essentially a scaled up version of Experiment 4 with three clusters. The results obtained confirm this and for the simulation runs with past positive results like $\{k_f = 3, k_r = 4, k_a = 5\}$ we obtained the correct clustering picture having small intra-cluster distances and large inter-cluster distances. Figure 6 shows us the result obtained for this simulation run.

Experiments with Larger Graphs: In real-life, a data-set is much larger and hence, two distinct experiments on graphs with $N = 300$, which are 30x larger graphs than before, have been carried out. The purpose of both experiments is the verification of the knowledge and comprehension gained from the previous experiments along with proving how a simple idea of simulation can also be implemented on large and complex graphs. Interestingly, what was observed was that both the experiments required different values of k_f, k_r, k_a than before. The reason behind this is simple and is due to the fact that the graph properties are way different than before. Essentially different force constants are required to account for the larger number of attractions and an even larger number of repulsions experienced by each node. The details of the two experiments are covered in the next two sub-sections.

3.6 Experiment 6

The predefined graph and expected outcome:

- Number of nodes: 300
- Number of nodes per cluster: 100
- Starting positions of nodes: Randomized
- Edge-Weights: Intra-Cluster edges with $E_w = 1$. Along with a 0.8 edge-density per cluster.
- Desired Outcome = 3 visually distinct clusters of 100 nodes each.

Correct clustering is observed at $k_f = 10, k_r = 1, k_a = 10$ which can be seen in Fig. 7, where the different coloured 'cross' symbols signify the nodes' starting positions, while the 'diamond' symbols signify the nodes' ending positions. The nodes belonging to the same cluster have the same colour. As we can see, all nodes cluster correctly at the end of the simulation, even though they were randomly spawned initially, essentially strengthening the idea of a physics based simulation approach to Node Clustering. Coming to the different values force constants, for k_r a lower value than before is required for the fact that each node now experiences repulsion from 299 other nodes which is almost 30x more than what observed in the smaller graphs. To overcome such a large number of repulsions, for k_a a larger value than

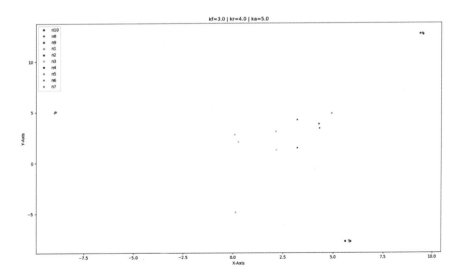

Fig. 6 Experiment 5 run with $k_f = 3, k_r = 4, k_a = 5$

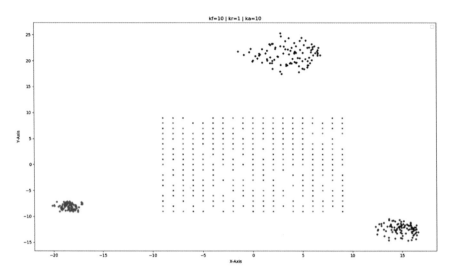

Fig. 7 Experiment 6 run with $k_f = 10$, $k_r = 1$, $k_a = 10$

before is required, while a larger value of k_f helps maintain the overall stability of system by countering the erratic movement of the nodes due to the larger number of forces of both attraction and repulsion.

3.7 Experiment 7

The predefined graph and expected outcome:

– Number of nodes: 300
– Number of nodes per cluster: 30
– Starting positions of nodes: Randomized
– Edge-Weights: Intra-Cluster edges with $E_w = 1$. Along with a 0.8 edge-density per cluster.
– Desired Outcome = 10 visually distinct clusters of 30 nodes each.

Correct clustering is observed at $k_f = 20$, $k_r = 1$, $k_a = 40$ which can be seen in Fig. 8, where the different coloured 'cross' symbols signify the nodes' starting positions, while the 'diamond' symbols signify the nodes' ending positions. The nodes belonging to the same cluster have the same colour. Just like experiment 6, all nodes cluster correctly at the end of the simulation, even though they were randomly spawned initially. Coming to the different values of force constants, the number of repulsions faced by a node does not change from Experiment 6, hence the value of k_r does not change. But as the number of attractions per node has reduced, a larger value of k_a is required for the clusters to form, while a larger value k_f helps balance the side-effects of having a larger k_a.

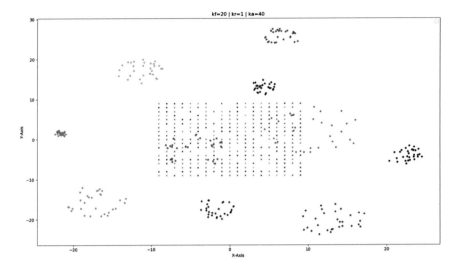

Fig. 8 Experiment 7 run with $k_f = 20, k_r = 1, k_a = 40$

4 Discussion

4.1 Time Complexity

If T represents the total number of time steps in a simulation run, N represents the total number of nodes in the graph and D represents the dimensionality of the system then a single simulation run has the time complexity $\sim O(TN^2D)$. This can be explained in the following steps:

1. As the position, velocity and acceleration vectors of all of the nodes are D dimensional, the three mathematical operations—addition of two vectors, subtraction of two vectors and multiplication of a vector with a scalar—have a time complexity $\sim O(D)$. This is because in all of the three operations, the math (addition/subtraction/multiplication) is carried out for each component of the vector. Using just these three operations—each with an overall time complexity $\sim O(D)$—we can update the positions and velocities of the nodes after every time step (Eqs. (4) and (5)), calculate the kinetic frictional force acting on a node at a particular time step (Eq. (3)), and also determine the attractive and repulsive forces present between any two nodes (Eqs. (1) and (2)).
2. To calculate the net acceleration experienced by a node, we calculate the net attractive force, the net repulsive force and the frictional forces being experienced by the node. This can be broken down into the following sub-operations:

   – To calculate the net attractive force experienced by a node we first determine all the attractive forces present between the concerned node and the remaining nodes of the system. Then, by adding all these attractive force vectors together, we calculate the net attractive force vector that is acting on the concerned node. This means that we need to carry out N-1 operations that have a time complexity $O(D)$ each. The addition of all these N-1 vectors, also has the overall time complexity $\sim O((N-1)*D)$. Therefore, calculating the net attractive force experienced by a node has a total time complexity $\sim O((N-1)*D) + O((N-1)*D) \sim O(ND)$.

- To calculate the net repulsive force acting on a node we follow the same procedure as done in calculating the net attractive force acting on a node. Hence, this operation also has a time complexity $\sim O(ND)$.
- To calculate the kinetic friction being applied on a node by the medium, we multiply the node's velocity vector with the scalar k_f, hence this operation has a time complexity $\sim O(D)$. On the other hand the static frictional force is a fixed D dimensional vector which is added to the node's net attractive and net repulsive force, having a time complexity $\sim O(D)$

Therefore the operation of calculating the net force and hence the net acceleration acting on a node, has an overall time complexity $\sim O(ND) + O(ND) + O(D) \sim O(ND)$

3. At every time step we need to update the position, velocity and acceleration vectors of each of the nodes in the graph. Therefore, executing a single time step has a time complexity $\sim O(N^2 D)$.
4. So, if a simulation is run for T time steps then the total time complexity of the run $\sim O(TN^2 D)$.

4.2 Tunable Parameters of the System

For running a simulation we rely on a number of parameters. These parameters have been kept tunable so as to have a generalized approach to running the simulation for a given data set. These parameters along with their effects on the simulation run have been mentioned below:

1. δt and $T \implies$ The size of the time-step and the number time-steps present in a simulation run: δt represents the time-step size. This quantity decides how the three characteristic quantities $(\vec{x}, \vec{v}$ and $\vec{a})$ of a node get updated with time. It represents the time-frame for which we consider the $\vec{v}$ and $\vec{a}$ of a node as constant quantities. Using this knowledge we then update $\vec{v}$ and $\vec{x}$ using Eqs. (4) and (5) respectively. A large value of δt will result in a faster progress in the simulation but will lead to abrupt values changes which could impact the accuracy of the simulation. A small value of δt will result in smoother but longer simulations to achieve the correct result. Whereas, the total number of time-steps, T, should be large enough so that the simulation reaches a steady state, but not so large that the simulation runs without a marked improvement in the final result.
2. $z \implies$ The softening constant: As seen in Eqs. (1) and (2) this parameter helps avoid singularities.

3. $m \implies$ The order of the ranged forces: This parameter as explained earlier ranges from $[0,\infty)$, controlling how far the effects of the attractive and repulsive forces can be felt. The larger the value of m chosen, the faster the force's magnitude decays with increasing distance.

4. k_f, k_r and $k_a \implies$ The three force constants: These three parameters have a linear relation to the magnitude of the frictional force, the repulsive force and the attractive force, respectively. The important point to note here is that we always take $k_a >= k_r$. This is done because if we have two nodes with the maximum edge-weight ($E_w = 1$) between them with $k_a < k_r$, then the two nodes will still repel each other which is incorrect. Hence we need $k_a >= k_r$.

5. $b \implies$ The initialization boundary: This parameter is responsible for deciding the maximum distance from origin within which the nodes can spawn. For a D Dimensional space, we restrict that each component of a node's starting position vector can have values only between $[-b,b]$ along any of the D axes. Sometimes two nodes can spawn very close to each other causing a singularity right at the start of the simulation, which could mess up the clustering result for the given time frame of the simulation run. Hence, in these situations, we give the system a shake, we re-initialize the starting position of one of the conflicting nodes until we have avoided the singularity. Therefore, we take b proportional to the number of nodes in the graph. Also to avoid any dependence of the clustering result on the initialization procedure, we can choose the starting location of each node as the average of multiple initializations.

6. $D \implies$ The dimensionality of the system: This parameter decides the number of dimensions of the model's universe. A smaller value of D increases the chances of a dissimilar node blocking the direct path between two similar nodes leading to incorrect results, for the given time frame, while with a larger value of m we run into problems associated with high-dimensional spaces [3].

7. $f_s \implies$ The static frictional force: This parameter represents the amount of static frictional force experienced by the nodes when in inertia. The larger the value of f_s, the larger the net resultant force is required to move a stationary node. This also results in a larger cutoff velocity.

8. $v_c \implies$ The cutoff velocity: This parameter plays the important role of avoiding the physically impossible situation of a particle moving in the direction of the kinetic frictional force experienced by it. When a node is continuously decelerating (due to kinetic friction) there will arise a situation when its velocity is so small in magnitude that in the next time step, its updated velocity vector will be in the direction of kinetic friction experienced by it in the current time step. So to avoid this, if at any point a node's velocity (magnitude) falls below a certain threshold, v_c, we equate its velocity vector for the next-step to a zero vector. This causes kinetic friction to be also 0, and hence avoids the above mentioned problem. We roughly can estimate $v_c = f_s * t$, as this is equal to the velocity change that a force of a magnitude same as static friction causes. This is done because, if a node's velocity is lesser than this then the static frictional force numerically has a significant impact on its velocity, to the extent that it will reverse the node's direction of movement to be along the acting frictional forces' vectors. Therefore,

if static friction is applicable to a node then its velocity essentially should be zero which is exactly what we do. This precaution is taken as the simulation procedure is being carried out numerically and not analytically.

5   Future Work

The next step in our research is to find optimum values for all the tunable parameters such that the unique values of $\{k_f, k_r, k_a\}$ found, help preprocess any data set for clustering, independent of the properties of the given data set. We aim to estimate the optimum values of k_f, k_r and k_a for a chosen value of $\{z, m, D, b, f_s, v_c, \delta t, T\}$ using supervised machine learning algorithms for regression tasks.

We aim to first roughly estimate good values of $\{z, m, D, b, f_s, v_c\}$ that will be applicable to a wide variety of large data sets. We then aim to estimate T such that the time complexity of the simulation runs for these large graphs is not very huge but is still significant to allow the system to achieve a steady state. To find δt we will study the simulation runs for a variety of given data sets in a controlled universe having only attractive and repulsive forces. The correct value of δt would be when the total energy of the N-body system, which is nothing but the sum of all the nodes' energies, is nearly constant throughout a simulation run. This is because the controlled universe has no presence of a dissipative force, and hence confirming that our updates to $\vec{x}$, $\vec{v}$ and $\vec{a}$ of the nodes are correct.

Only after this, we will find optimum values of k_f, k_r and k_a for the given values of other tunable parameters using supervised learning. We believe that for any given decent set of values of $\{z, m, D, b, f_s, v_c, \delta t, T\}$ there will be an optimum set of values of k_f, k_r and k_a such that we will obtain improved results for any density-based node clustering algorithm by preprocessing the dataset using the optimized *N-body simulation*. The cost function to be used in this supervised learning and to validate the clustering results at the end of a run will consist of two important metrics—first the proportion of the graph that has been clustered correctly and second the silhouette scores [11] of each of the nodes.

Finally we will also use the silhouette scores to evaluate the state of the simulation. We aim to stop a simulation run prematurely if we feel that the clustering picture is notgoing to improve any further, to reduce the time spent preprocessing a data set. We believe that the final clustering result is not going to change when the graph reaches a steady state during the run. A graph reaches a steady state when the silhouette scores of all its nodes has become somewhat constant for a considerable period of time during a simulation run.

6 Conclusion

We have presented a simulation based approach to node clustering. We believe that this idea is radically different to existing approaches to node clustering in recent literature. We have tested the idea against small and large inputs and it has faired well as a preprocessing step to existing density based clustering algorithms for the task of network community detection, and we believe that it can function well as a stand-alone clustering algorithm as well. We are aware that these observations are preliminary, and require further investigation. The theoretical issues of node clustering are intractable. All existing methods are approximation algorithms. In view of that, the apparent computationally demanding nature of simulation should not be a deterrent to exploring it as an alternative, as it provides great flexibility and scales well with graph sizes. Viewed as a computational approach in a traditional sense, the proposed approach is indeed very expensive. However, this approach would scale naturally in a radically different computing model, e.g. one based on actual physical particles. Such computing platforms do not exist today, but may become a reality in future. What manifests as separate computational steps—e.g. computation of inter-particle forces in our simulation—in a traditional computing platform would naturally take place in parallel in such a physics based computational system, leading to scale-up of the systems that are currently impossible to achieve even in the most massively parallelised computing platforms. However, there are still several questions to answer: what should be the values of tunable parameters? What should be the duration of simulation? Can the simulation parameters be machine-learned? These are themes we pursue in our future investigations.

References

1. Aggarwal, C.C., Wang, H.: A Survey of Clustering Algorithms for Graph Data, pp. 275–301. Springer US, Boston, MA (2010). https://doi.org/10.1007/978-1-4419-6045-0_9
2. Backstrom, L., Huttenlocher, D., Kleinberg, J., Lan, X.: Group formation in large social networks: Membership, growth, and evolution. In: Proceedings of the 12th ACM SIGKDD International Conference on Knowledge Discovery and Data Mining, pp. 44–54. Association for Computing Machinery (2006). https://doi.org/10.1145/1150402.1150412
3. Bellman, R., Bellman, R., Collection, K.M.R.: Adaptive Control Processes: A Guided Tour. Princeton Legacy Library, Princeton University Press (1961). https://books.google.co.in/books?id=POAmAAAAMAAJ
4. Campello, R.J.G.B., Moulavi, D., Sander, J.: Density-based clustering based on hierarchical density estimates. In: Pei, J., Tseng, V.S., Cao, L., Motoda, H., Xu, G. (eds.) Advances in Knowledge Discovery and Data Mining, pp. 160–172. Springer, Berlin, Heidelberg (2013)
5. Diestel, R.: Graph Theory, 4th Edition, Graduate texts in mathematics, vol. 173. Springer (2012)
6. Gold, V. (ed.): The IUPAC Compendium of Chemical Terminology: The Gold Book. International Union of Pure and Applied Chemistry (IUPAC), Research Triangle Park, NC, 4 edn. (2019). https://doi.org/10.1351/goldbook, https://goldbook.iupac.org/
7. Khan, B.S., Niazi, M.A.: Network community detection: a review and visual survey. CoRR **abs/1708.00977** (2017). http://arxiv.org/abs/1708.00977

8. Mahdi, O.A., Abdul Wahab, A.W., Idna Idris, M.Y., Abu znaid, A.M.A., Khan, S., Al-Mayouf, Y.R.B., Guizani, N.: A comparison study on node clustering techniques used in target tracking wsns for efficient data aggregation. Wirel. Commun. Mobile Comput. **16**(16), 2663–2676 (2016). https://doi.org/10.1002/wcm.2715

9. Radicchi, F., Castellano, C., Cecconi, F., Loreto, V., Parisi, D.: Defining and identifying communities in networks. Proc. National Acad. Sci. **101**(9), 2658–2663 (2004). https://doi.org/10.1073/pnas.0400054101, https://www.pnas.org/content/101/9/2658

10. Reddy, K.S.S., Bindu, C.S.: A review on density-based clustering algorithms for big data analysis. In: 2017 International Conference on I-SMAC (IoT in Social, Mobile, Analytics and Cloud) (I-SMAC), pp. 123–130 (2017). https://doi.org/10.1109/I-SMAC.2017.8058322

11. Rousseeuw, P.: Silhouettes: a graphical aid to the interpretation and validation of cluster analysis. J. Comput. Appl. Math. **20**(1), 53–65 (1987). https://doi.org/10.1016/0377-0427(87)90125-7

12. Schaeffer, S.E.: Survey: graph clustering. Comput. Sci. Rev., 27–64 (2007). https://doi.org/10.1016/j.cosrev.2007.05.001

13. Trenti, M., Hut, P.: Gravitational n-body simulations (2008)

14. Walker, J., Resnick, R., Halliday, D.: Halliday & Resnick Fundamentals of Physics. Wiley, Hoboken, NJ, 10th edition edn. (2014)

15. Yang, J., Leskovec, J.: Defining and evaluating network communities based on ground-truth. In: Proceedings of the ACM SIGKDD Workshop on Mining Data Semantics. MDS '12, Association for Computing Machinery, New York, NY, USA (2012). https://doi.org/10.1145/2350190.2350193

Isolated Bangla Spoken Digit and Word Recognition Using MFCC and DTW

Bachchu Paul, Rakesh Paul, Somnath Bera, and Santanu Phadikar

Abstract Digit recognition is one of the elegant research topics in modern world. Scientists had already got an excellent output in their research work on this topic for English and Chinese like languages. However, very few works exist on digit recognition for regional language. Moreover in case of regional language the pronunciation rapidly varies on the basis of area and the performance of their proposed recognition reasonably low. In this paper we have worked on digit recognition on a regional language, Bengali referred to as Bangla. In our proposed work of isolated digit and word recognition, we created a small speech database, containing ten Bangla digit zero to nine (pronounced as 'sunno' to 'noi') and four Bangla words ডান, বাম, উপর, নীচ (English equivalent Right, Left, Above, and Below respectively) with 100 samples for each class. We have done a pre-processing phase, followed by a 39 dimensional feature extraction procedure of Mel Frequency Cepstral Coefficients (MFCC), Δ MFCC and $\Delta \Delta$ MFCC. Finally, we used the Dynamic Time Warping (DTW) for classification of testing purpose. The system has achieved a highest accuracy of 93% for classification of Bangla words and digits, which is considerably satisfactory. The comparative analysis with the existing method is given in Sect. 5.

Keywords Pre-processing · ASR · Zero crossing · MFCC · Δ MFCC · DTW

1 Introduction

Speech is the most powerful communication medium. Human can express their views and thoughts through speech. From last two decades, researchers used speech as the technical equipment of development. Using speech as commands, they provided

B. Paul (✉) · R. Paul · S. Bera
Department of Computer Science, Vidyasagar University, Midnapore 721102, West Bengal, India
e-mail: ableb.paul@gmail.com

S. Phadikar
Department of Computer Science and Engineering, Maulana Abul Kalam Azad University of Technology, Kolkata 700064, India

P. Gyei-Kark et al. (eds.), *Engineering Mathematics and Computing*,
Studies in Computational Intelligence 1042,
https://doi.org/10.1007/978-981-19-2300-5_16

us voice dependent devices like computers, mobile, etc. During the time of using speech as commands, it is converted into text internally. This automatic speech to text conversion is treated as automatic speech recognition. In speech recognition a huge amount of speech are collected. Then it is used as voice sample and compared it with the existing speech database. The principle objective of this above process is to identify the speech itself [1]. To identify the speech, another main objective is speaker identification which always comes in front of first of this research work. It has also closed relation with speaker verification. Among the various research area, text-dependent speaker recognition and text independent speaker recognition are the widely used research area both in noisy and noiseless environment. Both techniques can mainly identify the voice of speaker. Using this techniques many voice controlled devices are made in our regular life [2]. There are different types of language have been used in different countries. Each country has multiple regional languages. Bangla is a regional language which has been considered in this paper. More than 215 million people all over the world speak in Bangla [3] as their native language. But very few research works have been done on Bangla ASR. So there is a good opportunity to us to do more research work in ASR of Bangla language to improve more. The proposed work based on spoken Bangla digit and four Bangla word recognition. This research work can help to those people who are interested to do their research in Bangla language. In our proposed work 1400 audio files of Bangla digit and words are taken as input samples. These collected samples are trained and finally tested individually with the reference template words for best match. Every language has different digit pronunciations from other languages. Here zero is pronounced as "Shunno", five as "Panch", etc. have been collected from ten speakers uttered by ten times for each sample.

There are many applications using numeric digit. It is used in ATM machines, biometric system, cellular phone, computer, smart wheel chair, etc. In railway system, announcement of train no of arriving or departure trains this system can be used. Besides who find difficult for calling using mobile an individual easily can do it [6]. The best application is automatic voice record and answering just speaks out numbers instead of pressing buttons to go through the menu options [13]. Though the DTW algorithm fails for large data set and noisy data, it has also benefit of faster recognize the word and digits. Thus where the number of class and dataset is small but faster recognition is important, there the proposed method of DTW algorithms can be adopted.

In our proposed work of isolated Bengali spoken work recognition, we have created small corpus, then we have done a preprocessing phase, followed by a MFCC, ΔMFCC and $\Delta\Delta$MFCC. Finally, the mean features are compared with for best matching using DTW algorithm.

2 Literature Review

Nahid et al. [4] worked on automatic Bangla real number recognition. In their research work, they used CMU Sphinx 4 for speech recognition. They took help of the Bengali digit writing software Avro in their work. In recognition phase, they used MFCC as a feature extractor and after recognition they used Net beans 8 to develop the whole system. Here statistical grammar is used for training purpose and the system is also tested with the help of personal computer and smart phone. Finally after testing they got 85% accuracy that gives the satisfactory result.

Abdullah-al-Mamun et al. [5] implemented Isolated Speech Recognition (ISR) combining MFCC and Hidden Markov Model (HMM) for recognition and training purpose. They recorded their data and extracted the feature for recognition and training purpose using a model. They have preferred HMM for the capability of non-linearity dependencies and estimation of model parameters. They have taken 5 male speakers and 5 female speakers of 56 Bangla alphabet which includes Bangla vowels, digits, and constants with both conditions (noisy and noiseless) and have received 85.714% accuracy.

In recognition of spoken Bangla digits, Gupta et al. [6] presented a method in both noisy and noiseless environment using 48 speakers. They have used MFCC for extracting features, PCA for feature summarization and SVM, Random Forest and Multi-Layer-Perceptron for classification purpose. In Multi-Layer-Perceptron input node were 39 for dimension of feature vector and 100 Hidden Layers and 10 output layer for 10 digit's separation. They got the accuracy of 87% in MLP, 91.67% in SVM and 84.06% in RF based classifier. The overall accuracy of their system was 90%.

Ahammad et al. [7] proposed connected Bangla digit speech recognition using Back Propagation Neural Network (BPNN) to train the network and have used MFCC for feature extraction. They have faced a problem at the time of first phase testing because of similarity of pronunciation for some digits. Besides they have seen that Artificial Neural Network (ANN) training time depends on hidden layers, epochs, and error threshold. Considering this they have reached 98.46% recognition accuracy.

For recognition of Tamil digits, Karpagavalli et al. [8] used vector quantization. For creation of Tamil codebook, they used Linde-Buzo-Gray (LBG) vector quantization algorithm. They have recorded 20 speakers' data with 6 times for each digit. They have used two-third data for training, and rest is for testing purpose. They used MFCC for feature extraction and pass it through vector quantizer with k parameter to generate codebook and for measuring the performance of vector quantizer they have used Word Error Rate (WER) and Word Recognition Rate (WRR) and finally they received 91.8% accuracy.

Hejazi et al. [9] introduced a new method for solving traditional problem in isolated digit recognition in "Persian" language. They introduced digit recognition procedure into three phases. In first phase word is decomposed into smaller parts using reliable algorithms to make HMM. After that best essential candidate is taken using Support Vector Machine (SVM). Finally recognition is done through SVM by

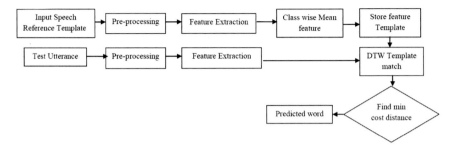

Fig. 1 Schematic diagram of the proposed method

counting the maximum number of similar parts of a word. In their work they have used 400 Persian male and 50 female's voice of different age in which 330 speech sample is used for training and 120 sample is used for testing purpose and got about 98% accuracy.

The proposed method is given in schematic diagram 1. Section 3 discussed the data set used and feature extraction phase. The DTW algorithm to find the distance matrix is given in Sect. 4. The result, analysis and comparative statement is discussed in Sect. 5 and finally Sect. 6 gives the conclusion (Fig. 1).

3   Data Set and Feature Extraction

3.1   Dataset Used

For the proposed work of isolated Bangla spoken digit recognition, we have taken a small data set of 10 Bangla digits zero to nine (pronounced as "sunno" to "noi"), uttered by 10 speakers, among them five male and five female with the age group from twenty to forty. Each word is uttered by ten times for each speaker. We have also taken four Bangla words ডান, বাম, উপর, নীচ (English equivalent Right, Left, Above, and Below respectively). Actually for DTW based word recognition works for small vocabulary and a smaller number of classes. But some of the applications where correct classification is important rather than the size of the data set, then DTW works fine. Each word is recorded in an ideal situation in a room with the sampling frequency of 16KHz and 32-bit mono channels with the help of audacity software. The data set contains 1400 audio samples of 14 classes. The training data contains 700 samples and test set having 700 samples considering the first 5 samples as the training set and last five as test data set for each speaker.

3.2 Pre-processing

During this phase, the signal is passed through a pre-emphasis filter to reduce the noise and to capture the smooth spectral information of the frequency components given in Eq. 1 [14].

$$y(n) = x(n) - ax(n - 1) \tag{1}$$

where 'a' lies between 0.9 and 1.

Since speech is a highly varying signal, the same word pronounced by two speakers varies in length. For better estimation, we find the most important portion of the signal where linguistic information carrying the identifiable word. This portion is selected by calculating the average short term energy and averages zero crossing of a frame, given by Eqs. 2 and 3 respectively. After that, the signal is fitted into one-second audio sample.

$$E_n = \sum_{m=-\infty}^{\infty} [X(m) - W(n - m)]^2 \tag{2}$$

where X(.) is the frame and W(.) is the windowing function of the same length.

$$ZCR = \frac{1}{2N} \sum_{j=i-N+1}^{i} |sgn(x(j) - sgn(x(j - 1))|w(i - j) \tag{3}$$

where,

$$sgn(x(j)) = \begin{cases} 1, & if(x(j) \geq 0. \\ 0, & if x(j) < 0. \end{cases}$$

where, N is the frame length.

3.3 Feature Extraction

We used a 39-dimensional feature in our proposed model for isolated word recognizer. We extracted a 13-dimensional MFCC, 13-dimensional ΔMFCC and 13-dimensional $\Delta\Delta$ MFCC from each of the signals. The calculation of MFCC feature from the audio signal is given in Fig. 2.

3.3.1 Framing

The voiced section for each audio sample detected in Sect. 3.2 is segmented into a 25ms frame with 15ms overlap. A single frame contains 400 samples i.e. 98 frames per second.

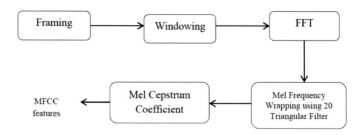

Fig. 2 Phases to find mel frequency cepstral coefficients

3.3.2 Windowing

Since speech is an aperiodic signal, to maintain the continuity at two extreme ends of a frame, the signal is multiplied by a Hamming window of the same size. The equation of a hamming window is given by Eq. 3.

$$w(n) = 0.54 - 0.46cos\left(\frac{2\pi n}{N-1}\right) \tag{4}$$

where, N is the number of samples in a single frame or simply the frame length.

3.3.3 Fast Fourier Transform (FFT)

The time domain into the frequency domain is converted using the FFT to measure the energy distribution over frequencies. The FFT is calculated using the Discrete Fourier Transform (DFT) formula given in Eq. 4.

$$S_i(k) = \sum_{n=1}^{N} s_i(n)e^{-\frac{j2\pi kn}{N}} \qquad 1 \le k \le K \tag{5}$$

K is the DFT length

3.3.4 Mel-Frequency Wrapping

In this step, the power spectrum is mapped onto the mel scale using 20 triangular band-pass filters. The relationship between frequency (f) and mel (m) is given in Eq. 5.

$$m = 2595log_{10}(1 + \frac{f}{700}) \tag{6}$$

3.3.5 Mel Cepstrum Coefficient

The frequency domain into time domain of the signal is converted by a Discrete Cosine Transform (DCT) [16] using Eq. 6.

$$C_m = \sum_{k=1}^{M} cos[m(k - \frac{1}{2})\frac{\pi}{M}]E_k \tag{7}$$

where M is the number of filter bank and 20 in our case, $1 \leq m \leq L$ is the number of MFCC coefficients and E_k is the mel power spectrum for the frame i.

The first 13 coefficients are the MFCC as our primary feature. The ΔMFCC is calculated by the Eq. 8 from the MFCC feature and similarly Δ ΔMFCC are computed from ΔMFCC parameters. Thus for a single frame the 39 numbers of features as our feature vector.

$$D(n) = \frac{1}{\sqrt{\sum_{i=-r}^{r} i^2}} \sum_{i=-r}^{r} i.C(n + i) \tag{8}$$

where D(n) is nth frame ΔMFCC , C(n) is the nth frame MFCC and r is a constant taken as 2 [15]. The Δ ΔMFCC is calculated by the same formula (8) where C(n+i) is replaced by D(n+i).

Now for a single audio sample with M frames, we obtained an M $\times$ 39 feature matrix. For N words we got N $\times$ M $\times$ 39 feature, which is large for analysis. Instead of storing separate feature for each of the uttered word, we just calculated the average of the feature for the same class and have been used as our training reference template. For testing purpose, all the previous stages have been followed with an individual audio sample. Then the DTW algorithm is used to take the decision.

4 Dynamic Time Warping (DTW)

DTW is dynamic programming problem. It is applied to find best warping path between temporal sequences to find similarity of audio which may changes in speed and time. It compares warped non-linear series pattern either stretching or shrinking with respect to time axis. DTW actually compares feature vector which are overlapped at beginning and end. Speech alignment time is a major problem for measuring distance. The pronunciation time of a word varies in time. Euclidean distance measure normal distance and gives poor performance with time varying feature vector patterns. DTW solves this alignment problem by performing linear mapping, considering two vectors P and V and it is represented by an nxn matrix. The distance between two vector of (ith, jth) element is represented by [10].

$$D(V_i, P_i) = \sqrt{\sum_{i=1}^{n}(V_i - P_i)^2} \qquad (9)$$

Best alignment between two vectors is a path along grid, called global distance. In DTW first we have used local constraint. To reach the point (ith, jth) from the earliest point. Here we have considered three predecessors(i−1,j),(i−1,j−1),(i,j−1) to find optimal path [11, 12]. The optimal value is calculated by given formula 10.

$$D(i, j) = |V(i) - P(j)| + min \begin{cases} D(i+1, j) \\ D(i+1, j+1) \\ D(i, j+1) \end{cases} \qquad (10)$$

It works as follows: Suppose we want to compare and estimate the difference between the following two signals:

(a) (Input) test signal x[t] : 2 2 3 2 3 2 0
(b) (Stored) reference signal y[t] : 1 2 2 3 2 3 2 1

Sample-by-sample difference x[t]-y[t] : 1 0 1 −1 1 −1 −2 undefined
Both signals are equivalence. However, the stored reference signal is longer than the test signal and not synchronized in time. To calculate the difference between them, considering every sample of x[t] and each sample of y[t]. The distance matrix D is shown in Fig. 3.

Now for each of the test audio sample, the feature is extracted as discussed in Sects. 3.2 and 3.3. We obtained an M × 39 feature matrix in a similar manner. This feature matrix is compared with all stored reference feature to find the best alignment or best match. This is done by calculating the distance from test feature matrix to all stored train feature matrix. The minimum distance is the winner for correct identification of the word.

5 Result and Discussion

As discussed earlier, the database of ten speakers with 1400 audio samples out of which 1000 Bangla numeric and 400 four Bangla words was created to show how the system performs. For each of the test audio sample, the 39 dimensional MFCC, ΔMFCC and Δ ΔMFCC has been computed. The distance between stored template and test sample has been computed using DTW. The distance obtained by applying DTW algorithm for every sample (one from each class) run one time from each of the reference template is given in Table 1.

Similarly the distance matrix for four Bangla words obtained from the proposed method is given by Table 2.

From each of the above table we observed that, the diagonal element has the lowest value in each row, which is the corresponding predicted spoken word. The test audio

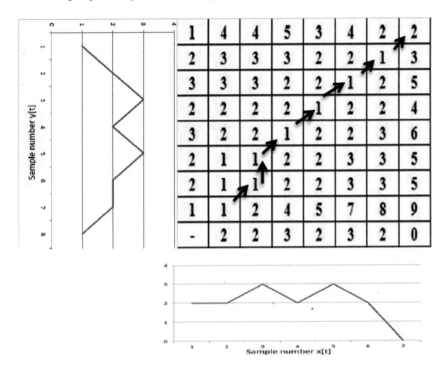

Fig. 3 Best matching path between two signal using DTW

Table 1 Distance from each class of test sample to reference template for number

Test Sample	Distance from reference template									
	"Sunno"	"Ek"	"Dui"	"Tin"	"Char"	"Panch"	"Choy"	"Sat"	"Aat"	"Noi"
"Sunno"	**387.64**	551.43	566.17	525.21	538.53	582.967	509.96	513.36	573.54	537.82
"Ek"	332.946	**221.31**	334.1	343.1	253.73	248.004	256.72	246.72	256.46	266.44
"Dui"	404.366	403.95	**283.77**	369.96	409.41	401.785	384.04	442.47	444.92	354.66
"Tin"	409.022	380.92	365.2	**264.99**	444.57	434.728	428.8	439.82	462.12	432.46
"Char"	365.541	354.79	437.61	433.07	**273.13**	337.941	286.49	325.75	313.32	302.8
"Panch"	378.157	339.12	409.54	444.47	316.97	**285.672**	324.19	331.3	296.61	316.39
"Choy"	337.479	349.46	379.25	398.45	286.07	350.277	**260.24**	317.58	336.58	291.52
"Sat"	324.952	324.99	441.98	360.6	297	345.377	301.43	**281.41**	338.04	354.69
"Aat"	419.691	361.26	449.18	489.87	321.71	332.343	323.61	351.05	**284.75**	320.53
"Noi"	396.97	369.03	327.19	404.29	337.62	354.335	319.75	385.71	341.25	**278.46**

Table 2 Distance from each class of test sample to reference template for words

Test Sample⬇	Distance from reference sample			
	"Bam"	"Dan"	"Nich"	"Upar"
"Bam"	**414.2**	453.127	813.184	506.367
"Dan"	457.533	**432.414**	787.271	520.051
"Nich"	755.321	748.69	**591.691**	797.203
"Upar"	623.37	602.193	709.378	**554.572**

Table 3 Confusion matrix for test samples

		Target Class								
		0	1	2	3	4	5	6	7	8	9
Output Class	0	**46**	0	1	0	0	0	1	1	1	0
	1	1	**47**	0	0	0	0	0	1	1	0
	2	0	1	**46**	0	1	0	1	0	0	1
	3	0	0	0	**49**	0	0	0	0	0	1
	4	0	1	0	0	**42**	1	0	4	2	0
	5	1	0	1	0	0	**45**	1	1	1	0
	6	0	0	0	1	0	0	**44**	1	1	3
	7	0	1	0	1	2	0	1	**43**	2	0
	8	0	1	0	0	0	0	0	1	**48**	0
	9	0	0	1	0	0	0	1	2	0	**46**

Table 4 Comparative accuracy with the existing method

Method	Feature	Classifier	Accuracy (%)
Abdullah-al-Mamun and Mahmud [5]	MFCC	Hidden Markov model (HMM)	85.71
Gupta and Sarkar [6]	MFCC and PCA for reduction	SVM, random forest and multi-layer-perceptron	91.67
Karpagavalli et al. [8]	MFCC	Linde-Buzo-Gray (LBG) and vector quantization	91.8
Proposed method	MFCC, ΔMFCC, ΔΔMFCC	DTW	93.5

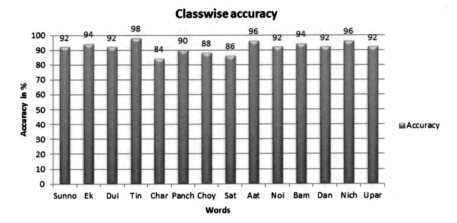

Fig. 4 Class wise percentage of accuracy

samples are run individually to predict the most accurate words. The confusion matrix is given by Table 3 for the Bangla digit (Fig. 4).

The percentage testing accuracy for the different Bangla digit and words is given in Fig. 3.

5.1 Comparative Statement

A very few research work done on Bangla spoken digit recognition. However, very few research work has been done on regional language for isolated digit or word recognition. Table 4 shows the accuracy obtained by the proposed method compared with the existing method given in Sect. 2.

6 Conclusion

Automatic Bangla digit recognition is a promising research area. This paper has focused on speech recognition of regional language Bangla. Speech recognizer is the main purpose of this research work. In our system MFCC is used in feature extraction and DTW is used as a classifier. Though this paper is based on Bangla speech recognition but it can also help to do research on same topic in other languages also for isolated word recognition. In this paper we have solved the Bangla digit recognition as well as Bangla word recognition. Now in future we want to innovate some new research arena using the Bangla speech recognition. The experimental analysis shows that this process has achieved 92% accuracy on digit recognition and 93% accuracy on words recognition of the regional language Bangla. This method is very simple and fruitful for small database and smaller number of class. The data set subject to noiseless also. But for a large data set and noisy data we should think some other methods. Too many calculations are to be done for the proposed method. Though, the DTW algorithm has some shortfalls, still it is an elegant method for some of the applications where smaller number of voiced based command are passed to the machine.

References

1. Mohanty, P., Nayak, A.K.: Isolated odia digit recognition using htk: an implementation view. In: 2018 2nd International Conference on Data Science and Business Analytics (ICDSBA), pp. 30–35. IEEE (2018)
2. Ghanty, S.K., Shaikh, S.H., Chaki, N.: On recognition of spoken Bengali numerals. In: 2010 International Conference on Computer Information Systems and Industrial Management Applications (CISIM), pp. 54–59. IEEE (2010)
3. Muhammad, G., Alotaibi, Y.A., Huda, M.N.: Automatic speech recognition for Bangladigits. In: 2009 12th International Conference on Computers and Information Technology, pp. 379–383. IEEE (2009)
4. Nahid, M.M.H., Islam, M.A., Islam, M.S.: A noble approach for recognizing Bangla real number automatically using CMU Sphinx4. In: 2016 5th International Conference on Informatics, Electronics and Vision (ICIEV), pp. 844–849. IEEE (2016)
5. Abdullah-al-Mamun, M.D., Mahmud, F.: Performance analysis of isolated Bangla speech recognition system using Hidden Markov Model
6. Gupta, A., Sarkar, K.: Recognition of spoken bengali numerals using MLP, SVM, RF based models with PCA based feature summarization. Int. Arab J. Inf. Technol. **15**(2), 263–269 (2018)
7. Ahammad, K., Rahman, M.M.: Connected bangla speech recognition using artificial neural network. Int. J. Comput. Appl. **149**(9), 38–41 (2016)
8. Karpagavalli, S., Rani, K.U., Deepika, R., Kokila, P.: Isolated Tamil digits speech recognition using vector quantization. Int. J. Eng. Res. Technol. **1**(4), 1–12 (2012)
9. Hejazi, S.A., Kazemi, R., Ghaemmaghami, S.: Isolated Persian digit recognition using a hybrid HMM-SVM. In: 2008 International Symposium on Intelligent Signal Processing and Communications Systems, pp. 1–4. IEEE (2009)
10. Hai, N.T., Van Thuyen, N., Mai, T.T., Van Toi, V.: MFCC-DTW algorithm for speech recognition in an intelligent wheelchair. In: 5th International Conference on Biomedical Engineering in Vietnam, pp. 417–421. Springer, Cham (2015)
11. Marković, B.G., Stevanović, G., Jovičić, S.T., Mijić, M., Galić, J.: Recognition of normal and whispered speech based on RASTA filtering and DTW algorithm. In: Proceedings of the International Conference IcETRAN-2017, pp. 8–2 (2017)
12. Permanasari, Y., Harahap, E.H., Ali, E.P.: Speech recognition using Dynamic Time Warping (DTW). In: Journal of Physics: Conference Series (Vol. 1366, No. 1, p. 012091). IOP Publishing (2019)
13. Shaikh, H., Mesquita, L.C., Araujo, S.D.C.S., Student, P.: Recognition of isolated spoken words and numeric using MFCC and DTW. Int. J. Eng. Sci., 10539 (2017)
14. Walid, M., Bousselmi, S., Dabbabi, K., Cherif, A.: Real-time implementation of isolated-word speech recognition system on raspberry Pi 3 Using WAT-MFCC. IJCSNS **19**(3), 42 (2019)
15. Zhang, L., Wu, D., Han, X., Zhu, Z.: Feature extraction of underwater target signal using Mel frequency cepstrum coefficients based on acoustic vector sensor. J. Sens. (2016)
16. Paul, B., Phadikar, S., Bera, S.: Indian regional spoken language identification using deep learning approach. In: Proceedings of the Sixth International Conference on Mathematics and Computing, pp. 263–274. Springer, Singapore (2021)

An EOQ Model for Deteriorating Items under Trade Credit Policy with Unfaithfulness Nature of Customers

Rituparna Mondal, Prasenjit Pramanik, Ranjan Kumar Jana, Manas Kumar Maiti, and Manoranjan Maiti

Abstract In this study, an EOQ model has been developed for a deteriorating item with time dependent deterioration with a fixed expiration date under partial trade credit policy considering the unfaithfulness nature of the base customers. The main purpose of this research work is two folds. First, the unethical behaviour of the base customers is considered. On the other hand, a non-trivial flaw has been rectified, considering all the inventory models with deteriorating items under trade credit policy, developed in the last two decades. The proposed model is illustrated with various numerical examples. Some managerial insights are also outlined.

Keywords Economic order quantity · Supply chain management · Deterioration · Partial trade credit · Unfaithful customers

R. Mondal · R. K. Jana (✉)
Department of Mathematics and Humanities (DoMH), Sardar Vallabhbhai National Institute of Technology, Surat, Gujarat, India
e-mail: rkjana2003@yahoo.com

P. Pramanik · M. Maiti
Department of Applied Mathematics with Oceanology and Computer Programming, Vidyasagar University, Paschim Medinipur, Midnapore, West Bengal, India

M. K. Maiti
Department of Mathematics, Mahishadal Raj College, Purba Medinipur, Mahishadal, West Bengal, India

© The Author(s), under exclusive license to Springer Nature Singapore Pte Ltd. 2023 247
P. Gyei-Kark et al. (eds.), *Engineering Mathematics and Computing*,
Studies in Computational Intelligence 1042,
https://doi.org/10.1007/978-981-19-2300-5_17

1 Motivation and Literature Review

In the recent competitive volatile marketing situation, it is very crucial to draw the optimal decisions for the decision maker/store manager/retailer. To enhance the demand of a product, the players of the inventory system/supply chain adopt various promotional activities. Among those, trade credit policy is the most effective activity. But in this policy there is a drawback, there is a possibility to cheat the credit amount, i.e., there is a default credit risk. However in a supply chain, this type of risk basically occurs in retailer-customers level as there is no such certified bonding between the retailer and the customers, credit is given based on the complete faith. In a local market, it is observed that there is an uncertain percentage of customers who are not comeback to repay the credit amount. In this circumstances, to draw an optimal decision it is mandatory to consider this phenomenon into the account.

With the introduction of trade credit policy by [13], there are so many research works found under different levels of trade credit policy: single level trade credit policy [1, 9, 19, 26, 27], two level trade credit policy [4–7, 14, 17, 18, 20, 23, 31–33, 35–38], three level trade credit policy [28, 29]. From the above studies, it is obvious that trade credit policy is one of the most important factor in any inventory control system. But trade credit policy occurs a default credit risk. To reduce the default credit risk, the practitioners and the researchers allow partial credit opportunity [16, 21, 22, 26, 28–30] to the customers. Also to reduce the default credit risk, Teng [35] considered different credit policies for different customers, i.e., full credit policy for good (faithful) customers and partial credit policy for the unfaithful customers. But in reality, it is really unpredictable to trace out the bad (unfaithful) customers, who will not obey the business ethics. Also, different credit opportunity for different customers is not only a malpractice for any type of business transaction but also effects negative impression to some customers. Though in a local market, it is unpredictable to trace out the good and bad customers, a retailer of the market roughly knows the percentage of the unfaithful customers. So it is more practical to consider partial trade credit policy for all customers to reduce the default credit risk assuming a percentage of total customers are unfaithful customers, who are not comeback to repay the credit amount. Though it is a real life phenomenon, it is overlooked by the researcher of this field.

On the other hand, proper maintenance of an inventory system of the deteriorated items is another crucial problem for the decision maker in any business sector. Ghare and Schrader [12] first developed an EOQ model for a deteriorating item. Afterwards, a large number of research articles found with this consideration [1, 8, 15, 16, 25]. In the above studies, they have considered that the deterioration rate of an item is constant. But there are so many products like fruits, vegetables, medicines, etc., those are not deteriorate continuously and also there is an expiration date for each product. There are very few research works [21, 22, 34] considering time dependent deterioration with a fixed expiration date. Baker [2] provided a research article covering all the models of inventory system with deterioration since 2001. In all the above studies it is observed that they have considered a deteriorating item to develop their models under trade credit policy, but unfortunately none has considered the interest paid part for the deteriorated units in the mathematical development of their model. In other words, if a wholesaler offers a credit period M to a retailer with retailer's replenishment time $T (> M)$ and some units are deteriorated during $(0, M)$, then the retailer has to pay the pricing value of those deteriorated units at the time M to the wholesaler by a bank loan and thus the retailer has to pay interest for those deteriorated units for a time duration $(T - M)$ (considered that the retailer's account will be settled at T). But none has considered this phenomena in the development of their models, they have only consider the pricing value of the deteriorated units in their model development section. In this study an attempt has been made to overcome this mentioned flaw.

Base demand or market demand is the key of any inventory control system/supply chain management. When the demand depends on the selling price of an item, then the inventory control model should have to consider the selling price as a decision variable. Then the optimal business strategy should be occur with an optimal pricing for the retailer. Several studies have been found with this consideration [3, 10, 22, 24, 26, 39].

In this study, for the first time an EOQ model has been developed for a deteriorating item having time dependent deterioration with a fixed expiration time under partial trade credit policy considering the unfaithfulness nature of the base customers. Here it is considered that a wholesaler offers a full credit period to a retailer, but in turns the retailer offers a partial credit period to his/her customers i.e., offers a credit period on a fraction of the total purchased amount. Also it is assumed that a fraction of total customers are unfaithful customers, who will not comeback for the repayment of the credit amount to the retailer. The goal of this investigation can be stated as below:

– In each and every business transaction with credit policy, it is observed that there are few customers, who does not obey the business ethics, in other words there are few customers who ran way without paying the credit amount to the retailer. This phenomenon is considered in this study.
– In all the inventory models for deteriorating items under trade credit policy, the deterioration of the item is considered but unfortunately the amount of charged interest paid for the same is overlooked. For instant, let us assume that a supplier offers a full credit period M to a retailer with retailer's replenishment time $T (> M)$. Now some units are deteriorated during $(0, M)$ and thus the retailer has to pay purchase cost of the units at the time M to the supplier by a bank loan whose

account is settled at the end of the return of all credit amount, i.e. at $(T + N)$. Since $T > M$ then the retailer has to pay interest for those deteriorated units for a time duration $(T - M)$. Moreover, some units will be deteriorated during $[M, T]$ and the retailer has to pay interest for those units also as payment of those units are made from bank loan and account of retailer will be settled with the bank at the time $(T + N)$. This phenomenon is incorporated in this study.

- Also most of the existing literature on trade credit policy assumed that the retailer earned interest from bank on the selling price of the sold units, which is unrealistic, because some expenditure (holding cost, ordering cost, etc.) is incurred to run the inventory and it comes from a portion of the profit part of the sell units. Normally the retailer receive bank interest from purchase cost portion of the selling units before the settlement of his account to the supplier. So the retailer actually earned interest from bank on the purchase cost of the sold units. This phenomenon is incorporated in this study.

The rest of this paper is organized as follows. Some notations and assumptions that are used throughout this paper are listed in Sect. 2. Mathematical formulation of the proposed model is given in Sect. 3. Theoretical results and the existence of optimal solutions of the proposed model are discussed in Sect. 4. Numerical results and analyses of the results are presented in Sect. 5. Finally, conclusions are drawn with few future research directions in Sect. 6.

2 Assumptions and Notations for the Proposed Models

To develop the model mathematically, following assumptions and notations are considered throughout the paper.

2.1 Assumptions

(i) Time horizon is infinite.
(ii) Replenishment is instantaneous.
(iii) In today's time-based competition, it is assume that shortages are not allowed to occur.
(iv) The market demand for the item is assumed to be sensitive to the customer's retail prices and is defined as $D(s) = as^{-b}$, which is decreasing function of the retail price s; where $a(> 0)$ is a scaling factor and $b(> 1)$ is a price elasticity coefficient. For notational simplicity, $D(s)$ and D will be used interchangeably in this paper.
(v) Here it is assumed that customers have to pay the α-fraction of the total purchased amount as a collateral deposit at the receiving time of the units of the

item and the retailer offers a credit period N to the customers on $(1 - \alpha)$-fraction of the total purchased amount.

(vi) It is assumed that β-fraction of the total customers are treated as bad or unfaithful customers and the $(1 - \beta)$-fraction of the total customers are good customers, obey the business ethics.

(vii) All units of the item have their expiration dates. The physical significance of the deterioration rate is the rate to be closed to 1 when time is approaching to the maximum lifetime m. The items deteriorates at a rate $\theta(t)$ which depends on time as follow:

$$\theta(t) = \frac{1}{1 + m - t}, 0 \leq t \leq T \leq m \tag{1}$$

Note that it is clear from Eq. (1) that the replenishment cycle time T must be less than or equal to m, and the proposed deterioration rate is a general case for non-deteriorating items, in which $m \to \infty$ and $\theta(t) \to 0$.

(viii) During the credit period offered by the supplier, the retailer earn interest at a rate I_e on the sold units. At the end of the permissible delay period, the retailer pays off the purchase cost to the supplier by a bank loan and pays off the interest to the bank at a rate I_c for the items in stock or the items already sold but have not been paid for yet and for the deteriorated units.

2.2 Notations

(i) $I(t)$, retailer's inventory level at any time t.
(ii) D, marketing demand for the item.
(iii) Q, retailer's order quantity per cycle.
(iv) T, retailer's replenishment time, a decision variable.
(v) c, unit purchasing cost in $.
(vi) s, unit selling price in $ with $(s > c)$, a decision variable.
(vii) A, ordering cost per order in $.
(viii) h, inventory holding cost in $ per year excluding interest charges.
(ix) I_e, earned interest rate per $ per year.
(x) I_c, payable interest rate per $ per year.
(xi) M, retailer's trade credit period offered by the supplier per cycle.
(xii) N, customer's trade credit period offered by the retailer per cycle.
(xiii) $\theta(t)$, deterioration rate of the item at the time t, where $0 \leq \theta(t) < 1$.
(xiv) m, expiration time or maximum lifetime of the item.
(xv) α, a positive real number between 0 and 1.
(xvi) β, a positive real number between 0 and 1.
(xvii) $TP(T, s)$ retailer's total profit in $ per year.

3 Model Development and Analysis

To develop the model mathematically, it is assumed that the inventory starts with Q units, i.e., at time $t = 0$, the retailer receives Q units of the item and then the inventory level gradually depletes to zero due to the combined effect of demand and deterioration. Thus the instantaneous state of the inventory level is given by:

$$\frac{dI(t)}{dt} + \theta(t)I(t) = -D; 0 \le t \le T. \tag{2}$$

with boundary condition $I(0) = Q, I(T) = 0$. Solving the above differential Eq. (2) with the boundary condition $I(T) = 0$, we get the retailer's inventory level at time t as

$$I(t) = D(1 + m - t) \ln\left(\frac{1 + m - t}{1 + m - T}\right) \qquad \text{for } 0 \le t \le T \tag{3}$$

As a result, the retailer's order quantity is

$$Q = I(0) = D(1 + m) \ln\left(\frac{1 + m}{1 + m - T}\right) \tag{4}$$

The relevant costs of the inventory system are as below:

1. **Annual sales revenue,** $SR = sD\{\alpha + (1 - \alpha)(1 - \beta)\}$
2. **Annual ordering cost,** $OC = \frac{A}{T}$.
3. **Annual purchase cost,** $PC = \frac{c}{T}I(0) = \frac{cD(1+m)}{T} \ln\left(\frac{1+m}{1+m-T}\right)$
4. **Annual stock holding cost (AHC) (excluding interest charges) is given by**

$$AHC = \frac{h}{T} \int_0^T I(t)dt = \frac{hD}{T}\left[\frac{(1+m)^2}{2} \ln\left(\frac{1+m}{1+m-T}\right) + \frac{T^2}{4} - \frac{(1+m)T}{2}\right]$$

5. **Annual interest charge (payable/earning) for the retailer:**

Depending upon the values of M (retailer's credit period) and N (customer's credit period), there are two cases may arise: **Case 1**: $N < M$ and **Case 2**: $N \ge M$.

Case 1: $N < M$
In this case, based on the values of M (retailer's delay period), T (retailer's replenishment time) and $(T + N)$ (the time at which the retailer receives the payment from the last customer), three situations may arise, those are:

- **Situation-1.1** $0 < T + N < M$
- **Situation-1.2** $T < M \le T + N$ and
- **Situation-1.3** $M \le T$

For each situation of this case, earned and payable interest of the retailer are presented below:

Situation 1.1: $0 < T + N < M$ (i.e., $0 < T < M - N$)
In this situation, the retailer receives all the returns from the customers before the delay period M. Therefore the retailer not to pay any charged interest whereas the retailer can earn interest from the customers' payment.

Thus the annual interest paid by the retailer is given by

$$IP = 0 \tag{5}$$

On the other hand, interest earned per year is

$$IE = \frac{1}{T}[IE_1 + IE_2], \text{ where} \tag{6}$$

$IE_1 = $ interest earned due to instant payment of sold units during $(0, T]$

$$= I_e \int_0^T \alpha c D(M - t)dt = I_e \alpha c D \left[MT - \frac{T^2}{2} \right] \tag{7}$$

$IE_2 = $ interest earned due to credit payment of sold units during $(0, T]$

$$= I_e \int_0^T (1 - \alpha)(1 - \beta)c D(M - N - t)dt$$

$$= I_e(1 - \alpha)(1 - \beta)c DT (M - N - \frac{T}{2}) \tag{8}$$

Situation 1.2: $T < M \leq T + N$(i.e., $M - N \leq T < M$)
In this situation, the retailer has to pay interest for the customers' credit amount for the sold units during $[M - N, T]$.

Again there are some units, those are deteriorated during $(0, T]$, but the retailer has to pay the full purchased amount to the supplier at the time M by a bank loan and the retailer's account will be settled with the bank at the time $t = (T + N)$. Thus the retailer has to pay interest for those deteriorated units for a time duration $(T + N - M)$. But this phenomenon is overlooked by the researchers.

Thus the retailer has to pay interest per year is

$$IP = (IP_1 + IP_2)/T, \text{ where} \tag{9}$$

$IP_1 = $ interest charged due to credit amount of sold units during $[M - N, T]$

$$= I_c \int_{M-N}^T (1 - \alpha)c D(t - M + N)dt = I_c \frac{(1 - \alpha)c D}{2}(T + N - M)^2 \tag{10}$$

$IP_2 = $ interest charged due to deteriorated units during$(0, T]$

$$= I_c c(T + N - M) \int_0^T \theta(t) I(t) dt$$

$$= I_c c D(T + N - M) \left[(1 + m) \ln \left(\frac{1 + m}{1 + m - T} \right) - T \right] \tag{11}$$

On the other hand, the retailer can earn interest for the collateral deposit of the customers for the sold units during $(0, T]$ and for the repayment of the customers' credit amount for the sold units during $(0, M - N]$.

Thus the retailer's earned interest per year is

$$IE = (IE_1 + IE_2)/T], \quad \text{where} \tag{12}$$

$IE_1 = $ interest earned due to instant payment of sold units in time $(0, T]$

$$= I_e \int_0^T \alpha c D(M - t) dt = I_e \alpha c D \left[MT - \frac{T^2}{2} \right] \tag{13}$$

$IE_2 = $ interest earned due to credit payment of sold units in time $(0, M - N]$

$$= I_e \int_0^{M-N} (1 - \alpha)(1 - \beta) c D(M - N - t) dt$$

$$= I_e (1 - \alpha)(1 - \beta) c D \frac{(M - N)^2}{2} \tag{14}$$

Situation 1.3: $M \leq T$

In this situation the retailer has to pay interest due to

  (i) the stocked units during $[M, T]$.
 (ii) the customers' credit amount for the sold units during $(M - N, M]$
(iii) the customers' credit amount for the sold units during $[M, T]$
 (iv) the deteriorated units during $(0, M]$ and
  (v) the deteriorated units during $(M, T]$.

Thus the retailer has to pay interest per year is

$$IP = (IP_1 + IP_2 + IP_3 + IP_4 + IP_5)/T, \quad \text{where} \tag{15}$$

$IP_1 = $ interest charged due to stock units during $[M, T] = I_c c \int_M^T I(t) dt$

$$= I_c c D \left[\frac{(1 + m - M)^2}{4} \left\{ 2 \ln \left(\frac{1 + m - M}{1 + m - T} \right) - 1 \right\} + \frac{(1 + m - T)^2}{4} \right] \tag{16}$$

$IP_2 = $ interest charged due to the customers' credit of sold units during

$$[M - N, M] = I_c \int_{M-N}^{M} (1 - \alpha)cD(t - M + N)dt = I_c(1 - \alpha)cD\frac{N^2}{2} \quad (17)$$

$I P_3$ = interest charged due to the customers' credit amount of sold units

$$\text{during } [M, T] = I_c \int_{M}^{T} (1 - \alpha)cDNdt = I_c(1 - \alpha)cDN(T - M) \quad (18)$$

$I P_4$ = interest charged due to deteriorated units during $(0, M]$

$$= I_c c(T + N - M) \int_{0}^{M} \theta(t)I(t)dt$$

$$= I_c cD(T + N - M) \left[(1 + m) \ln \left(\frac{1 + m}{1 + m - M} \right) - M \left\{ 1 - \ln \left(\frac{1 + m - M}{1 + m - T} \right) \right\} \right]$$

$$(19)$$

$I P_5$ = interest charged due to deteriorated units during $(M, T]$

$$= I_c c \int_{M}^{T} \theta(t)I(t)(T + N - t)dt$$

$$= I_c cD[\left\{ \frac{M^2}{2} - \frac{(1 + m)^2}{2} + (T + N)(1 + m - M) \right\} \ln \left(\frac{1 + m - M}{1 + m - T} \right)$$

$$+ \frac{T^2 - M^2}{4} - (T + N)(T - M) + \frac{(1 + m)(T - M)}{2}] \quad (20)$$

On the other hand, retailer can earn interest on the initial payment of the customers for the sold units during $(0, M]$ and on the credit payment of the customers for the sold units during $(0, M - N]$.

Thus the retailer's earned interest per year is

$$I E = (I E_1 + I E_2)/T, \quad \text{where} \quad (21)$$

$I E_1$ = interest earned due to instant payment of sold units during $(0M]$

$$= I_e \int_{0}^{M} \alpha cD(M - t)dt = I_e \alpha cD\frac{M^2}{2} \quad (22)$$

$I E_2$ = interest earned due to credit payment of sold units during $(0, M - N]$

$$= I_e \int_{0}^{M-N} (1 - \alpha)(1 - \beta)cD(M - N - t)dt = \frac{I_e}{2}(1 - \alpha)(1 - \beta)cD(M - N)^2$$

$$(23)$$

Therefore, the total annual profit of the retailer when $N < M$ is given by

$$TP_1(T, s) = \begin{cases} TP_{11}(T, s), & \text{if } 0 < T < M - N \\ TP_{12}(T, s), & \text{if } M - N \leq T < M \\ TP_{13}(T, s), & \text{if } M \leq T \end{cases} \tag{24}$$

where

$$TP_{11}(T, s) = sD[\alpha + (1 - \alpha)(1 - \beta)] - \frac{cD(1 + m)}{T} \ln\left(\frac{1 + m}{1 + m - T}\right) - \frac{A}{T}$$

$$- \frac{hD}{T}\left[\frac{(1 + m)^2}{2} \ln\left(\frac{1 + m}{1 + m - T}\right) + \frac{T^2}{4} - \frac{(1 + m)T}{2}\right]$$

$$+ \frac{cDI_e}{2}[T + 2\alpha(M - T) - (1 - \alpha)\beta T + 2(1 - \alpha)(1 - \beta)(M - N - T)] \tag{25}$$

$$TP_{12}(T, s) = sD[\alpha + (1 - \alpha)(1 - \beta)] - \frac{cD(1 + m)}{T} \ln\left(\frac{1 + m}{1 + m - T}\right) - \frac{A}{T}$$

$$- \frac{hD}{T}\left[\frac{(1 + m)^2}{2} \ln\left(\frac{1 + m}{1 + m - T}\right) + \frac{T^2}{4} - \frac{(1 + m)T}{2}\right]$$

$$+ \frac{cDI_e}{2T}\left[\alpha T^2 + 2\alpha T(M - T) + (1 - \alpha)(1 - \beta)(M - N)^2\right] - \frac{cDI_c}{T}(T + N - M)$$

$$\left[\frac{(1 - \alpha)(T + N - M)}{2} + (1 + m)\ln\left(\frac{1 + m}{1 + m - T}\right) - T\right] \tag{26}$$

$$TP_{13}(T, s) = sD[\alpha + (1 - \alpha)(1 - \beta)] - \frac{cD(1 + m)}{T} \ln\left(\frac{1 + m}{1 + m - T}\right) - \frac{A}{T}$$

$$- \frac{hD}{T}\left[\frac{(1 + m)^2}{2} \ln\left(\frac{1 + m}{1 + m - T}\right) + \frac{T^2}{4} - \frac{(1 + m)T}{2}\right]$$

$$+ \frac{cDI_e}{2T}\left[\alpha M^2 + (1 - \alpha)(1 - \beta)(M - N)^2\right] - \frac{cDI_c}{T}\left[(1 + m)(T + N - M)\right.$$

$$\left.\ln\left(\frac{1 + m}{1 + m - T}\right) + (1 - \alpha)N\left(T - M + \frac{N}{2}\right) + \frac{T^2 + M^2}{2} - T(T + N)\right] \tag{27}$$

Case 2: $N \geq M$

In this case, based on the values of M and T, here two situations arise:

- **Situation-2.1** $T < M$ and
- **Situation-2.2** $T \geq M$

Situation 2.1: $T < M$

In this situation, the retailer has to pay interest per year is

$$IP = (IP_1 + IP_2)/T, \quad \text{where} \tag{28}$$

$IP_1 = $ interest charged due to credit amount of sold units during $(0, T]$

$$= I_c \int_0^T (1 - \alpha)cD(t - M + N)dt = \frac{1}{2}I_c(1 - \alpha)cDT\{T + 2(N - M)\} \quad (29)$$

$IP_2 = $ interest charged due to deteriorated units during $(0, T]$

$$= I_c c(T + N - M) \int_0^T \theta(t)I(t)dt$$

$$= I_c c D(T + N - M) \left\{ (1 + m) \ln \left(\frac{1 + m}{1 + m - T} \right) - T \right\} \quad (30)$$

On the other hand, retailer can earn interest for the initial payment of the customers during $(0, T]$.

Thus the retailer's earned interest per year is

$$IE = IE_1/T, \text{ where} \quad (31)$$

$IE_1 = $ interest earned due to instant payment of sold units during $(0, T]$

$$= I_e \int_0^T \alpha c D(M - t)dt = I_e \alpha c D \left[MT - \frac{T^2}{2} \right] \quad (32)$$

Situation 2.2: $M \leq T$ (i.e., $T \geq M$)

In this situation the retailer has to pay interest due to

  (i) the stocked units during $[M, T]$
 (ii) the customers' credit amount for the sold units during $(0, M]$
(iii) the customers' credit amount for the sold units during $[M, T]$
 (iv) the deteriorated units during $(0, M]$ and
  (v) the deteriorated units during $[M, T]$.

Thus the retailer has to pay interest per year is

$$IP = (IP_1 + IP_2 + IP_3 + IP_4 + IP_5)/T, \text{ where} \quad (33)$$

$IP_1 = $ interest charged due to stocked units during $[M, T] = I_c c \int_M^T I(t)dt$

$$= \frac{1}{4}I_c c D \left[(1 + m - M)^2 \left\{ 2 \ln \left(\frac{1 + m - M}{1 + m - T} \right) - 1 \right\} + (1 + m - T)^2 \right] \quad (34)$$

$IP_2 = $ interest charged due to the customers' credit amount of sold units

during $(0, M] = I_c \int_0^M (1 - \alpha)cD(t - M + N)dt = \frac{1}{2}I_c(1 - \alpha)cDM(2N - M)$

(35)

$IP_3 = $ interest charged due to the customers' credit amount of sold units

$$\text{during } [M, T] = I_c \int_M^T (1 - \alpha)cDNdt = I_c(1 - \alpha)cDN(T - M) \qquad (36)$$

$IP_4 = $ interest charged due to deteriorated units during $(0, M]$

$$= I_c c(T + N - M) \int_0^M \theta(t)I(t)dt$$

$$= I_c cD(T + N - M)\left[(1 + m)\ln\left(\frac{1 + m}{1 + m - M}\right) - M\left\{1 - \ln\left(\frac{1 + m - M}{1 + m - T}\right)\right\}\right]$$

(37)

$IP_5 = $ interest charged due to deteriorated units during $[M, T]$

$$= I_c c \int_M^T \theta(t)I(t)(T + N - t)dt$$

$$= I_c cD\left[\left\{\frac{M^2}{2} - \frac{(1 + m)^2}{2} + (T + N)(1 + m - M)\right\}\ln\left(\frac{1 + m - M}{1 + m - T}\right) + \right.$$
$$\left. \frac{T^2 - M^2}{4} - (T + N)(T - M) + \frac{(1 + m)(T - M)}{2}\right]$$

(38)

On the other hand, the retailer's earned interest per year is

$$IE = IE_1/T, \quad \text{where} \qquad (39)$$

$IE_1 = $ interest earned due to the customers' instant payment of sold units

$$\text{during } (0, M] = I_e \int_0^M \alpha cD(M - t)dt = I_e \alpha cD\frac{M^2}{2} \qquad (40)$$

Therefore, the retailer's annual total profit for $N \geq M$ is given by

$$TP_2(T, s) = \begin{cases} TP_{21}(T, s), & \text{if } T < M \\ TP_{22}(T, s), & \text{if } T \geq M \end{cases} \qquad (41)$$

where

$$TP_{21}(T, s) = sD[\alpha + (1 - \alpha)(1 - \beta)] - \frac{cD(1 + m)}{T}\ln\left(\frac{1 + m}{1 + m - T}\right) - \frac{A}{T}$$

$$-\frac{hD}{T}\left[\frac{(1+m)^2}{2}\ln\left(\frac{1+m}{1+m-T}\right)+\frac{T^2}{4}-\frac{(1+m)T}{2}\right]$$

$$+\frac{\alpha cDI_e}{2}(2M-T)-\frac{cDI_c}{T}\left[(1-\alpha)T\left(\frac{T}{2}+N-M\right)\right.$$

$$+(T+N-M)\left\{(1+m)\ln\left(\frac{1+m}{1+m-T}\right)-T\right\}\right] \tag{42}$$

$$TP_{22}(T,s)=sD\left[\alpha+(1-\alpha)(1-\beta)\right]-\frac{cD(1+m)}{T}\ln\left(\frac{1+m}{1+m-T}\right)-\frac{A}{T}$$

$$-\frac{hD}{T}\left[\frac{(1+m)^2}{2}\ln\left(\frac{1+m}{1+m-T}\right)+\frac{T^2}{4}-\frac{(1+m)T}{2}\right]+\frac{cDI_e}{2T}\alpha M^2$$

$$-\frac{cDI_c}{T}\left[(T+N-M)(1+m)\ln\left(\frac{1+m}{1+m-T}\right)+(1-\alpha)\left(NT-\frac{M^2}{2}\right)\right.$$

$$+\frac{T^2+M^2}{2}-T(T+N)\right] \tag{43}$$

Hence the problem is to

$$\text{maximize } TP(T,s)=\begin{cases} TP_1(T,s), \text{ if } N<M \\ TP_2(T,s), \text{ if } N\geq M \end{cases} \tag{44}$$

where $TP_i(T,s), i=1,2$ is defined in Eqs. (24) and (41) respectively.

It is to be noted that $TP_{11}(M-N,s)=TP_{12}(M-N,s)$, $TP_{12}(M,s)=TP_{13}(M,s)$ and $TP_{21}(M,s)=TP_{22}(M,s)$. Hence $TP_i(T,s)$ is a continuous function $\forall\ s$ and $T>0$ for $i=1,2$.

4 Theoretical Discussion for Existence of a Solution

In this section, a theoretical discussion is made for each of the situations of the proposed model to find the retailer's optimal replenishment time (T) and the optimal selling price (s) in such a way the retailer's annual profit is maximum.

4.1 Discussion for an Optimal Solution to the Situation-1.1

For a fixed selling price (s), the first order partial derivative of $TP_{11}(T,s)$ with respect to T is

$$\frac{\partial TP_{11}(T,s)}{\partial T}=\frac{F_{11}(T)}{T^2} \tag{45}$$

where

$$F_{11}(T) = A + (1 + m)D\left\{c + \frac{h(1 + m)}{2}\right\}\left\{\ln\left(\frac{1 + m}{1 + m - T}\right) - \frac{T}{1 + m - T}\right\}$$

$$-\frac{DT^2}{4}\left\{h + 2cI_e\{1 - (1 - \alpha)\beta\}\right\} \tag{46}$$

Differentiating $F_{11}(T)$ with respect to $T \in (0, M - N]$ we have

$$F'_{11}(T) = \frac{dF_{11}(T)}{dT}$$

$$= -DT\left[\frac{(1 + m)}{(1 + m - T)^2}\left\{c + \frac{h(1 + m)}{2}\right\} + \frac{h}{2} + cI_e\{1 - (1 - \alpha)\beta\}\right] < 0 \tag{47}$$

Thus $F_{11}(T)$ is strictly decreasing function with respect to $T \in (0, M - N]$. Moreover, $\lim_{T \to \infty} F_{11}(T) = -\infty$, $F_{11}(0) = A > 0$ and

$$F_{11}(M - N) = A + (1 + m)D\left\{c + \frac{h(1 + m)}{2}\right\}\left\{\ln\left(\frac{1 + m}{1 + m - M + N}\right) - \frac{M - N}{1 + m - M + N}\right\}$$

$$- \frac{h}{4}D(M - N)^2 - \frac{DcI_e}{2}(M - N)^2\{1 - (1 - \alpha)\beta\} = \Delta_1(say) \tag{48}$$

If $\Delta_1 \leq 0$ then by Intermediate Value Theorem there exists a unique value of T (say $T_{11} \in (0, M - N]$) such that $F_{11}(T_{11}) = 0$.

Again, $\left[\frac{\partial^2 T P_{11}(T, s)}{\partial T^2}\right]_{T = T_{11}} = \left[-\frac{2}{T^3}F_{11}(T) + \frac{1}{T^2}\frac{dF_{11}(T)}{dT}\right]_{T = T_{11}}$

$$= -\frac{2}{T_{11}^3}F_{11}(T_{11}) - \frac{D}{T_{11}}\left[\frac{(1 + m)}{(1 + m - T_{11})^2}\left\{c + \frac{h(1 + m)}{2}\right\} + \frac{h}{2} + cI_e\{1 - (1 - \alpha)\beta\}\right]$$

$$= -\frac{D}{T_{11}}\left[\frac{(1 + m)}{(1 + m - T_{11})^2}\left\{c + \frac{h(1 + m)}{2}\right\} + \frac{h}{2} + cI_e\{1 - (1 - \alpha)\beta\}\right] < 0 \tag{49}$$

Thus $T P_{11}(T, s)$ has a global maxima at the point $T = T_{11}$.

Conversely, if $\Delta_1 > 0$ we have $F_{11}(T) > 0 \ \forall T \in (0, M - N]$ which implies $T P_{11}(T, s)$ is a strictly increasing function of T for a fixed s. Hence $T P_{11}(T, s)$ has a maximum value at the boundary point $T = (M - N)$.

From this discussion, it is seen that for a fixed s, there exist an optimal $T = T_{11}^*$ such that $T P_{11}$ is maximized, where T_{11}^* is given by the Eq. (50).

$$T_{11}^* = \begin{cases} T_{11}, & \text{if } \Delta_1 \leq 0 \\ (M - N), & \text{if } \Delta_1 > 0 \end{cases} \tag{50}$$

Again for the fixed T_{11}^* defined in Eq. (50), the first order partial derivative of $TP_{11}(T_{11}^*, s)$ with respect to s gives

$$\frac{\partial TP_{11}(T_{11}^*, s)}{\partial s} = -\frac{(b-1)}{s^b}\eta_1 + \frac{b}{s^{b+1}}\eta_2 + \frac{b}{s^{b+1}}\eta_3 - \frac{b}{s^{b+1}}\eta_4 \qquad (51)$$

where

$$\eta_1 = a\{\alpha + (1-\alpha)(1-\beta)\} \qquad (52)$$

$$\eta_2 = \frac{ac}{T}(1+m)log(\frac{1+m}{1+m-T}) \qquad (53)$$

$$\eta_3 = \frac{ah}{T}[\frac{(1+m)^2}{T}log(\frac{1+m}{1+m-T}) + \frac{T^2}{4} - \frac{(1+m)}{2}T] \qquad (54)$$

$$\eta_4 = \frac{acI_e}{2}[T + 2\alpha(M-T) - (1-\alpha)\beta T + 2(1-\alpha)(1-\beta)(M-N-T)] \qquad (55)$$

Now, $\frac{\partial TP_{11}(T_{11}^*, s)}{\partial s} = 0$ implies that

$$s = \frac{b(\eta_2 + \eta_3 - \eta_4)}{(b-1)\eta_1} = s_{11}^* \text{ (say)} \qquad (56)$$

Again at $s = s_{11}^*$,

$$\frac{\partial^2 TP_{11}(T_{11}^*, s)}{\partial s^2} = -\frac{(b-1)}{s_{11}^{*b+1}}\eta_1 < 0 \qquad (57)$$

Thus from the above discussion it is evident that for an optimal replenishment time (T_{11}^*) there is an optimal retail selling price (s_{11}^*) for which annual retailer's profit $(TP_{11}(T, s))$ is maximum and hence there is an optimal solution of the proposed model for the situation-1.1.

4.2 Discussion for an Optimal Solution to the Situation-1.2

For a fixed selling price (s), the first order partial derivative of $TP_{12}(T, s)$ with respect to T is

$$\frac{\partial TP_{12}(T, s)}{\partial T} = \frac{F_{12}}{T^2} \qquad (58)$$

where

$$F_{12}(T) = A + (1+m)D\left\{c + \frac{h(1+m)}{2}\right\}\left\{\ln\left(\frac{1+m}{1+m-T}\right) - \frac{T}{1+m-T}\right\}$$

$$-\frac{hDT^2}{4} - \frac{cDI_e}{2}\{\alpha T^2 + (1-\alpha)(1-\beta)(M-N)^2\}$$

$$+cDI_cT^2 - \frac{cDI_c(1-\alpha)}{2}\{T^2 - (M-N)^2\}$$

$$-cDI_c(1+m)\left\{(M-N)\ln\left(\frac{1+m}{1+m-T}\right) + \frac{T(T+N-M)}{1+m-T}\right\} \quad (59)$$

Differentiating $F_{12}(T)$ with respect to $T \in [M-N, M]$ we have

$$F'_{12}(T) = \frac{dF_{12}(T)}{dT}$$

$$= -DT\left[\frac{(1+m)}{(1+m-T)^2}\left\{c + \frac{h(1+m)}{2}\right\} + \frac{h}{2} + cI_e\alpha + \right.$$

$$\left. cI_c\left\{(1-\alpha) + \frac{2T}{1+m-T} + \frac{(1+m)(T+N-M)}{(1+m-T)^2}\right\}\right] < 0 \quad (60)$$

Thus $F_{12}(T)$ is strictly decreasing function with respect to $T \in [M-N, M]$. Moreover, $\lim_{T\to\infty} F_{12}(T) = -\infty$. Now,

$$F_{12}(M-N) = A + (1+m)D\left\{c + \frac{h(1+m)}{2}\right\}\left\{\ln\left(\frac{1+m}{1+m-M+N}\right) - \right.$$

$$\left. \frac{M-N}{1+m-M+N}\right\} - (M-N)^2[\frac{hD}{4} + \frac{cDI_e}{2}\{1-\beta(1-\alpha)\}]$$

$$-cDI_c(M-N)\left\{(1+m)\ln\left(\frac{1+m}{1+m-M+N}\right) - (M-N)\right\} \text{ and} \quad (61)$$

$$F_{12}(M) = A + (1+m)D\left\{c + \frac{h(1+m)}{2}\right\}\left\{\ln\left(\frac{1+m}{1+m-M}\right) - \frac{M}{1+m-M}\right\}$$

$$-\frac{hDM^2}{4} - \frac{cDI_e}{2}\left\{\alpha M^2 + (1-\alpha)(1-\beta)(M-N)^2\right\}$$

$$+\frac{cDI_c}{2}\left\{2M^2 - (1-\alpha)N(2M-N)\right\}$$

$$-cDI_c(1+m)\left\{(M-N)\ln\left(\frac{1+m}{1+m-M}\right) + \frac{MN}{1+m-M}\right\} \quad (62)$$

Here, three scenarios may arise:

Scenario-1: $F_{12}(M) < F_{12}(M-N) < 0$

Then, $F_{12}(T) < 0$ for all $T \in [M-N, M]$ and hence $TP_{12}(T,s)$ is a decreasing function of $T \in [M-N, M]$ for a fixed s and hence $TP_{12}(T,s)$ attain its optimal value at $T = (M-N)$.

Scenario-2: $F_{12}(M) < F_{12}(M-N) > 0$

Then, $F_{12}(T) > 0$ for all $T \in [M - N, M]$ and hence $TP_{12}(T, s)$ is a increasing function of $T \in [M - N, M]$ for a fixed s and hence $TP_{12}(T, s)$ attain its optimal value at $T = M$.

Scenario-3: $F_{12}(M) < 0$ but $F_{12}(M - N) > 0$

Then, $\exists$ a $T = T_{12}(say) \in [M - N, M]$ such that $F_{12}(T_{12}) = 0$.

$$
\begin{aligned}
\text{Now, } \left[\frac{\partial^2 TP_{12}(T, s)}{\partial T^2}\right]_{T=T_{12}} &= \left[-\frac{2}{T^3}F_{12}(T) + \frac{1}{T^2}\frac{dF_{12}(T)}{dT}\right]_{T=T_{12}} \\
&= -\frac{2}{T_{12}^3}F_2(T_{12}) - \frac{D}{T_{12}}\left[\frac{(1+m)}{(1+m-T_{12})^2}\left\{c + \frac{h(1+m)}{2}\right\} + (\frac{h}{2} + cI_e\alpha) + \right. \\
&\quad \left. cI_c\left\{(1-\alpha) + \frac{2T}{1+m-T_{12}} + \frac{(1+m)(T_{12}+N-M)}{(1+m-T_{12})^2}\right\}\right]
\end{aligned}
\tag{63}
$$

$$
\begin{aligned}
&= -\frac{D}{T_{12}}\left[\frac{(1+m)}{(1+m-T_{12})^2}\left\{c + \frac{h(1+m)}{2}\right\} + (\frac{h}{2} + cI_e\alpha) + \right. \\
&\quad \left. cI_c\left\{(1-\alpha) + \frac{2T}{1+m-T_{12}} + \frac{(1+m)(T_{12}+N-M)}{(1+m-T_{12})^2}\right\}\right] < 0.
\end{aligned}
\tag{64}
$$

Thus for a fixed s, there exist an optimal $T = T_{12}^*$ such that $TP_{12}(T, s)$ is maximized, where T_{12}^* is given by the Eq. (65).

$$
T_{12}^* = \begin{cases} (M - N), & \text{if } F_{12}(M) < F_{12}(M - N) < 0 \\ M, & \text{if } F_{12}(M) < F_{12}(M - N) > 0 \\ T_{12}, & \text{if } F_{12}(M) < 0 \text{ but } F_{12}(M - N) > 0 \end{cases}
\tag{65}
$$

Again for the fixed T_{12}^* defined in Eq. (65), the first order partial derivative of $TP_{12}(T_{12}^*, s)$ with respect to s gives

$$
\frac{\partial TP_{12}(T_{12}^*, s)}{\partial s} = -\frac{(b-1)}{s^b}\eta_1 + \frac{b}{s^{b+1}}\eta_2 + \frac{b}{s^{b+1}}\eta_3 - \frac{b}{s^{b+1}}\eta_4 + \frac{b}{s^{b+1}}\eta_5,
\tag{66}
$$

where, η_1, η_2, η_3 are same as Eqs. (52), (53), (54) respectively and

$$
\eta_4 = \frac{acI_e}{2T}[\alpha T^2 + 2\alpha T(M - T) + (1 - \alpha)(1 - \beta)(M - N)^2]
\tag{67}
$$

$$
\eta_5 = \frac{acI_c}{T}(T + N - M)[\frac{(1-\alpha)}{2}(T + N - M) + (1 + m)\log(\frac{1+m}{1+m-T}) - T]
\tag{68}
$$

Now, $\frac{\partial TP_{12}(T_{12}^*, s)}{\partial s} = 0$ implies that

$$
s = \frac{b(\eta_2 + \eta_3 - \eta_4 + \eta_5)}{(b-1)\eta_1} = s_{12}^* \text{ (say)}
\tag{69}
$$

Again at $s = s_{12}^*$,

$$\frac{\partial^2 T P_{12}(T_{12}^*, s)}{\partial s^2} = -\frac{(b-1)}{s_{12}^{*b+1}} \eta_1 < 0 \tag{70}$$

Thus from the above discussion it is observed that there is an optimal solution (T_{12}^*, s_{12}^*) of the proposed model for the situation-1.2, for which the retailer's annual profit $(T P_{12}(T, s))$ should be maximum.

4.3 Discussion for an Optimal Solution to the Situation-1.3

Likewise, for a fixed s the first order partial derivative of $T P_{13}(T, s)$ with respect to T is given by

$$\frac{\partial T P_{13}(T)}{\partial T} = \frac{F_{13}}{T^2} \tag{71}$$

where

$$F_{13}(T) = A + (1+m)D\left\{c + \frac{h(1+m)}{2}\right\}\left\{\ln\left(\frac{1+m}{1+m-T}\right) - \frac{T}{1+m-T}\right\}$$

$$-\frac{hDT^2}{4} - \frac{cDI_e}{2}\left\{\alpha M^2 + (1-\alpha)(1-\beta)(M-N)^2\right\}$$

$$+\frac{cDI_c}{2}\left\{T^2 + M^2 - (1-\alpha)N(2M-N)\right\}$$

$$-cDI_c(1+m)\left\{(M-N)\ln\left(\frac{1+m}{1+m-T}\right) + \frac{T(T+N-M)}{1+m-T}\right\} \tag{72}$$

Differentiating $F_{13}(T)$ with respect to $T \in [M, \infty)$ we have

$$F_{13}'(T) = \frac{dF_{13}(T)}{dT}$$

$$= -DT\left[\frac{(1+m)}{(1+m-T)^2}\left\{c + \frac{h(1+m)}{2}\right\} + \frac{h}{2} - cI_c\right.$$

$$\left.+\frac{cI_c(1+m)}{(1+m-T)^2}\left\{(1+m-T) + (1+m-M+N)\right\}\right] < 0 \tag{73}$$

Thus $F_{13}(T)$ is strictly decreasing function with respect to $T \in [M, \infty)$. Moreover, $\lim_{T \to \infty} F_{13}(T) = -\infty$. Now,

$$F_{13}(M) = A + (1+m)D\left\{c + \frac{h(1+m)}{2}\right\}\left\{\ln\left(\frac{1+m}{1+m-M}\right) - \frac{M}{1+m-M}\right\}$$

$$-\frac{hDM^2}{4} - \frac{cDI_e}{2}\left\{\alpha M^2 + (1-\alpha)(M-N)^2\right\} + \frac{cDI_c}{2}\left\{2M^2 - (1-\alpha)\right.$$

$$N(2M - N)\Big\} - cDI_c(1 + m)\Big\{(M - N)\ln\left(\frac{1 + m}{1 + m - M}\right) + \frac{MN}{1 + m - M}\Big\} \quad (74)$$

If $F_{13}(M) < 0$ we have $F_{13}(T) < 0$ for all $T \in [M, \infty)$, which implies that $TP_{13}(T, s)$ is strictly decreasing function of $T \in [M, \infty)$. Hence $TP_{13}(T, s)$ has a maximum value at the boundary point $T = M$.

Conversely if $F_{13}(M) \geq 0$, then by Intermediate Value Theorem there exists a unique value of $T = T_{13}(say) \in [M, \infty)$ such that $F_{13}(T_{13}) = 0$. Again at $T = T_{13}$,

$$[\frac{d^2 TP_{13}(T)}{dT^2}]_{T=T_{13}} = \frac{F_{13}'(T_{13})}{T_{13}^2} - \frac{2F_{13}(T_{13})}{T_{13}^3} < 0, \text{ as } F_{13}'(T) < 0 \forall T \in [M, \infty) (75)$$

Thus for a fixed s, there exist an optimal $T = T_{13}^*$ such that $TP_{13}(T, s)$ is maximized, where T_{13}^* is given by the Eq. (76).

$$T_{13}^* = \begin{cases} M, \text{ if } F_{13}(M) < 0 \\ T_{13}, \text{ if } F_{13}(M) > 0 \end{cases} \quad (76)$$

Again for the fixed T_{13}^* defined in Eq. (76), the first order partial derivative of $TP_{13}(T_{13}^*, s)$ with respect to s gives

$$\frac{\partial TP_{13}(T_{13}^*, s)}{\partial s} = -\frac{(b - 1)}{s^b}\eta_1 + \frac{b}{s^{b+1}}\eta_2 + \frac{b}{s^{b+1}}\eta_3 - \frac{b}{s^{b+1}}\eta_4 + \frac{b}{s^{b+1}}\eta_5, \quad (77)$$

where η_1, η_2, η_3 are respectively given by Eqs. (52), (53), (54) and

$$\eta_4 = \frac{acI_e}{2T}[\alpha M^2 + (1 - \alpha)(1 - \beta)(M - N)^2] \quad (78)$$

$$\eta_5 = \frac{acI_c}{T}[(1 + m)(T + N - M)\log(\frac{1 + m}{1 + m - T}) + (1 - \alpha)N(T - M + \frac{N}{2})$$

$$+ \frac{T^2 + M^2}{2} - T(T + N)] \quad (79)$$

Now, $\frac{\partial TP_{13}(T_{13}^*, s)}{\partial s} = 0$ implies that

$$s = \frac{b(\eta_2 + \eta_3 - \eta_4 + \eta_5)}{(b - 1)\eta_1} = s_{13}^* \text{ (say)} \quad (80)$$

Again at $s = s_{13}^*$,

$$\frac{\partial^2 TP_{12}(T_{12}^*, s)}{\partial s^2} = -\frac{(b - 1)}{s_{13}^{*b+1}}\eta_1 < 0 \quad (81)$$

Thus for the situation-1.3 also, it is seen that there exists an optimal solution (T_{13}^*, s_{13}^*) for which the retailer's total annual profit $(TP_{13}(T, s))$ is maximum.

4.4 Discussion for an Optimal Solution to the Situation-2.1

For a fixed s, the 1st order partial derivative of $TP_{21}(T, s)$ with respect to T is

$$
\frac{\partial TP_{21}(T)}{\partial T} = \frac{1}{T^2}\left[A + (1+m)D\left\{ c + \frac{h(1+m)}{2} \right\}\left\{ \ln\left(\frac{1+m}{1+m-T} \right) - \right. \right.
$$
$$
\left. \frac{T}{1+m-T} \right\} - \frac{hDT^2}{4} - \frac{cDI_e}{2}\alpha T^2 - cDI_c\left\{ \frac{1}{2}(1-\alpha)T^2 + \frac{T+N-M}{1+m-T}T^2 \right.
$$
$$
\left. \left. -(N-M)\left\{ (1+m)\ln\left(\frac{1+m}{1+m-T} \right) - T \right\} \right\} \right] \tag{82}
$$

Proceeding similarly as in situation-1.1, there exists an unique value $T = T_{21}(say)$ $\in (0, M]$ at which $\frac{\partial TP_{21}(T,s)}{\partial T} = 0$.

Hence for a fixed s, the profit function $TP_{21}(T, s)$ is maximum at an optimal value $T = T_{21}^*(say)$ given by the Eq. (83).

$$
T_{21}^* = \begin{cases} M, & \text{if } \Delta_2 < 0 \\ T_{21}, & \text{if } \Delta_2 \geq 0 \end{cases} \tag{83}
$$

where

$$
\Delta_2 = A + (1+m)D\left\{ c + \frac{h(1+m)}{2} \right\}\left\{ \ln\left(\frac{1+m}{1+m-M} \right) - \frac{M}{1+m-M} \right\} -
$$
$$
\frac{hDM^2}{4} - \frac{cDI_e}{2}\alpha M^2 - cDI_c\left\{ \frac{1}{2}(1-\alpha)M^2 + \frac{NM^2}{1+m-M} - \right.
$$
$$
\left. (N-M)\left\{ (1+m)\ln\left(\frac{1+m}{1+m-M} \right) - M \right\} \right\} \tag{84}
$$

Again for the fixed T_{21}^* defined in Eq. (83), the first order partial derivative of $TP_{21}(T_{21}^*, s)$ with respect to s gives

$$
\frac{\partial TP_{21}(T_{21}^*, s)}{\partial s} = -\frac{(b-1)}{s^b}\eta_1 + \frac{b}{s^{b+1}}\eta_2 + \frac{b}{s^{b+1}}\eta_3 - \frac{b}{s^{b+1}}\eta_4 + \frac{b}{s^{b+1}}\eta_5, \tag{85}
$$

where η_1, η_2, η_3 are respectively given by Eqs. (52), (53), (54) and

$$
\eta_4 = \frac{ac\alpha I_e}{2}(2M - T) \tag{86}
$$
$$
\eta_5 = \frac{acI_c}{T}\left[(1+m)(\frac{T}{2} + N - M) + (T + N - M) \right.
$$
$$
\left. \left\{ (1+m)\ln\left(\frac{1+m}{1+m-T} \right) - T \right\} \right] \tag{87}
$$

Now, $\frac{\partial T P_{21}(T_{21}^*,s)}{\partial s} = 0$ implies that

$$s = \frac{b(\eta_2 + \eta_3 - \eta_4 + \eta_5)}{(b-1)\eta_1} = s_{21}^* \text{ (say)} \tag{88}$$

Again at $s = s_{21}^*$,

$$\frac{\partial^2 T P_{21}(T_{21}^*, s)}{\partial s^2} = -\frac{(b-1)}{s_{21}^{*b+1}}\eta_1 < 0 \tag{89}$$

Hence from the above discussion it is evident that there is an optimal solution of the proposed model for the situation-2.1.

4.5 Discussion for an Optimal Solution to the Situation-2.2

Similarly for a fixed s, the first order partial derivative of $T P_{22}(T, s)$ with respect to T is

$$\frac{\partial T P_{22}(T, s)}{\partial T} = \frac{1}{T^2}\left[A + (1+m)D\left\{c + \frac{h(1+m)}{2}\right\}\left\{\ln\left(\frac{1+m}{1+m-T}\right) - \right.\right.$$
$$\left.\frac{T}{1+m-T}\right\} - \frac{hDT^2}{4} - \frac{cDI_e}{2}\alpha M^2 - cDI_c(1+m)\left\{(M-N)\ln\left(\frac{1+m}{1+m-T}\right)\right.$$
$$\left.+\frac{T(T+N-M)}{1+m-T}\right\} + \frac{cDI_c}{2}\left\{T^2 + M^2 - (1-\alpha)M^2\right\}\right] \tag{90}$$

Proceeding similarly as in situation-1.3, there exists an unique value $T = T_{22}$(say) $\in [M, \infty)$ at which $\frac{\partial T P_{22}(T,s)}{\partial T} = 0$.

Hence for a fixed s, the profit function $T P_{22}(T, s)$ is maximum at an optimal value $T = T_{22}^*$(say), given by the Eq. (91).

$$T_{22}^* = \begin{cases} M, & \text{if } \Delta_3 < 0 \\ T_{22}, & \text{if } \Delta_3 \geq 0 \end{cases} \tag{91}$$

where

$$\Delta_3 = A + (1+m)D\left\{c + \frac{h(1+m)}{2}\right\}\left\{\ln\left(\frac{1+m}{1+m-M}\right) - \frac{M}{1+m-M}\right\}$$
$$-\frac{hDM^2}{4} - \frac{cDI_e}{2}\alpha M^2 - cDI_c$$
$$\left[(1+m)\left\{\frac{NM}{1+m-M} - (N-M)\ln\left(\frac{1+m}{1+m-M}\right)\right\} - \frac{(1+\alpha)}{2}M^2\right] \tag{92}$$

Again for the fixed T_{22}^* defined in Eq. (91), the first order partial derivative of $T P_{22}(T_{22}^*, s)$ with respect to s gives

$$\frac{\partial T P_{22}(T_{22}^*, s)}{\partial s} = -\frac{(b-1)}{s^b}\eta_1 + \frac{b}{s^{b+1}}\eta_2 + \frac{b}{s^{b+1}}\eta_3 - \frac{b}{s^{b+1}}\eta_4 + \frac{b}{s^{b+1}}\eta_5, \quad (93)$$

where, η_1, η_2, η_3 are same as Eqs. (52), (53), (54) respectively and

$$\eta_4 = \frac{acI_e}{2}\alpha M^2 \quad (94)$$

$$\eta_5 = \frac{acI_c}{T}\left[(T + N - M)(1 + m)\ln\left(\frac{1+m}{1+m-T}\right) + (1-\alpha)\left(NT - \frac{M^2}{2}\right)\right.$$
$$\left. + \frac{1}{2}(T^2 + M^2) - T(T + N)\right] \quad (95)$$

Now, $\frac{\partial T P_{22}(T_{22}^*, s)}{\partial s} = 0$ implies that

$$s = \frac{b(\eta_2 + \eta_3 - \eta_4 + \eta_5)}{(b-1)\eta_1} = s_{22}^* \text{ (say)} \quad (96)$$

Again at $s = s_{22}^*$,

$$\frac{\partial^2 T P_{22}(T_{22}^*, s)}{\partial s^2} = -\frac{(b-1)}{s_{22}^{*b+1}}\eta_1 < 0 \quad (97)$$

Thus from the above discussion it is seen that there exists an optimal replenishment time (T_{22}^*) and an optimal selling price (s_{22}^*) for which annual retailer's profit $(T P_{22}(T, s))$ is maximum and hence there is a feasible solution of the proposed model for the situation-2.2.

5 Numerical Experiments

From Section-4, it is evident that there exists an optimal solution of the proposed model. To find the optimal solution of the proposed model, a soft computing technique, Particle Swarm Optimization (PSO) [11] algorithm is developed using C-language. In this section PSO is used to run several numerical examples in order to illustrate the model as well as to gain some managerial insights.

Here five numerical examples are taken with different parametric values, results are obtained and presented in Table 1.

Example-1: The following parametric values are considered for Example-1:
$a = 5 \times 10^6$, $b = 2.3$, $A = 120$, $h = 7$, $c = 15$, $I_c = 0.15$, $I_e = 0.12$, $m = 1$ year, $M = 60/365$, $N = 10/365$, $\alpha = 0.5$ and $\beta = 0.05$.

Table 1 Results of the proposed model for the Examples 1-5

Example	T^*	s^*	Q^*	$TP^*(T^*, s^*)$	Result belongs to the
1	0.0777	27.90	187.39	26437.10	Situation-1.1
2	0.0624	22.96	235.07	33570.64	Situation-1.2
3	0.0974	34.40	145.95	19796.96	Situation-1.3
4	0.0531	18.77	316.68	44658.81	Situation-2.1
5	0.0395	14.72	410.86	62822.47	Situation-2.2

Example-2: In the Example-2, assumed values of the parameters are given below:
$a = 5 \times 10^6$, $b = 2.3$, $A = 100$, $h = 6$, $c = 12$, $I_c = 0.10$, $I_e = 0.08$, $m = 1$ year, $M = 45/365$, $N = 30/365$, $\alpha = 0.3$ and $\beta = 0.07$.

Example-3: The assumed parametric values of Example-3 are as follows:
$a = 5 \times 10^6$, $b = 2.3$, $A = 150$, $h = 10$, $c = 18$, $I_c = 0.12$, $I_e = 0.10$, $m = 1$ year, $M = 25/365$, $N = 15/365$, $\alpha = 0.6$ and $\beta = 0.06$.

Example-4: The assumed parametric values of Example-5 are as follows:
$a = 5 \times 10^6$, $b = 2.3$, $A = 130$, $h = 9$, $c = 10$, $I_c = 0.14$, $I_e = 0.11$, $m = 1$ year, $M = 30/365$, $N = 45/365$, $\alpha = 0.5$ and $\beta = 0.04$.

Example-5: The assumed parametric values of Example-4 are as follows:
$a = 5 \times 10^6$, $b = 2.3$, $A = 80$, $h = 5$, $c = 8$, $I_c = 0.10$, $I_e = 0.08$, $m = 1$ year, $M = 10/365$, $N = 15/365$, $\alpha = 0.6$ and $\beta = 0.04$.

5.1 Parametric Study on M and N

To gain some managerial insights for the retailer, a parametric study with respect to retailer's credit period (M) and customers' credit period (N) is performed here. The results are obtained with the parametric values of the Example-1 and presented in Table 2.

Based on the results of Table 2, some managerial insights are found, which are listed below:

- From Table 2, it is observed that for a fixed M, when the customers' credit period increases then the retail selling price increases. With the increase of selling price demand decreases and for that retailer's order quantity decreases as well as retailer's annual profit decreases.
- Also from Table 2, it is observed that for a fixed N, when retailer's credit period increases then the retail selling price decreases and with the less selling price demand increases as well as retailer's annual profit increases.

Table 2 Parametric study on M and N

M	N	T^*	s^*	Q^*	TP^*
10/365	10/365	0.0777	28.37	180.39	25838.24
	15/365	0.0775	28.40	179.61	25801.80
	30/365	0.0779	28.49	179.11	25692.10
	45/365	0.0782	28.59	178.47	25584.96
	60/365	0.0784	28.67	177.62	25477.70
15/365	10/365	0.08	28.35	182.36	25882.46
	15/365	0.0775	28.35	180.23	25870.27
	30/365	0.0776	28.43	179.38	25760.97
	45/365	0.0781	28.53	179.06	25652.46
	60/365	0.0782	28.62	178.02	25544.73

Table 3 Effect of α

α	T^*	s^*	Q^*	TP^*
0.00	0.0788	28.99	174.26	24417.25
0.10	0.0785	28.82	176.02	24761.91
0.20	0.0781	28.63	177.60	25109.57
0.30	0.0777	28.45	179.33	25460.24
0.40	0.0772	28.28	180.55	25813.93
0.50	0.0770	28.10	182.79	26170.67
0.60	0.0767	27.94	184.42	26530.43
0.70	0.0761	27.75	185.94	26893.29
0.80	0.0758	27.58	187.78	27259.20
0.90	0.0755	27.41	189.70	27628.19
1.00	0.0751	27.25	191.31	28000.28

5.2 Effect of Customers' Collateral Deposit (α) on the Proposed Model

Here a study is made for the proposed model with respect to customers' collateral deposit (α) with the same parametric values as taken in Example-1 except $M = 45/365$ and $N = 30/365$. Results are obtained and presented in Table 3, form which some managerial insights are outlined.

From Table 3, it is seen that if the customers' collateral deposit increases then the optimal selling price decreases. Since the selling price decreases so the base demand increases and for that the retailer's annual profit increases, which lead to a realistic phenomenon.

Table 4 A sensitivity analysis with respect to m

m	T^*	s^*	Q^*	$TP^*(T^*, s^*)$
0.6	0.0728	28.16	172.52	25995.36
0.8	0.0748	28.13	177.51	26091.61
1.0	0.0770	28.10	182.79	26170.67
1.2	0.0786	28.07	186.69	26236.85
1.4	0.0802	28.05	190.75	26293.04

5.3 A Sensitivity Analysis with Respect to Expiration Time of the Item (m)

Here a sensitivity analysis is performed with respect to the expiration time (m) of the item for the proposed model with the parametric values as taken in Example-1 except $M = 45/365$ and $N = 30/365$. Results are obtained and presented in Table 4. From the results of Table 4, it is observed that if the expiration time of an item increases then the retailer's replenishment time (T) also increases and for that the selling price (s) decreases. Due to the less selling price, base demand (D) increases as well as the retailer's profit increases. This observation lead to a natural phenomenon.

6 Conclusion

In this study for the first time an EOQ model has developed for a deteriorating item under two level partial trade credit period incorporating the interest paid due to the deteriorated units of the item in the profit function. Moreover in this study the unfaithfulness nature of the customers are taken into account to reduce the default credit risk of the inventory system. To solve the proposed model a soft computing technique is developed and the results are taken for the numerical illustration of the model. Also from the numerical study some managerial insights are outlined, which help a decision maker to draw a optimal business strategy. Due to the generalized nature of the proposed model, it can be extended in various circumstances, such as under inflation, retailer's discount policy, with various forms of demand etc.

References

1. Aggarwal, S.P., Jaggi, C.K.: Ordering policies of deteriorating items under permissible delay in payments. J. Oper. Res. Soc. **46**, 658–662 (1995)
2. Bakker, M., Riezebos, J., Teunter, R.H.: Review of inventory systems with deterioration since 2001. Europ. J. Oper. Res. **221**(2), 275–284 (2012)

3. Chao, X., Gong, X and Zheng, S.: Optimal pricing and inventory policies with reliable and random-yield suppliers: Characterization and comparison. Ann. Oper. Res. https://doi.org/10.1007/s10479-014-1547-0 (2014)
4. Chung, K.J.: Comments on the EOQ model under retailer partial trade credit policy in supply chain. Int. J. Prod. Econ. **114**, 308–312 (2008)
5. Chung, K.J.: The simplified solution procedures for the optimal replenishment decisions under two level trade credit policy depending on the order quantity in a supply chain system. Expert Syst. Appl. **38**, 13482–13486 (2011)
6. Chung, K.J.: The EPQ model under conditions of two levels of trade credit and limited storage capacity in supply chain management. Int. J. Syst. Sci. **44**, 1675–1691 (2013)
7. Chung, K.J., Cardenas-Barron, L.E., Ting, P.S.: An inventory model with non-instantaneous receipt and exponentially deteriorating items for an integrated three layer supply chain system under two levels of trade credit. Int. J. Prod. Econ. **155**, 310–317 (2014)
8. Chung, K.J., Huang, T.S.: The optimal retailer's ordering policies for deteriorating items with limited storage capacity under trade credit financing. Int. J. Prod. Econ. **106**, 127–145 (2007)
9. Chung, K.J., Liao, J.J.: Lot-size decisions under trade credit depending on the ordering quantity. Comput. Oper. Res. **31**, 909–928 (2004)
10. Dye, C.Y.: Joint pricing and ordering policy for a deteriorating inventory with partial backlogging. Omega **35**, 184–189 (2007)
11. Engelbrecht, A.P.: Fundamentals of Computational Swarm Intelligence. Wiley (2005)
12. Ghare, P.M., Schrader, G.H.: A model for exponentially decaying inventory system. Int. J. Prod. Econ. **21**, 449–460 (1963)
13. Goyal, S.K.: Economic order quantity under conditions of permissible delay in payment. J. Oper. Res. Soc. **36**, 335–338 (1985)
14. Guchhait, P., Maiti, M.K., Maiti, M.: Multi-item inventory model of breakable items with stock-dependent demand under stock and time dependent breakability rate. Comput. Indus. Eng. **59**, 911–920 (2010)
15. Guchhait, P., Maiti, M.K., Maiti, M.: Inventory model of a deteriorating item with price and credit linked fuzzy demand?: A fuzzy differential equation approach. OPSEARCH **51**(3), 321–353 (2014)
16. Guchhait, P., Maiti, M.K., Maiti, M.: Two storage inventory model of a deteriorating item with variable demand under partial credit period. Appl. Soft Comput. **13**, 428–4489 (2013)
17. Huang, Y.F.: Optimal retailer's ordering policies in the EOQ model under trade credit financing. J. Oper. Res. Soc. **54**, 1011–1015 (2003)
18. Huang, Y.F.: An inventory model under two levels of trade credit and limited storage space derived without derivatives. Appl. Math. Model. **30**, 418–436 (2006)
19. Huang, Y.F.: Economic order quantity under conditionally permissible delay in payments. Europ. J. Oper. Res. **176**, 911–924 (2007)
20. Kundu, A., Guchhait, P., Pramanik, P., Maiti, M.K., Maiti, M.: A production inventory model with price discounted fuzzy demand using an interval compared hybrid algorithm. Swarm Evolu. Comput. **34**, 1–17 (2016)
21. Mahata, G.C.: Partial trade credit policy of retailer in economic order quantity models for deteriorating items with expiration dates and price sensitive demand. J. Math. Model. Algorithms Oper. Res. **14**(4), 363–392 (2015)
22. Mahata, G.C., De, S.K.: Supply chain inventory model for deteriorating items with maximum lifetime and partial trade credit to credit risk customers. Int. J. Manag. Sci. Eng. Manag. **12**(1), 21–32 (2016)
23. Maiti, M.K.: A fuzzy genetic algorithm with varying population size to solve an inventory model with credit-linked promotional demand in an imprecise planning horizon. Europ. J. Oper. Res. **213**, 96–106 (2011)
24. Maiti, M.K., Maiti, M.: Two-storage inventory model with lot-size dependent fuzzy lead-time under possibility constraints via genetic algorithm. Europ. J. Oper. Res. **179**, 352–371 (2007)
25. Min, J., Zhou, Y.W., Zhao, J.: An inventory model for deteriorating items under stock dependent demand and two-level trade credit. Appl. Math. Model. **34**, 3273–3285 (2010)

26. Ouyang, L.Y., Teng, J.T., Goyal, S.K., Yang, C.-T.: An economic order quantity model for deteriorating items with partially permissible delay in payments to order quantity. Europ. J. Oper. Res. **194**, 418–431 (2009)
27. Pramanik, P., Maiti, M.K., Maiti, M.: An appropriate business strategy for a sale item. OPSEARCH. https://doi.org/10.1007/s12597-017-0310-0 (2017)
28. Pramanik, P., Maiti, M.K., Maiti, M.: A supply chain with variable demand under three level trade credit policy. Comput. Indus. Eng. **106**, 205–221 (2017)
29. Pramanik, P., Maiti, M.K., Maiti, M.: Three level partial trade credit with promotional cost sharing. Appl. Soft Comput. **58**, 553–575 (2017)
30. Pramanik, P., Maiti, M.K.: An inventory model for deteriorating items with inflation induced variable demand under two level partial trade credit: a hybrid ABC-GA approach. Eng. Appl. Artif. Intell. **85**, 194–207 (2019)
31. Pramanik, P., Maiti, M.K.: An inventory model with variable demand incorporating unfaithfulness of customers under two-level trade credit. Europ. J. Indus. Eng. **13(4)**, 461–488 (2019)
32. Pramanik, P., Das, S.M., Maiti, M.K.: Note on: Supply chain inventory model for deteriorating items with maximum lifetime and partial trade credit to credit risk customers. J. Indus. Manag. Optim. **15(3)**, 1289–1315 (2019)
33. Pramanik, P., Maiti, M.K.: Trade credit policy of an inventory model with imprecise variable demand: an ABC-GA approach. Soft Comput. https://doi.org/10.1007/s00500-019-04502-5 (2019)
34. Tayal, S., Singh, S.R., Sharma, R.: Multi item inventory model for deteriorating items with expiration date and allowable shortages. Indian J. Sci. Technol. **7(4)**, 463–471 (2014)
35. Teng, J.T.: Optimal ordering policies for a retailer who offers distinct trade credits to its good and bad customers. Int. J. Prod. Econ. **119**, 415–423 (2009)
36. Teng, J.T., Chang, C.T.: Optimal manufacture's replenishment policies in the EPQ model under two level trade credit policy. Europ. J. Oper. Res. **195**, 358–363 (2009)
37. Teng, J.T., Cheng, C.T., Chern, M.S., Chan, Y.L.: Retailer's optimal ordering policies with trade credit financing. Int. J. Syst. Sci. **38**, 269–278 (2007)
38. Teng, J.T., Yang, H.L., Chern, M.S.: An inventory model for increasing demand under two levels of trade credit linked to order quantity. Appl. Math. Model. **37**, 7624–7632 (2013)
39. You, P.S.: Inventory policy for product with price and time-dependent demands. J. Opera. Res. Soc. **56**, 870–873 (2005)

Analysis of 2PADCL Energy Recovery Logic for Ultra Low Power VLSI Design for SOC and Embedded Applications

Samik Samanta, Rajat Mahapatra, and Ashis Kumar Mal

Abstract Here in this scope we have proposed a noble adiabatic dynamic CMOS logic circuit known as two phase adiabatic dynamic logic. The proposed two 2 phase ADCL uses two complementary sinusoidal power supply clocks known as power clock. As a result, the propagation delay of the 2PADCL is smaller than that of the conventional ADCL circuits and conventional CMOS circuits. The simulation results also show that the power dissipation of the 2PADCL circuit is lower than those of other conventional adiabatic logic circuits and conventional CMOS circuits. The adiabatic performance parameter which is known as energy saving factor is also estimated in lower and higher frequency ranges. We have also estimated the power delay product of ADCL and proposed 2PADCL logic.

Keywords VLSI · Energy recovery · CMOS · Adiabatic · Power dissipation · MOSFET

1 Introduction

In conventional CMOS circuits, power dissipation primarily occurs during circuit node switching. Zimmermann et al. had shown that the dynamic power of digital circuits depends on supply voltage, clock frequency, and node switching activities. A sudden flow of current through channel resistive elements results in half of the supplied energy being dissipated as heat at each transition. Low-power based systems are designed by implementing the concept of adiabatic switching and energy recovery principle have been applied to various circuits with adiabatic circuitry for ultra-low power portable applications. In the circuit-level adiabatic technique, charge transfer occurs without the generation of heat. During adiabatic switching, all the nodes are

S. Samanta (✉)
ECE Department, Management and Science, Neotia Institute of Technology, Kolkata,
West Bengal, India
e-mail: samik.nitmas@gmail.com

R. Mahapatra · A. K. Mal
ECE Department, National Institute of Technology, Durgapur, India

P. Gyei-Kark et al. (eds.), *Engineering Mathematics and Computing*,
Studies in Computational Intelligence 1042,
https://doi.org/10.1007/978-981-19-2300-5_18

275

charged or discharged at a constant current to minimize power dissipation. This is achieved by using a ramped-step voltage or an AC power supply to initially charge the circuit during specific adiabatic phases and then discharging the circuit to recover the supplied charge of the operation in the next phase. In many earlier research papers we can find use of four phase clocking schemes for adiabatic logic circuits or systems.4-phase clocking suffers from various drawbacks or hazards like complexity of the hardware. Pipelining cannot be done in these circuits which results lower propagation delay [1]. Moreover the clock driver is very difficult to design and implement in case of 4-phase clocking circuits. ADCL design is also very complex, they also have higher propagation delays, and an extra hardware is required to store or hold the output in these circuits. Many researchers have already presented 2PADCL circuits. PADCL is one type of quasi adiabatic circuit [2]. Dickinson et al. and Ye et al. have presented 2PADCL circuits earlier but these systems were needed complex circuitry and complex power clock hardware requirements. PADCL circuits have improved delay characteristics and less hardware requirements over ADCL circuits. Moreover they have less power dissipation over ADCL circuits. Here in this paper we will present the design and power dissipation of 2PADCL circuit and compare the simulation results with other adiabatic circuits and conventional CMOS circuit. From the simulation results it can be clearly shown that 2PADCL have better performance over other adiabatic circuits. These circuits realize the advantage of energy efficiency through the use of gate over drive and reduced switching power. These low power circuits can be applied in system on chip (SOC) and embedded wearable devices.

2   Adiabatic Logic

Conventional CMOS is very useful technology for low power digital circuit design due to its negligible static power. Generally, CMOS has two design networks which are known as pull up and pull down networks. Dynamic power dissipation of CMOS circuits takes place due to charging and discharging of nodes. This power is proportional to load capacitance, square of the supply voltage, and switching frequency. Capacitance and frequency factors increase the power consumption. Therefore, the regular complementary MOS design needs to be changed in order to satisfy the market demand of ultra-low power. In Fig. 1 we have shown the equivalent circuit of

Fig. 1 Principle of adiabatic logic

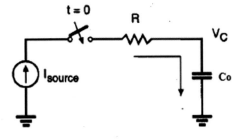

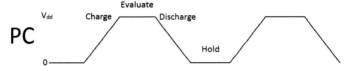

Fig. 2 Power clock of adiabatic logic

adiabatic logic [3]. Output load capacitance is charged by a constant current source instead of a constant voltage source used in conventional CMOS structures. This circuit is same as the equivalent model used in charging process in conventional CMOS. On resistance of pull up PMOS network is represented by R, and CO is the output capacitance [4]. It is noted that constant charging current corresponds to a linear voltage ramp. Energy dissipated through adiabatic logic is given as

$$E(\text{diss}) = \left(C_o / V_{dd}\right)^2 . RT = RC_o / T.C_o V_{dd}^2 \tag{1}$$

Adiabatic process ideally operates as a reversible thermodynamic process. This is done without loss or gain of energy. Adiabatic computation works by making very small changes in energy levels in circuits. This is done by sufficiently slow, ideally resulting process with no energy dissipation. Figure 2 shows the power clock of adiabatic circuits. It has four phases. They are charge, evaluate, discharge, and hold.

3 Two Phase Adiabatic Dynamic Logic

The output of Normal ADCL circuit is properly synchronized with the power supply voltage. The frequency of the power clock determines the operating speed of ADCL circuits. As the size of the gates or stages of the gate increase lower will be the speed of the circuit. The circuit diagram of ADCL can be shown in Fig. 3.

Fig. 3 Circuit diagram of ADCL

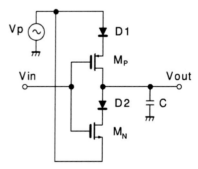

In 2PADCL, the circuit is operated with complementary phases of power supply signals. The power supply signal consists of evaluation and hold mode [5]. Generally adiabatic circuit has power clocks with evaluation, hold, and recovery and wait phases. In 2PADCL, we are using only two phases namely evaluation and hold. The 2PADCL inverter having output transition 0 to 1 or 1 to 0 has only shown in the simulation. The inverter is compatible with adiabatic mode under the condition of the phase difference between 0 rad and $\pi/2$ rad. The minimum energy consumption in a single inverter is reached when the phase difference between the power supply, and the input data is $\pi/2$ rad so that the data is lacking the power supply. Generally, 2PADCL achieves ultra-low energy dissipation by restricting current to flow across devices with low voltage drop and by recycling the energy stored in internal capacitors. The 2PADCL inverter having output transition low to high or high to low has only shown in the simulation (Figs. 4 and 5).

Figure 2 shows the circuit diagram of 2PADCL, and Fig. 3 shows the power clock of 2PADCL.When Vp and complement of Vp are in evaluation mode there will be a conducting path in NMOS devices or PMOS devices, and the output may change from low to high or vice versa. The output can also be remained unchanged. No extra circuit is required to hold the output at every phase.

When Vp and complement of Vp are in hold phase, output will hold its value. In every clock cycle circuit nodes need not change its value. This mean in every clock cycle it will not charge or discharge. There the switching activity will be reduced. This will improve the speed of 2PADCL circuit [6]. These circuits can maintain

Fig. 4 Power clock of 2PADCL

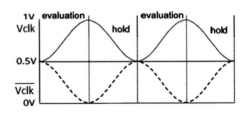

Fig. 5 Circuit diagram of 2PADCL

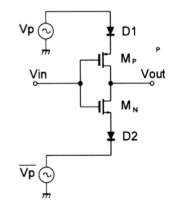

the output without load capacitor or extra capacitance. This results simplification of hardware complexity.

The energy dissipation in this logic family occurs due to threshold voltage of MOSFET, diode cut in potential and energy dissipated in resistance of MOS devices.

The energy dissipation of inverter per cycle is

$$E_d = 2C_{Gs}(V_p - 2V_d)V_d + C_{Gs}(V_t - V_d)^2 \tag{2}$$

As diodes are constructed from MOSFETs, we can write $V_t = V_d$. Therefore, energy dissipation of inverter per cycle is

$$E_d = 2C_{Gs}(V_p - 2V_d)V_d \tag{3}$$

CGS is the gate source capacitance of next stage, Vp is the peak value of power clock, and Vd the is cut-in potential of the diode in use. Here, we have ignored the energy dissipation through ON resistance of MOSFET as this component is very negligible (Fig. 6).

Fig. 6 Simulation results of 2PADCL

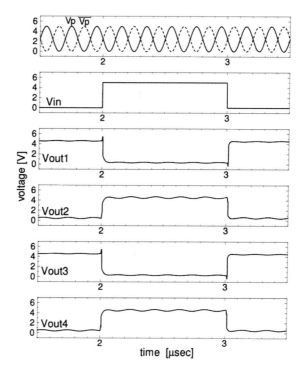

4   Simulation Results

We have simulated CMOS, ADCL, and 2PADCL inverters. The power dissipations and delays are calculated for the various inverters. Table1 shows the power dissipations, and Table 2 shows the delay variations of various inverters (Fig. 7).

2PADCL NOR circuit is presented in Fig. 4. It is operated with complementary phases of power clock signals. Supply waveform has two modes as discussed previously namely evaluation and hold. The assumption made for adiabatic mode in which clk and complementary clk signals are in evaluation mode, then there is conducting path(s) in either PMOS devices or NMOS devices.

Evaluation of output node made from low to high or from high to low or remain unchanged, which resembles to the conventional CMOS circuit. Thus, there is no need to restore the node voltage to 0 (or V_{dd}) every cycle. On the other side when clk and complementary clk signals are in hold mode, output node holds its value in spite of the fact that clk and complementary clk signals are changing their values. By observing the function of diodes and the fact that the inputs of a gate have a different phase with the output we can find the above mentioned operations. The reduction in node switching activities subsequently takes place due to not necessarily charging and discharging of circuit nodes with every clock cycle [7]. This phenomena decreases the energy dissipation and increases the speed of operation (Fig. 8 and Tables 3, 4, 5, and 6).

Table 1 Power dissipation of various inverters

Frequency (MHz)	Power dissipation (μW)		
	CMOS	ADCL	PADCL
1	6.00	2.2	1.2
10	15.2	4.5	1.4
20	25.2	11.3	2.6
50	27.9	12.7	3.1
70	56.1	13.4	5.3
100	92.6	15.2	7.6

Table 2 Propagation delays of various inverters

Frequency (MHz)	Propagation delay (μS)		
	CMOS	ADCL	PADCL
1	11.2	4.2	2.2
10	14.5	5.6	3.6
20	16.3	7.2	4.9
50	24.9	8.1	5.1
70	29.7	10.5	8.7
100	57.9	23.4	10.2

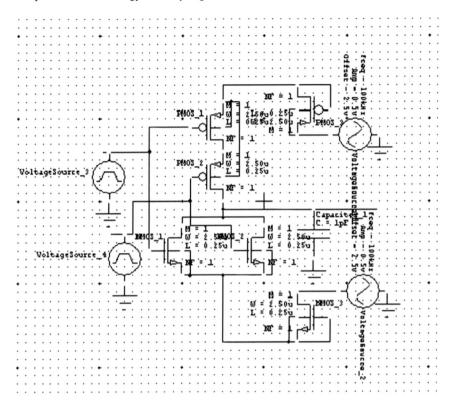

Fig. 7 2PADCL based NOR gate

5 Conclusion

Based on the simulation results it is clearly found the 2PADCL logic has better performance over conventional CMOS and conventional ADCL logic. We have observed power dissipation, propagation delay, and PDP parameters of various adiabatic logics, based on the results it is found that 2PADCL logic has better power performance and better speed performance over conventional CMOS. The energy saving factors in various frequency ranges have been observed. It can be clearly found that 2PADCL has very good energy saving factor in lower and higher frequencies. Generally embedded systems, system on chip, and biochips require very less power and minimum area. For this reason the present PADCL circuits are well suited.

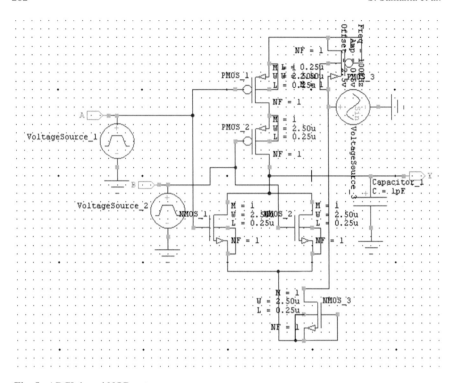

Fig. 8 ADCL based NOR gate

Table 3 Power dissipation in various frequencies

Frequency (MHz)	Power dissipation (μW)	
	ADCL	PADCL
1	5.2	0.7
10	6.6	1.7
20	7.8	2.8
50	8.6	3.1
70	11.2	5.1
100	13.3	7.3

Table 4 Propagation delays of various inverters

Frequency (MHz)	Power dissipation (nS)	
	ADCL	PADCL
1	3.2	0.2
10	6.8	0.7
20	8.3	1.4
50	9.5	3.1
70	11.4	5.3
100	14.4	8.1

Table 5 Power delay products in various frequencies

Frequency (MHz)	PDP	
	ADCL	PADCL
1	16.6	0.14
10	44.8	1.19
20	64.7	3.92
50	81.7	9.61
70	127.6	27
100	191.5	59.3

Table 6 ESF of various inverters

Frequency (MHZ)	ESF
1	8.5
10	8.9
20	9
50	9
70	11
100	12.6

References

1. Kanungo, J., Dasgupta, S.: Study of scaling trends in energy recovery logic: an analytical approach. IOP Sci. J. Semicond. **34**(8), 085001-1-085001-5 (2013)
2. Kanungo, J., Dasgupta, S.: Single phase energy recovery logic and conventional CMOS logic: a comparative analysis, Special Issue on "emerging device and circuit techniques for ultra low power design in the Nano scale technologies". J. Microelectron. Solid State Electron. Scientific Acad. Publishing (SAP) (2013)
3. Hu, J., Dai, J., Zhang, W., Wu, Y.: Pre-settable adiabatic flip-flops and sequential circuits. IEEE International Conference on Communications, Circuits and Systems (2016)
4. Takahashi, Y., Fukuta, Y., Sekine, T., Yokoyama, M.: 2PADCL: two phase drive adiabatic dynamic CMOS logic. Proceedings IEEE APCCAS, Dec 2006, pp. 1486–1489 (2006)
5. Bhattacharjee, A., Bandyopadhyay, C,. Wille, R., Drechsler, R., Rahaman, H.: Improved look-ahead approaches for nearest neighbor synthesis of 1D quantum circuits. 32nd International Conference on VLSI Design, VLSID 2019 & 18th International Conference on Embedded Systems, ES 2019 (2019)
6. Zhao, P., Darwish, T., Bayoumi, M.: High performance and low power conditional discharge flip-flop. IEEE Trans. Very Large Scale Integr. (VLSI) Syst. **12**(5), 477–484 (2004)
7. Samanta, S., Mahapatra, R., Mal, A.K.: Parameter analysis of ECRL and PFAL comparators. Int. J. Appl. Innov. Engg. Manag. **8**(7) (2019)
8. Kramer, A., Denker, J.S. et al.: 2nd order adiabatic computation with2N-2P and 2N-2N2P logic circuits. Proceedings International Symposium. Low Power Design, pp. 191–196 (1995)
9. Denker, J.S.: A review of adiabatic computing. In: Proceedings of the Symposium on Low Power Electronics, pp. 94–97 (1994)
10. Samanta, S.:Power efficient VLSI inverter design using adiabatic logic and estimation of power dissipation using VLSI-EDA tool. Spec. Issue Int. J. Comput. Commun. Technol. **2**(2,3,4), 300–303 (2010)

11. Mal, S., Podder, A., Chowdhury, A., Chanda, M.: Comparative analysis of ultra-low power adiabatic logics in near-threshold regime. IEEE EDS Sponsored Conference on Devices for Integrated Circuit (DevIC), pp. 664–669 (2017)
12. Samanta, S., Mahapatra, R., Mal, A.K.: Analysis of adiabatic flip-flops for ulta low power applications. 3rd International Conference on Devices for Integrated Circuits 2019. organized by KGEC, India
13. Fornaciari, W., Gubian, P., Sciuto, D., Silvano, C.:High level power estimation of VLSI systems. IEEE International Symposium on Circuits and Systems, Hong Kong, June 1997, pp. 1804–1807 (1997)
14. Maheshwari, S., Bartlett, V.A., Kale, I.: 4-phase resettable quasi-adiabatic flip-flops and sequential circuit design. 12th International Conference on Ph.D. Research in Microelectronics and Electronics (PRIME), Portugal, Lisbon, June 27–30, 2016

Investigating the Impact of Social Media Marketing on Business Performance of Different Brands in Indian Cosmetics Market: An Empirical Study

Jayeeta Majumder, Arunangshu Giri, and Sourav Gangopadhyay

Abstract The Indian cosmetic product is becoming a booming sector. The Indian cosmetic market is growing rapidly with the help of social media over the last few decades. Indian cosmetics industries are always trying to invent and use different ways of marketing techniques to enhance the sale. Social media marketing is an intriguing and effective way of marketing. It is a kind of marketing that is unique, and it is a way of unpaid advertisement on behalf of the brand which is conducted by loyal customers. In this paper, we are trying to understand how social media marketing has impacted the Indian cosmetic industry. The aim of this paper is to study the factors related to social media marketing and their impact on the purchase intention of customers which deals with business performance of a wide range of Indian cosmetic products including branded low or medium price category. In this study, Structural Equation Modeling and Exploratory Factor Analysis (EFA) by SPSS-23 and AMOS-23 were used to analyze the data and to establish a research model for assessing influences of Social Media Marketing toward business performance of different Brands in Indian cosmetics Market.

Keywords Social media marketing · Business performance · Indian cosmetics market · Brands

1 Introduction

Traditional marketing is changing day by day new technologies, and concepts are propelling the process of development of new ways of marketing. Now marketing is becoming more personal, more target-oriented, and more opportunities for enhancement of customer involvement in the process [7]. Building a relationship with the

J. Majumder · S. Gangopadhyay
Haldia Institute of Management, Haldia, West Bengal, India

A. Giri (✉)
Haldia Institute of Technology, Haldia, West Bengal, India
e-mail: arunangshu.giri@hithaldia.ac.in

© The Author(s), under exclusive license to Springer Nature Singapore Pte Ltd. 2023
P. Gyei-Kark et al. (eds.), *Engineering Mathematics and Computing*,
Studies in Computational Intelligence 1042,
https://doi.org/10.1007/978-981-19-2300-5_19

customer is needed most that is why the relation is understood by the time when it is seen that the consumers are becoming the co-creator of the band [6]. Customers nowadays are participating in the process of brand promotion. Customers always love to act independently; they love to promote those brands which they love. Social media marketing is an effective tool that enhances business performance for all industries including the cosmetics industry. Social media, as networking and interactive platform, shares and exchanges information which facilitates to promote cosmetic products and deliver customized services to millions of people. The cosmetic market is a fast-growing area. They are always trying to focus on the new way of marketing strategies. The marketing strategies of cosmetic products have changed dramatically due to influence of the social media. The public is sharing their knowledge through social media. In India in the year 2010, it was reported that nearly 20,000 crores were spent on media marketing, of which 40% was on the newspapers, 50% for television, 5% on the outdoor, 2% radio, and 1% internet [34]. Many advertisers concluded that the following web is a medium in their advertising media mix strategy. Many researchers have found that online advertising is increasing. The global online advertising industry will grow even more [40]. Internet helps to understand the consumer's self-concept, attitude, purchase intention, behavior, satisfaction, and loyalty [9]. Buying the cosmetics product is depending largely upon the social media they are using. Nowadays the people find cosmetics as a way of looking attractive [29]. For the greater business performance of the cosmetic industry, building customer trust and loyalty are very important components. Various social media platforms can help this industry to come closer to all types of customers by providing smooth access and faster services facilities. In India, a trend of putting more attention to beauty care and using cosmetic products by the customers has been observed for their increasing income [37].

2   Literature Review

Social media is helping together to participate in an information exchange program. Also by social media invitation in the profile can be done. Social media is used by many corporate sectors to make communication with customers more effective. Many organizations are using social media pages to confirm attachment between brands and consumers [38]. A large number of people are using social networking sites through smartphones, personal computers, and new technologies that make the companies change their attitude toward marketing [39]. Social media became popular among people like about 1.32 billion people are using Facebook every day. 93% of the organization use social media for marketing. A study shows that people are more spending time on Facebook than on websites. It is now become necessary to reach the consumer through social media [25]. Through social media, the consumer and the company can communicate easily. Social media helps to build a solid and mutual relationship with the customer. Companies that are ignoring the utilities of social media marketing and social media strategies, lose the opportunities of taking full advantage of social media marketing. Without social media marketing, as an alternative distribution channel, business of cosmetic companies can lose their competitive advantage in Indian market [22]. Social media platform can give opportunities to engage in cruising, questioning, and extolling a product or service which is helpful for the organization [4], and also it gives opportunities for conversation regarding every aspect of product with other customers [28]; in that way, social media acts as a potential platform to exchange ideas and also it increases the customer expectation and brand experience. Social media gives immense opportunity to circulate information so the people act as impelling factor in this context. Loyal customers are creating content of a brand they like, and it is called social money, the social money when the customer shares a brand or data about a brand they like [41]. The customers portray different kinds of media content which is accessible easily and freely. So it can be said the social cash is the influencing factor of brand execution. For the increasing popularity of green organic cosmetic products, consumer purchase intention is also increasing and ecological concerns of customers affect the marketing efficiency of cosmetic products [15, 23]. Customers can access varieties of products from online portals as per their convenience [24]. Social media is playing a good platform of word of mouth. Virtual Brand Community (VBC) is a unique term in this context. This community is containing customers who are amalgamated on the internet and show their enthusiasm for some brand or item [32]. Some researchers have found that trust and securities are the keys of VBC, the customer became loyal to the brand when they have a positive experience in the investment process of that brand [8]. Virtual brand communities have a significant impact on online purchasing behavior. Inside the Virtual Brand Community, a social connection can be seen such as Electronic Consumer to Consumer Interaction to facilitate the cooperation between consumers and retailers by sharing experience. The important part of technology acceptance is perceived usefulness [11]. Some researchers have found perceived usefulness has an indirect effect on the acceptance of the system, but some other researchers have found it has indirect effect on the acceptance of a system [14]. Some other researchers have found that the perceived usefulness is significantly depending upon the quality of the website, quality of information, and quality of service the product [1]. Sometimes the relationship acts as an important factor of purchase [17]. Shopping is more than only purchasing the product; it is about enjoyment, experience, and entertainment [13]. Social media is fulfilling the requirement, which changes the shopping behavior drastically. According to a research paper, social media promotes online shopping which mimics the actual market. Social media is a useful platform to share information [10]. People are getting interested to exchange information and getting social support. Review of the product and services are available which are useful for both the customer and the organization [33]. The companies are always encouraging the customer to rate and give comments on the product or the service [5] which is becoming a deciding factor of purchasing the product or avail the particular service [35]. Apart from sharing views, the sharing recommendation is also useful for both the customer and the organization because the consumers are more attracted to the recommendation of other users. In this way, trust is built. Learning and doing is the main essence of the virtual world [31], but the quality of shared information by anonymous consumers in social media sometimes creates challenges for the organization. Social networking site is able to provide information and emotional support to the customers as the customer who are engaged in the online group they need social, emotional support, and friendship from the group. Moreover it draws attraction of the new customers toward this social group as it is useful to get information directly from other customers [19]. Online forearms, communities, ratings, reviews, recommendations, all are useful tools of social interaction. Trust between the consumer and the organization is important for every organization [30]. Trust is built up upon empathy, competency, integrity, ability, and benevolence; some researchers are seeing trust depending upon competence and benevolence [26]. If we sum up all the dimensions of trust we can find benevolence and credibility are the main pillars of trust. Credibility trust is built upon the situation when both the parties trust each other, and the benevolence is a repeated seller-buyer relationship [2]. Social networking sites are playing important role in the building up the trust and perceived security [21]. Social networking sites are helping the customers to exchange their experience and by that, they are able to produce trust and ultimately increase the desire to buy [20]. Trust and perceived-usefulness have a positive correlation with each other. Social networking sites are playing important role in building trust as connected with user interactions. Electronic word of mouth takes over traditional word of mouth because interaction with consumers is more effective in social media platform [15]. Some researchers have identified some drawbacks of social media, as it can cause the uncontrolled spreading of negative publicities. Social media gives opportunity for self-expression and self-presentation. Social media gives immense opportunities to reach mass audience by sharing pictures, videos, and text. That is how the customers are more attracted toward social media to get information about a product [3].

3 Objectives of the Study

- To study the factors related to Social Media Marketing that affects business performance of different brands in Indian cosmetics market.
- To assess the impact of the factors that affect marketing efficiency and business performance by studying consumer behavior.

4 Hypotheses and Research Model

H1: 'Perceived Usefulness of Social Media' positively influences 'Marketing Efficiency' of different Brands in Indian Cosmetic Market.

H2: 'Experiencing Enjoyment' positively influences 'Marketing Efficiency' of different Brands in Indian Cosmetic Market.

H3: 'Trustworthiness' positively influences 'Marketing Efficiency' of different Brands in Indian Cosmetic Market.

H4: 'Perceived Security' positively influences 'Marketing Efficiency' of different Brands in Indian Cosmetic Market.

H5: 'Virtual Platform for Social Support' positively influences 'Marketing Efficiency' of different Brands in Indian Cosmetic Market.

H6: 'Quick E-Word of Mouth Circulation' positively influences 'Marketing Efficiency' of different Brands in Indian Cosmetic Market.

H7: 'Marketing Efficiency' positively influences 'Business Performance' of different Brands in Indian Cosmetic Market.

5 Research Methodology

Secondary and Primary data were used in this study for analysis to achieve the research objective and to establish the hypothesized research model (Fig. 1). This hypothesized research model was developed by using 8 constructs (Perceived Usefulness, Experiencing Enjoyment, Trustworthiness, Perceived Security, Virtual Platform for Social Support, Quick E-Word of Mouth Circulation, Marketing Efficiency, and Business Performance). Survey questionnaire was prepared using 5 Point Likert Scale (Strongly Agree-5; Agree-4; Neutral-3; Disagree-2; and Strongly Disagree-1) for the purpose of collecting primary data. We collected the feedback regarding the impact of Social Media Marketing on business performance of different brands in Indian

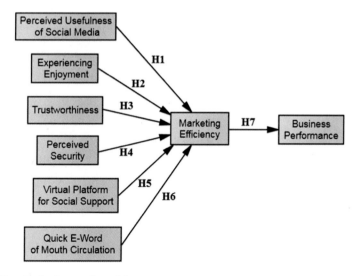

Fig. 1 Hypothesized research model

Table 1 Reliability statistics

Cronbach's Alpha	No. of variables
0.835	16

Table 2 KMO Measureand Bartlett's test

Kaiser–Meyer–Olkin Measure		**0.738**
Bartlett's test of Sphericity	Chi-square	995.318
	P-value	**<0.01**

Cosmetics Market from 200 social media users through Convenience Sampling technique. Primary data collection period was from September, 2019 to November, 2019. Structural Equation Modeling and Exploratory Factor Analysis (EFA) by SPSS-23 and AMOS-23 were used in this study as statistical tools.

6 Data Analysis and Findings

Reliability and Validity Testing

For checking the reliability of collected primary data, we evaluated Cronbach's Alpha value which should be more than 0.70 for maintaining acceptable range. Here, alpha value (0.835) showed the acceptable range of reliability (Table 1). On the other hand, for validity testing of primary data, we performed Exploratory Factor Analysis [27]. Also, KMO value (0.738) and Bartlett's Test of Sphericity (<0.001) indicated good sampling adequacy and aptness of executing EFA (Table 2). EFA extracted 8 separate factors with 'factor loadings' larger than 0.5 using Varimax Rotation Method (Table 3).

On the other hand, we examined all 'Variance Inflation Factor' (VIF) values which proved that all independent factors were free from Multi-co-linearity error (Acceptable range of VIF values are less than 3) (Table: 4).

The following table showed different fitness indices (Table 5) to judge the structural appropriateness of the research model [16, 36].

Path analysis (Table 6) through Structural Equation modeling (Fig. 2) has been executed for finding out the impact of Social Media Marketing on business performance of different brands in Indian Cosmetics Market.

7 Implication

In the present scenario, cosmetic companies are investing more money in social media marketing to draw the attention of the customers who can easily access their required cosmetic products [12]. Promoting cosmetics through social media helps customers

Table 3 Result of EFA (Exploratory Factor Analysis)*

Factors	Questions/variables	Factor loading	% of variance
Perceived usefulness of social media	q1	0.950	12.204
	q2	0.938	
Experiencing enjoyment	q12	0.930	11.635
	q11	0.913	
Trustworthiness	q4	0.935	11.507
	q3	0.923	
Perceived security	q15	0.925	10.931
	q16	0.901	
Virtual platform for social support	q10	0.911	10.925
	q9	0.877	
Quick E-word of mouth circulation	q6	0.921	10.909
	q5	0.867	
Marketing efficiency	q8	0.891	10.314
	q7	0.856	
Business performance	q14	0.874	9.837
	q13	0.827	

*PCA Extraction Method
Varimax and Kaiser Normalization
Rotation converged in 6 iterations

Table 4 Co-linearity statistics

'Marketing Efficiency' as dependent factor		
	Tolerance	VIF values
Perceived usefulness of social media	0.545	**1.682**
Experiencing enjoyment	0.564	**1.614**
Trustworthiness	0.631	**1.242**
Perceived security	0.522	**1.876**
Virtual platform for social support	0.559	**1.698**
Quick E-word of mouth circulation	0.620	**1.256**

to get detailed information regarding product and price and several recommendations regarding those products. Social media marketing has a huge impact on any brand. The cosmetics industry is not an exception. If any organization wants to promote the sale of cosmetic products, then they can use social media for that purpose. The employment structure is an important issue for any organization, and the skilled and experienced employee has a positive influence in the market of any product including the cosmetic product. People nowadays are becoming more and more quality savvy, and they are always looking for quality products and services, and in this regard trustworthiness and Security are playing a major role in it. Indian

Table 5 Fit indices for research model

Fit Index	Fitness values (tolerable range)
$\chi 2$/df	1.963 (<3)
RMSEA	0.041 (<0.06)
GFI	0.986 (>0.90)
AGFI	0.915 (>0.90)
NFI	0.984 (>0.90)
CFI	0.992 (>0.90)

$\chi 2$/df:Chi-square/degree of freedom; **RMSEA :** Root mean-square error of approximation
GFI: Goodness of fit index; **AGFI:** Adjusted goodness of fit index; **NFI:** Normed fit index
CFI: Comparative fit index

Table 6 Path analysis for Hypothesis Testing

Measurement path			Hypothesis testing	Regression estimate	S.E	P- value
Marketing efficiency	←	Experiencing enjoyment	**H2 (S)**	0.242	0.042	**<0.01***
Marketing efficiency	←	Trustworthiness	**H3 (S)**	0.115	0.041	**0.005***
Marketing efficiency	←	Perceived usefulness of social media	**H1 (S)**	0.202	0.035	**<0.01***
Marketing efficiency	←	Virtual platform for social support	**H5 (S)**	0.231	0.043	**<0.01***
Marketing efficiency	←	Quick E-word of mouth circulation	**H6 (S)**	0.242	0.038	**<0.01***
Marketing efficiency	←	Perceived security	**H4 (S)**	0.164	0.044	**<0.01***
Business performance	←	Marketing efficiency	**H7 (S)**	0.599	0.083	**<0.01***

Note (*) → 1percent level of significance
(S) → Hypothesis Supported
In this study, all hypotheses were accepted

Cosmetic product's marketing efficiency is largely depending upon trustworthiness and Security. People nowadays are becoming more and more tech-savvy. They are attracted to technology. The virtual platform is becoming a popular platform for everyone. Every organization including the cosmetics industry should have a virtual platform for social media that facilitates the marketing efficiency of any brand. Many researchers including this research have shown that the word of mouth is one of the most important ways of brand promotion. According to different researches, in the case of cosmetics products, the word of mouth plays a pivotal role in brand promotion. After social media is being most popular and useful ways of word of mouth, it has

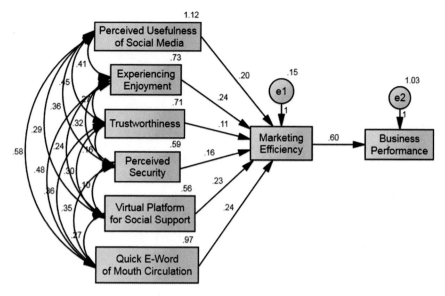

Fig. 2 Structural equation modeling (SEM)

a huge impact on the incrementing the marketing efficiency of Indian cosmetics products. This marketing efficiency has the power to mold the customer perception regarding purchasing cosmetics in a positive direction. So, marketing efficiency has a direct influence on business performance. So if any organization wants to enhance the business performance, they have to uplift the marketing division by augmenting the marketing efficiency.

8 Conclusion

In recent days, business dealing is shifting from transactional to social relationships. Social media marketing for cosmetics products can cover the customers with separate age groups, lifestyles, and affordability. Variety cosmetic products with discounted prices can be accessed through social media. The social media marketing process can easily increase the awareness level of the customers regarding cosmetics products in the fastest way. Indian cosmetics market has huge demand all over the world from the ancient period of time. People are more fascinated with using herbal cosmetics products. India has a huge number of herbal plants for that reason the Indian cosmetics market gets huge benefits from the availability of raw material resources. In the present scenario, if the Indian cosmetic market wants to promote the sale of cosmetics products, they need to adopt the latest ways of brand promotion like the use of social media, e-word of mouth, experiencing the enjoyments, trustworthiness, and perceived security.

References

1. Ahn, T., Ryu, S., Han, I.: The impact of web quality and playfulness on user acceptance of online retailing. Inf. Manag. **44**(3), 263–275 (2007)
2. Ba, S., Pavlou, P.A.: Evidence of the effect of trust building technology in electronic markets: price premiums and buyer behavior. MIS Q. **26**(3), 243–268 (2002)
3. Baird, C.H., Parasnis, G.: From social media to social customer relationship management. Strategy Leadership. **39**(5), 30–37 (2011)
4. Belch, G.E.: Advertising and Promotion: An Integrated Marketing Communications Perspective. McGraw-Hill, Sydney, Australia (2009)
5. Bronner, F., Hoog, R.: Consumer-generated versus marketer-generated websites in consumer decision making. Int. J. Mark. Res. **52**(2), 231 (2010)
6. Brown, S., Robert, V.K., John, F.S.: Teaching old brands new tricks: retro branding and the revival of brand meaning. J. Mark. **67**(7), 19–33 (2003)
7. Cappo, J.: The Future of Advertising: New Media, NewClients, New Customers in a Post Television Age. McGraw-Hill, New York (2003)
8. Cha, J.: Shopping on social networking websites: attitudes towards real versus virtual items. J. Interact. Advert. **10**(1), 77–93 (2009)
9. Chakraborty, S., Giri, A., Biswas, S., Bag, M.: Measuring the impact of celebrity endorsement on consumer purchase intention of beauty soap in Indian context. Int. J. Sci. Technol. Res. **9**(2), 1019–1022 (2020)
10. Chen, J., Xu, H., Whinston, A.B.: Moderated online communities and quality of user-generated content. J. Manag. Inf. Syst. **28**(2), 237–268 (2011)
11. Davis, F.D.: Perceived usefulness, perceived ease of use, and user acceptance of information technology. MIS Q. **13**(3), 319–340 (1989)
12. De Vries, L., Gensler, S., Leeflang, P.S.: Popularity of brand posts on brand fan pages: an investigation of the effects of social media marketing. J. Interact. Market. **26**(2), 83–91 (2012)
13. Dennis, C., Jayawardhena, C., Merrilees, W., Wright, L.T.: E-consumer behaviour. Eur. J. Mark. **43**(10), 1121–1139 (2009)
14. Gefen, D.: The relative importance of perceived ease of use in is adoption: a study of e-commerce adoption. J. Assoc. Inf. Syst. **1**(8), 1–30 (2000)
15. Giri, A., Biswas, W., Biswas, D.: The impact of social networking sites on college students: a survey study in West Bengal. Indian J. Market. **48**(8), 7–23 (2018)
16. Giri, A., Chatterjee, S., Biswas, S., Aich, A.: Factors influencing consumer purchase intention of daily groceries through b2c websites in metro-cities of India. Int. J. Sci. Technol. Res. **9**(1), 719–722 (2020)
17. Giri, A., Chatterjee, S., Paul, P., Chakraborty, S.: Determining the impact of artificial intelligence on 'developing marketing strategies' in organized retail sector of West Bengal, India. Int. J. Eng. Adv. Technol. **8**(6), 3031–3036 (2019)
18. Giri, A., Gangopadhyay, S., Majumder, J., Paul, P.: Model development for employee retention in Indian construction industry using structural equation modeling (SEM). Int. J. Manag. **10**(4), 196–204 (2019)
19. Gruzd, A., Wellman, B., ScTakhteyev, Y.: Imagining Twitter as an imagined community. Am. Behav. Sci. **55**(10), 1294–1318 (2011)
20. Han, B.O., Windsor, J.: Users' willingness to pay on social network sites. J. Comput. Inf. Syst. **51**(4), 31–40 (2011)
21. Jiyoung, C.: Shopping on social networking web sites: attitudes toward real versus virtual items. J. Interact. Advert. **10**(1), 77–93 (2009)
22. Kiang, M.Y., Chi, R.T.: A framework for analyzing the potential benefits of internet marketing. J. Electron. Commerce Res. **2**(4), 157–163 (2001)
23. Kim, H.Y., Chung, J.E.: Consumer purchase-intention for organic personal care products. J. Consum. Mark. **28**(1), 40–47 (2011)

24. Krishnan, G.A., Koshy, L., Mathew, J.: Factors affecting the purchasing behaviour of customers towards male grooming products: A descriptive study conducted at Ernakulam, Kerala, India. ZENITH Int. J. Multidiscip. Res. **3**(7), 48–60 (2013)
25. Kumar, V., Mirchandani, R.: Increasing the ROI of social media marketing. MIT Sloan Manag. Rev. **54**(1), 55–56 (2012)
26. Li, F., Zhou, N., Kashyap, R., Yang, Z.: Brand trust as a second-order factor: an alternative measurement model. Int. J. Mark. Res. **50**(6), 817 (2008)
27. Majumder, J., Giri, A., Gangopadhyay, S.: Factors affecting work life balance of employees in Indian manufacturing companies: An empirical analysis using structural equation modeling (SEM). Int. J. Innov. Technol. Exploring Eng. **8**(7), 1551–1555 (2019)
28. Mangold, W.G., Faulds, D.J.: Social media: The new hybrid element of the promotion mix. Bus. Horiz. **52**(4), 357–365 (2009)
29. Mansor, N., Abidin, A.F.A.: The application of ecommerce among Malaysian small medium enterprises. Eur. J. Sci. Res. **41**(4), 590–604 (2010)
30. McCole, E., Ramsey, E., Williams, J.: Trust considerations on attitudes towards online purchasing: the moderating effect of privacy and security concerns. J. Bus. Res. **63**(10), 1018–1024 (2010)
31. Mueller, J., Hutter, K., Fueller, J., ScMatzler, K.: Virtual worlds as knowledge management platform—a practice-perspective. Inf. Syst. J. **21**(6), 479–501 (2011)
32. Muniz, A.M., Lawrence, O.H.: Us versus them: oppositional brand loyalty and the cola wars, advances in consumer research, Provo, Utah: association for consumer. Research **28**(1), 355–361 (2001)
33. Nambisan, S.: Designing virtual customer environments for new product development: toward a theory. Acad. Manag. Rev. **27**(3), 392–413 (2002)
34. Oghabian, F., Babu, N.: Influence of advertisements on impressionable young minds. Southern Economist. **49**(12), 29–30 (2010)
35. Pan, L.Y., Chiou, J.S.: How much can you trust online information? Cues for perceived trustworthiness of consumer-generated online information. J. Interact. Mark. **25**(2), 67–74 (2011)
36. Paul, P., Giri, A., Chatterjee, S., Biswas, S.: Determining the effectiveness of 'cloud computing' on human resource management by structural equation modeling (SEM) in manufacturing sector of West Bengal, India. Int. J. Innov. Technol. Exploring Eng. **8**(10), 1937–1942 (2019)
37. Rani, K.: A study on consumer awareness, attitude and preference towards herbal cosmetic products. Int. J. Interdiscip. Res. Arts Humanities (IJIRAH) **1**(1), 64–70 (2016)
38. Rossiter, J.R.,. Percy, L.: Advertising Communication & Promotion Management. McGraw-Hill Companies, New York, NY (1997)
39. Saravanakumar, M., SuganthaLakshmi, T.: Social media marketing. Life Sci. J. **9**(4), 4444–4451 (2012)
40. Vekat, M.: The new marketing order—the rules of the game are changing. Indian Manag. **49**(10), 84–86 (2010)
41. Zinnbauer, M., Honer, T.: How brands can create social currency—a framework for managing brands in a network Era. Market. Rev. **28**(1), 50–55 (2011)

Printed in the United States
by Baker & Taylor Publisher Services